"十二五"普通高等教育本科规划教材
包装工程本科专业系列教材

包装工程概论
（双语教学用）
第二版

陈满儒　主　编
孙德强　副主编

·北京·

本书是在第一版同名教材的基础上修订而成的。全书由包装概述、包装材料和容器、包装动力学与运输包装、包装工艺与设备和包装研发共5个单元20课组成。修订后的教材内容补充了国外包装工程学科专业的最新资料，使之更能体现包装工程的专业特色。同时，为体现双语教学的内涵和特点，方便组织教学和自主学习，教材修订中也注重了内容的易于阅读和理解性。一是在每篇课文之后列出了专业词汇、疑难句注释和复习题；二是全部课文原文配有中文翻译和教学用演示文稿ppt；三是在全书最后附了词汇总表及索引。本教材既适合包装工程本科专业开展双语教学，也可作为从事包装及相关专业的技术人员提高业务及其专业英语应用水平的学习参考书。

图书在版编目（CIP）数据

包装工程概论（双语教学用）/陈满儒主编．—2版．—北京：化学工业出版社，2014.7
"十二五"普通高等教育本科规划教材．包装工程本科专业系列教材
ISBN 978-7-122-20680-0

Ⅰ.①包… Ⅱ.①陈… Ⅲ.①包装-工程技术-高等学校-教材-汉、英 Ⅳ.①TB48

中国版本图书馆CIP数据核字（2014）第098505号

责任编辑：杨　菁　　　　　　　　　文字编辑：李　瑾
责任校对：徐贞珍　　　　　　　　　装帧设计：史利平

出版发行：化学工业出版社（北京市东城区青年湖南街13号　邮政编码100011）
印　　装：三河市延风印装有限公司
787mm×1092mm　1/16　印张20¾　字数505千字　2016年4月北京第2版第1次印刷

购书咨询：010-64518888（传真：010-64519686）　　　售后服务：010-64518899
网　　址：http://www.cip.com.cn
凡购买本书，如有缺损质量问题，本社销售中心负责调换。

定　价：49.00元　　　　　　　　　　　　　　　　　　　　　　　　版权所有　违者必究

本书编写人员名单

主　　编：陈满儒
副 主 编：孙德强
编写人员（按姓名笔画排序）：
　　　　　邢月卿　巩桂芬　孙德强　李国志　陈满儒
　　　　　苟进胜　赵郁聪

本科护理人员守则

二版前言
Preface to the Second Edition

自 2005 年 1 月本书第一版首印以来，有许多读者和大学选其作为包装工程本科专业双语教学（或专业英语）教材和提高专业英语应用水平的自学读物，为拓展其国际视野发挥了积极作用。同时，使用原教材的专业人员和读者也提出了一些好的意见与建议。意见集中在，一些课文的阅读理解偏难，篇幅偏长，学起来难度较大；专业词汇及难句注释少，教师不便于备课，读者不便于自学。建议精选精编包装组分之经典内容；丰富包装学科专业的特色内容；编写符合学生和读者基本特征，利于面向国际化包装专业人才培养的双语教学改革的辅助配套教材。为此，该教材第二版的修订上着重在以下几个方面：

一是在教材内容上，更新了包装的经典内容。如包装的历史沿革、包装功能、包装材料和容器等。替换增添了 6 节课，如测定缓冲曲线的应力-能量法、实用包装技术、包装研究与开发等，使得包装颇具特色的专业知识得到加强，也使得包装工程专业的人才培养更符合国家规范和行业标准。同时，教材内容也适宜于有效开展包装专业双语教学及读者自学。

二是在编写结构上，注意教材的易于理解性。首先，每篇课文增加了专业生词与词组（New Words and Expressions）中英文对照表，并在书后增加了词汇总表和词汇索引（Glossary）；其次，就跨文化背景、惯用型表达和语法等对疑难句作了注释（Notes）；第三，增加了课后习题；第四，为便于教学和自学，新增了所有课文的中文翻译和 ppt 文稿，供读者从出版社索取使用。

三是教材修订符合了专业课双语教育教学改革的要求。本教材的修订较好地贯彻了双语课程内涵式教学的 4Cs 核心要素：学科内容（Content）、交流表达（Communication）、认知思维（Cognition）和文化习俗（Culture）。保证了双语教材与只偏重外语语言再学习的专业英语书籍的不同，换句话说，双语教材更注重专业性与知识性，外语语言只充当专业知识的学习工具。

本书由 5 大单元 20 课组成。第 1 和第 11 课由陈满儒编写；第 2、第 14 和第 15 课由李国志编写；第 3～5 和第 9 课由赵郁聪编写；第 6～8 课由邢月卿编写；第 10、第 18 和第 19 课由孙德强编写；第 12 和第 20 课由北京林业大学苟进胜编写；第 13、第 16 和第 17 课由巩桂芬编写。编写人员中除署名单位外，其他人员均属陕西科技大学。

本书既适合包装工程本科专业开展双语教学，也可作为从事包装及相关专业的技术人员提高业务及其专业英语应用水平的学习参考书。

在修订本书的过程中，再次得到了北美包装教育资深教授 Walter Soroka 博士的大力支持，也得到了美国威斯康星-斯陶特大学教授 Louis Moegenburg 博士等的帮助，在此一并表示谢意。

由于编者水平所限，书中疏漏之处在所难免，敬请读者批评指正。

主编

2014 年 8 月于陕西科技大学

一版前言
Preface to the First Edition

20世纪80年代初期，依照发达国家包装工程高等教育的发展模式，我国的一些大学相继创建了包装工程本科专业。20多年来，我国包装工程专业的办学规模、层次及其办学水平都有了很大的提高。进入21世纪以来，随着我国包装科学技术的迅速发展，随着我国从包装大国向包装强国迈进政策的实施，迫切要求从事包装的专业技术人员努力提高自身的能力与素质，更有效地学习和应用国外先进的包装技术。

国内各大学的包装工程专业自建立以来都开设有专业英语类课程，这对巩固学生的基础英语水平，拓宽知识面，提高对国外技术资料的阅读应用能力起到很好的促进作用。为了贯彻教育部关于"本科教育要创造条件使用英语等外语进行公共课和专业课教学……"的文件精神，一些学校在包装工程专业教学中开始用双语教学替代原有的专业英语教学。这种不是就英语学英语，而是通过英语来学习专业知识，将专业英语学习与专业知识学习融合一起的方法有利于学生第一时间用英语阅读和吸收英语原版专业科技信息打下基础，有利于增强他们用英语进行专业交流的实际能力。然而，国内尚缺少能全面反映包装工程学科特点、内容适当、系统性强、可供学生和专业技术人员选用的包装工程双语教材或读本，影响了本专业双语教学的水平提高及其规范化。

包装工程是一门综合性的交叉学科，内容非常丰富，涉及面广。本书的选材力求在有限的篇幅内尽可能涵盖包装工程的学科领域。

本书由五个方面的内容组成，即：透视包装；包装材料和容器；包装印刷与装潢；包装动力学和运输包装；包装机械。

为方便学习理解，书后列出了专业词汇与术语的中英文对照表。本书既可作为包装工程专业本科生技术基础课双语教学或专业英语教学用书，也可作为从事包装工程及相关专业的技术人员提高业务及其专业英语应用水平的学习参考书。

在编写该书的过程中，得到了加拿大莫哈克学院包装设计专业Walter Soroka教授、美国威斯康星-斯陶特大学包装专业Robert Berkemer教授、美国密歇根州立大学包装学院Bruce Harte教授和美国罗切斯特理工学院包装科学系Daniel Goodwin教授等的大力支持，在此一并致谢。

由于编者的水平所限，本书疏漏之处在所难免，敬请读者批评指正。

编者
2004年10月于陕西科技大学

目录

Contents

UNIT 1   An Overview Of Packaging ·· 1
  Lesson 1   A Historical Perspective On Packaging ····························· 1
    1   What Is Packaging? ·· 1
    2   Primitive Packaging ··· 2
    3   From Rome To The Renaissance ······································· 3
    4   The Industrial Revolution ·· 5
    5   The Evolution Of New Packaging Roles ······························ 6
    6   Packaging In The 20th Century ··· 8
    7   Modern Packaging ··· 9
      7.1   Changing Needs and New Roles ································· 9
      7.2   Packaging and the Modern Industrial Society ················ 9
      7.3   World Packaging ·· 11
    8   Waste Management And Environmental Issues ····················· 12
    9   The Modern Packaging Industry ·· 16
  Lesson 2   Basic Functions of Packaging ·· 22
    1   Introduction ·· 22
    2   The Containment Function ·· 23
    3   The Protection/Preservation Function ································· 23
    4   Food Preservation ··· 24
      4.1   The Nature of Food ·· 24
      4.2   Meat products ··· 26
      4.3   Fish ··· 26
      4.4   Produce ·· 26
      4.5   Barrier Packaging ·· 27
      4.6   Microorganisms ·· 28
    5   The Transport Function ·· 35
    6   The Information/Sales Function ·· 36
UNIT 2   Packaging Materials And Containers ····································· 42
  Lesson 3   Paper and Paperboard ··· 42

1	Introduction	42
2	Representative Papermaking Machines	42
	2.1 Fourdrinier machines	42
	2.2 Cylinder Machines	43
	2.3 Twin-Wire Machines	45
3	Machine Direction And Cross Direction	45
4	Surface Or Dry-End, Treatments And Coatings	46
5	Paper Chracterization	47
	5.1 Caliper and Weight	47
	5.2 Brightness	48
	5.3 Paper and Moisture Content	48
	5.4 Viscoelasticity	49
6	Paper Types	49
	6.1 Newsprint and Related Grades	49
	6.2 Book Papers	49
	6.3 Commercial Papers	49
	6.4 Greaseproof Papers	50
	6.5 Natural Kraft Paper	50
	6.6 Bleached Krafts and Sulfites	50
	6.7 Tissue Paper	50
	6.8 Label Paper	50
	6.9 Pouch Papers	51
	6.10 Containerboards (linerboard and medium)	51
7	Paperboard Grades	51
	7.1 Chipboard, Cardboard, Newsboard	51
	7.2 Bending Chipboard	51
	7.3 Lined Chipboard	51
	7.4 Single White-Lined (SWL) Paperboard	52
	7.5 Clay-Coated Newsback (CCNB)	52
	7.6 Double White-Lined (DWL) Paperboard	52
	7.7 Solid Bleached Sulfate (SBS)	52
	7.8 Food Board	52
	7.9 Solid Unbleached Sulfate (SUS)	52
Lesson 4	Corrugated Fiberboard Boxes	57
1	Historical Perspective	57
2	Corrugated Board	58
	2.1 Construction	58
	2.2 Flutes	59
	2.3 Fiberboard Grades	60
	2.4 Corrugating Adhesives	61

	2.5	Board Manufacture	62
3	Fiberboard Characterization Tests	63
	3.1	Mullen Burst Test (TAPPI T 810)	63
	3.2	Edgewise Compression Test (TAPPI T 811)	63
	3.3	Flat Crush Test (TAPPI T 808)	64
	3.4	Combined Weight of Facings	64
	3.5	Thickness of Corrugated Board (TAPPI T 411)	64
	3.6	Gurley Porosity (TAPPI T 460 and T 536)	64
	3.7	Flexural Stiffness (TAPPI T 820)	64
	3.8	Water Take-up Tests (TAPPI T 441)	64
	3.9	Puncture Test (TAPPI T 803)	64
	3.10	Pin Adhesion (TAPPI T 821)	65
	3.11	Ply Separation (TAPPI T 812)	65
	3.12	Coefficient of Friction (TAPPI T 815 and ASTM 04521)	65
4	Corrugated Boxes	65
	4.1	Selecting the Correct Flute	65
	4.2	Box Styles	67
	4.3	Manufacturer's joint	68
	4.4	Dimensioning	68
5	Carrier Rules	68
	5.1	Application	68
	5.2	Summary of Rules for Corrugated Box Construction	69

Lesson 5 Metal Containers — 75

1	Background	75
2	Common Metal Container Shapes	76
3	Three-Piece Steel Cans	77
4	Two-Piece Cans	79
	4.1 Draw Processes	79
	4.2 Draw-and-Redraw Process	79
	4.3 Draw-and-Iron Process	80
5	Impact Extrusion	81
6	Aerosols	83
	6.1 Aerosol Propellants	83
	6.2 Other Pressurized Dispensers	84

Lesson 6 Glass Containers — 89

1	Glass Types And General Properties	89
2	Bottle Manufacture	90
	2.1 Blowing the Bottle or Jar	90
	2.2 Annealing	92
	2.3 Surface Coatings	92

  2.4 Inspection and Packing ································· 93
 3 Bottle Design Features ······································· 93
  3.1 Bottle Parts and Shapes ································ 93
  3.2 Finish and Closures ···································· 94
  3.3 Neck and Shoulder Areas ······························· 94
  3.4 Sides ··· 94
  3.5 Heel and Base ··· 95
  3.6 Stability and Machinability ······························ 95
  3.7 Vials and Ampoules ···································· 95
  3.8 Carbonated Beverages ································· 96
Lesson 7 Plastics in Packaging ································· 98
 1 Introduction To Plastics ····································· 98
 2 Thermoplastic And Thermoset Polymers ······················ 100
 3 Shaping Plastics ·· 101
 4 Plasticating Extruders ······································· 101
 5 Extrusion ··· 102
  5.1 Profile Extrusion ······································ 102
  5.2 Sheet and Film Extrusion ······························· 103
  5.3 Blown-Film Extrusion ·································· 103
  5.4 Orientation ··· 104
  5.5 Co-Extrusion ··· 105
 6 Injection Molding ·· 105
  6.1 Injection Molding Machines ····························· 105
  6.2 Co-injection Molding ··································· 106
 7 Extrsion Blow Molding ······································ 106
 8 Injection Blow Molding ····································· 107
 9 Thermoforming ·· 109
  9.1 Principle and Applications ······························ 109
  9.2 Thermoforming Methods ······························· 109
  9.3 Billow forming ·· 110
 10 Other Forming Methods ···································· 111
  10.1 Pressblowing ·· 111
  10.2 Rotational Molding ··································· 111
  10.3 Compression Molding ································ 111
  10.4 Blow-Fill-Seal Molding ································ 112
 11 Recognizing Molding Methods ······························ 112
Lesson 8 Flexible Packaging Laminates ························· 116
 1 Laminates ·· 116
 2 Aluminum Foil ·· 116
  2.1 Chemical Characteristics ······························· 117

  2.2 Aluminum Foil in Flexible Packaging ……………………… 117
  2.3 Foil Coatings ………………………………………………… 118
 3 Vacuum Metallizing ……………………………………………… 119
  3.1 The Metallizing Process …………………………………… 119
  3.2 Vacuum-Metallizing Paper ………………………………… 120
  3.3 Vacuum-Metallizing Films ………………………………… 120
 4 Other Inorganic Coating ………………………………………… 121
  4.1 Silicone oxides ……………………………………………… 121
  4.2 Carbon coatings ……………………………………………… 121
  4.3 Nanocomposites ……………………………………………… 121
 5 Laminate Structural And Physical Properties ……………… 121
  5.1 Coefficient of Friction ……………………………………… 122
  5.2 Body and Dead-Fold Properties …………………………… 123
  5.3 Tear Properties ……………………………………………… 123
  5.4 Thermoformability …………………………………………… 123
  5.5 Use Environments …………………………………………… 123
 6 Flaxible Bags, Pouches And Sachets ………………………… 123
 7 Sealability ………………………………………………………… 126
 8 Barrier Properties ……………………………………………… 127
 9 Laminating Processes ………………………………………… 129
  9.1 Bonding Methods …………………………………………… 129
  9.2 Laminating Machines ……………………………………… 129
  9.3 Coating Stations …………………………………………… 130
 10 Specifying Laminates ………………………………………… 132
Lesson 9 Closures ………………………………………………… 138
 1 Selection Considerations ……………………………………… 138
 2 Tamper Evident Closures ……………………………………… 139
 3 Tamper Evident Systems ……………………………………… 140
 4 Child-Resistant Closures ……………………………………… 142
UNIT 3 Packaging Dynamics and Distribution Packaging ……… 145
Lesson 10 Shock, Vibration, and Compression ………………… 145
 1 Shock ……………………………………………………………… 145
  1.1 Shock Resulting from Drops ……………………………… 145
  1.2 Shock During Rail Transport ……………………………… 147
  1.3 Other Shock Conditions …………………………………… 147
  1.4 Quantifying Shock Fragility ……………………………… 148
  1.5 Cushioning Against Shock ………………………………… 151
 2 Vibration ………………………………………………………… 152
  2.1 Vibration Damage Due to Relative Motion ……………… 153
  2.2 Vibration Resonance ……………………………………… 153

    2.3   Stack Resonance ·· 154
    2.4   Isolating Vibration ·· 155
  3 Compression ··· 156
    3.1   Static and Dynamic Compression ·· 156
    3.2   Compression Strength and Warehouse Stack Duration ························ 157
    3.3   Compression Strength and Humidity ·· 157
    3.4   Other Factors Influencing Box Stack Strength ································ 157
    3.5   Contents' Effect on Compression Strength ····································· 159
    3.6   Plastic Bottle Stacking Factors ··· 161
    3.7   Stacking and Compression ·· 161
    3.8   Distribution Environment and Container Performance ······················· 162
    3.9   Estimating Required Compression Strength ···································· 163

Lesson 11 Mechanical Shock ·· 169
  1 Introduction ··· 169
  2 The Freely Falling Package ··· 169
  3 Mechanial Shock Theory ·· 171
  4 Shock Duration ·· 176
  5 Shock Amplification And The Critical Element ································· 177
  6 Horizontal Impacts ··· 179

Lesson 12 *Lansmont* Six Step Method ·· 184
  1 Step 1 Define The Environment ··· 185
    1.1   Shock Environment ·· 185
    1.2   Vibration Environment ·· 187
  2 Step 2 Product Fragility Analysis ··· 188
    2.1   Shock: Damage Boundary ·· 188
    2.2   Vibration: Resonance Search & Dwell ··· 191
  3 Step 3 Product Improvement Feedback ··· 192
  4 Step 4 Cushion Material Performance Evaluation ······························ 192
    4.1   Shock Cushion Performance ··· 192
    4.2   Vibration Cushion Performance ·· 193
  5 Step 5 Package Design ·· 194
    5.1   Shock: Package Design ··· 194
    5.2   Vibration: Package Design ··· 195
    5.3   Package Design Considerations ··· 196
  6 Step 6 Test The Product/package System ······································· 196
    6.1   Shock: Package Testing ··· 196
    6.2   Vibration: Package Testing ··· 197

Lesson 13 Distribution Packaging ··· 200
  1 Short History Of Distribution Packaging In The USA ························· 200
  2 Functions And Goals Of Distribution Packaging ································ 200

  2.1 Containment ··· 200
  2.2 Protection ·· 200
  2.3 Performance ·· 200
  2.4 Communication ·· 201
  2.5 Product Protection ··· 201
  2.6 Ease of Handling and Storage ·· 201
  2.7 Shipping Effectiveness ·· 201
  2.8 Manufacturing Efficiency ··· 201
  2.9 Ease of Identification ··· 201
  2.10 Customer Needs ·· 201
  2.11 Environmental Responsibility ·· 201
 3 The Cost Of Packaging ··· 201
 4 The Package Design Process ··· 202
 5 Taking A Total System Approach To Package Design ················ 202
 6 The Protective Package Concept ··· 203
 7 The 10-Step Process Of Distribution Packaging Design ·············· 205
  7.1 Identify the Physical Characteristics of the Product ············ 205
  7.2 Determine Marketing and Distribution Requirements ·········· 205
  7.3 Learn About the Environmental Hazards Your Packages
    will Encounter ·· 205
  7.4 Consider Packaging and Unitizing Alternatives ·················· 205
  7.5 Design the Distribution Package ····································· 205
  7.6 Determine Quality of Protection Through Performance-Testing ············ 206
  7.7 Redesign Package (and Unit Load) Until It Successfully
    Passes All Tests ··· 206
  7.8 Redesign the Product if Indicated and Feasible ················ 206
  7.9 Develop the Packaging Methods ···································· 207
  7.10 Document All Work ··· 207
 8 A Final Check ··· 207
 9 The Warehouse ··· 207
 10 Unit Loads ··· 208
  10.1 Pallets ··· 208
  10.2 Unit Load Efficiency ··· 210
  10.3 Stabilizing Unit Loads ·· 211
Lesson 14 Test Method for Product Fragility ································· 215
 1 Shock: Damage Boundary ··· 215
  1.1 Conducting A Fragility Test ··· 217
 2 Vibration: Resonance Search & Dwell ····································· 219
Lesson 15 Stress-Energy Method for Determining Cushion Curves ········ 223
 1 Current Practice ·· 223

	2	Something New	223
	3	Stress-Energy Method	224
	4	Procedure To Find The Stress-Energy Equation	225
	5	Using The Stress-Energy Equation To Generate Cushion Curves	226
	6	Test Procedure	226
	7	Conclusion	228

UNIT 4 Packaging Technology and Machinery 230

Lesson 16 Liquid Filling 230
1. Introduction 230
2. Type Of Filling Machine 230
3. Use Of Filling Machines 231
4. Rotary And Straight-Line Fillers 231
5. Liquid Volumetric Filling 231
   - 5.1 Introduction 231
   - 5.2 Piston Volumetric Filling 232
   - 5.3 Diaphragm Volumetric Filling 235
   - 5.4 Timed Flow Volumetric Filling 236
   - 5.5 Rotating Metering Discs 236
   - 5.6 Rotary Pumps 237
   - 5.7 Augers 238
6. Liquid Constant Level Filling 238
   - 6.1 Introduction 238
   - 6.2 Pure Gravity Filling 238
   - 6.3 Pure Vacuum Filling 240
   - 6.4 Gravity Vacuum Filling 241
   - 6.5 Pure Pressure Filling 242
   - 6.6 Level Sensing Filling 242
   - 6.7 Pressure Gravity Filling 243

Lesson 17 Dry Product Filling 248
1. Introduction 248
2. Type Of Dry Products 248
3. Type Of Dry Filling Operations 248
4. Product Delivery 249
5. Dry Volumetric Filling 249
   - 5.1 Introduction 249
   - 5.2 Cup or Flask Fillers 249
   - 5.3 Flooding or Constant Stream Fillers 251
   - 5.4 Auger Fillers 251
   - 5.5 Vacuum Fillers 252
6. Dry Filling By Weight 253

	6.1	Introduction	253
	6.2	Net Weight Filling	253
	6.3	Gross Weight Filling	254
	6.4	Types of Scales	255
7		Filling By Count	256
	7.1	Introduction	256
	7.2	Board or Disc Counters	256
	7.3	Slat Counters	257
	7.4	Column Counters	258
	7.5	Electronic Counters	259
	7.6	Maintenance	259

Lesson 18   Applied Packaging (I) ········ 262
  1 Carded Display Packages ········ 262
  2 Blister Packages ········ 263

Lesson 19   Applied Packaging (II) ········ 266
  1 Carded Skin Packaging ········ 266
  2 Pharmaceutical Packaging ········ 267
    2.1 Drug Properties ········ 267
    2.2 Packaging Emphases ········ 268
    2.3 Regulations ········ 269
    2.4 Manufacturing Practice ········ 270

UNIT 5   Packaging Development ········ 273

Lesson 20   Packaging Development Process ········ 273
  1 Managing The Packaging Project ········ 273
  2 Project Scope ········ 274
  3 The Package Development Process ········ 275
    3.1 An Overview of the Package Development Process ········ 275
    3.2 Generating Ideas ········ 276
    3.3 The Package Design Brief ········ 277
    3.4 The Development Timetable ········ 280
    3.5 Development and Testing of Alternatives ········ 280
  4 Specifications ········ 281

Glossary ········ 285
Resources ········ 316

UNIT 1

An Overview Of Packaging

Lesson 1 A Historical Perspective On Packaging

1 What Is Packaging?

① Packaging is best described as a coordinated system of preparing goods for transport, distribution, storage, retailing, and use of the goods. It is a complex, dynamic, scientific, artistic, and controversial business function, which in its most fundamental form contains, protects/preserves, transports, and informs/sells. Packaging is a service function that cannot exist by itself; it needs a product. If there is no product, there is no need for a package.

② Packaging functions range from those that are technical in nature to those that are marketing oriented (Figure 1.1). Technical packaging professionals need science and engineering skills, while marketing professionals need artistic and motivational understanding. Packaging managers need a basic understanding of both marketing and technical needs, mixed with good business sense. This unusual skill spread makes the packaging industry a unique career choice.

Technical Functions		Marketing Functions	
contain	measure	communicate	promote
protect	dispense	display	sell
preserve	store	inform	motivate

Figure 1.1 Packaging encompasses functions ranging from the purely technical to those that are marketing in nature

③ Packaging is not a recent phenomenon. It is an activity closely associated with the evolution of society and, as such, can be traced back to human beginnings. The nature, degree, and amount of packaging at any stage of a society's growth reflect the needs, cultural patterns, material availability and technology of that society. A study of packaging's changing roles and forms over the centuries is, in a very real sense, a study of the growth of civilization.

④ From an individual perspective, change often seems to be that which has already happened, but society is changing daily—meeting new challenges, integrating new knowledge, accommodating new needs and rejecting systems proven to be unacceptable. These changes are inevitably reflected in the way we package, deliver and consume goods.

⑤ Because the science of packaging is closely connected to everything we do as a society, it should come as no surprise that the packaging industry is always in a state of change. Entire sectors can become obsolete, or new industries generated by the discovery of a new material, process or need. For example, a whole new packaging sector was born with a single tragic tampering incident (the Tylenol episode of October 1982). Society suddenly required **tamper-evident closure** systems.

⑥ Until the 1950s, motor oil was delivered in bulk to service stations, which in turn measured it into 1-quart glass jars. The advantages of premeasured oil in metal cans swung the entire trade into metal cans. By the late 1960s, **foil/fiber composite cans** had replaced metal cans, and by the late 1970s, plastic bottles had replaced fiber cans.

⑦ Similarly, milk delivery went from glass bottles to today's variety of plain and **aseptic** paper cartons, plastic bottles and flexible bags, each packaging method offering its own particular advantages.

⑧ How oil or milk will be delivered tomorrow is open to speculation. Packaging choice can probably reflect an increasing need for **environmentally acceptable packaging** that will generate minimal waste. The relative costs of petrochemicals, **wood pulp**, and metal will likely govern choices. And finally, the way we buy and consume oil or milk will have a significant impact. No option can be ignored; it is not difficult to imagine a scenario where milk is delivered in **refillable aluminum cans.**

2 Primitive Packaging

⑨ We don't know what the first package was, but we can certainly speculate. Primitive humans were nomadic hunter/gatherers; they lived off the land. Such an existence has severe limitations. It takes considerable land area to support the wild animals and vegetation needed to feed a single person. Social groupings were therefore small, probably restricted to family units.

⑩ These early humans would have been subject to the geographical migrations of animals and the seasonal availability of plant food. This meant that humans followed their food sources around and quite often went hungry. Such an extreme nomadic existence does not encourage property accumulation beyond what can be carried on one's back.

⑪ Nonetheless, primitive people needed containment and carrying devices, and *out of this need came the first "package"*. It was most likely a wrap of leaves, an animal skin, the shell of a nut or gourd, or a naturally hollow piece of wood. Fire was carried from camp to camp, and evidence suggests that the role of **fire-bearer** and the "packaging" of fire carried a mystical significance.

⑫ Let's jump ahead to 5000 B.C., a time of some domesticated plants and animals. While the forage or hunt was still important, a reasonable food supply was available in a given vicinity. This evolutionary stage, which supported larger social groups, gave birth to small tribal villages. Storage and transport containers were needed for milk, honey, seed grains, nuts, and dried meat. Villages with access to different resources traded with their neighbors, requiring transport containers.

⓷ Fabricated **sacks**, baskets, and bags, made from materials of plant or animal origin, were added to the primitive packaging list. Wood boxes replaced hollow logs. Clay from a riverbank would have initially been shaped into containers and allowed to dry in the sun. This was fine for **dry products**, but wet products quickly converted such containers to mud. Some impatient Neolithic genius, probably trying to hurry the slow process of sun-drying, placed a clay bowl in a fire. Much to his or her pleasure, the fire-dried clay pots were more durable and held their shape when filled with water. Thus was born the pottery and ceramic trade.

⓸ Legend has it that Phoenician sailors who used salt blocks to protect their fire from wind on the sandy Mediterranean coast discovered a hard inert substance in the fire's remains. By 2500 B.C., **glass beads** and figures were being made in Mesopotamia (today's Iraq). The earliest hollow glass objects appeared in Mesopotamia and Egypt in about 1500 B.C.

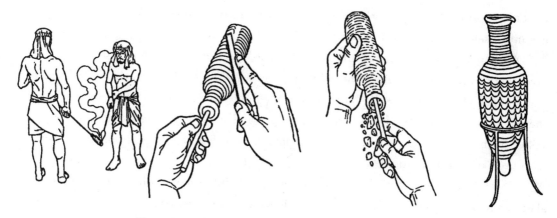

Figure 1.2 Forming a hollow glass vessel around a core

⓹ Ancient Egyptian glass containers were core-formed. Hot strands of glass were wrapped around a core of clay and dung (Figure 1.2). Wavy patterns could be introduced by dragging a stick across the soft hot glass. Rolling the glass against a smooth surface flattened and smoothed the strand lines. When the glass was cool, the core was dug out of the container.

⓺ Along with metal, these glass containers were the ancient packaging materials. Many centuries would pass before modern materials such as paper and plastics expanded the packager's portfolio.

⓻ While the printing arts and extensive packaging laws were still in the distant future, law that affected packaging were being enacted as early as the Greek city-state period (about 250 B.C.). For example, olive oil, at that time packaged in **amphora** (large clay jugs with elongated or pointed bottoms), was marked with a stamp identifying the city-state where it was produced, the time of pressing and the person responsible for it (Figure 1.3).

3 From Rome To The Renaissance

⓼ As time went on, cities were established, trade flourished across the European and

Figure 1.3 A portion of a Greek amphora handle dated 220~189 B.C. (The stamped image shows a rose, indicating that the olive oil was pressed on the island of Rhodes)

Asian continents, and conquering armies frequently sallied forth to plunder some other region's wealth. While the world witnessed many societal changes, the corresponding changes in packaging related mostly to the quality and quantity of existing packaging practices.

⑲　An important packaging event, attributed to the Romans in about 50 B.C., was the invention of the **glass blowpipe**. The blowpipe was a hollow steel rod on the end of which was placed **a gob of** molten glass. By blowing into the opposite end, the glassblower could inflate the gob into a hollow vessel in a variety of shapes and sizes. The glassblower could shape the vessel freehand by alternately blowing and shaping, or blow the glass bubble into a cup mold with pre-existing decorations.

⑳　The blowpipe's invention brought glass out of noble households and temples. Roman **glass beakers** decorated with chariots and gladiator contests—apparently sold as souvenirs and mementos of such events—are reasonably common.

㉑　The origin of the first **wooden barrel** is not clear, but it also probably had its start at this time, possibly in the Alpine regions of Europe. The barrel was destined to become one of the most common packaging forms for many centuries.

㉒　With the Roman Empire's collapse in about 450 A.D., Europe was reduced to minor city-states. Many established **arts and crafts** were forgotten or became stagnant. The 600 years following the fall of Rome were so devoid of significant change that historians refer to them as **the Dark Ages.**

㉓　Any progress came from the Far East and from Arabic nations newly inspired by the Muslim faith. In China, Ts'ai Lun is credited with making the first true paper from the inner bark of mulberry trees. When the Muslims sacked Samarkand in about 950, they carried the secret back to Europe. The Egyptians had been making a similar sheet product by weaving together the split stalks of **papyrus reeds.** By pounding, pressing and drying the woven strips, they created a useful **sheet material.** Centuries later the name "paper" was given to the Chinese invention made of matted plant fibers.

㉔　Printing from **woodcuts**—the ancient parent of the printing process known as **flexography**—also originated in the Far East. The oldest existing printed objects are Japanese Buddhist charms dated to 768. The oldest existing book is the *Diamond Sutra*, found in Turkistan and printed in 868.

25   The European world awoke in about 1100. Neglected crafts were revitalized, learning and the arts were revived and trade increased, and by the 1500s, the great age of exploration was well under way. The art of printing was born in this period.

26   Fundamental social structures had not changed significantly. Most of the population lived off the land, sometimes as freeholders, but more typically as serfs who owed their existence and part of everything they produced to a higher power. For the most part they ate what they raised, found or caught. At this level, consumer needs were nonexistent.

27   Shops and stores where a person could buy goods did not exist as we know them. Although money as an exchange medium was available, much of the population never saw any. *Manufacturing was strictly a* **custom business**, *and what we have called packages to this point was personally crafted, as were most goods*. Packages, where they existed, were valuable utensils, and were rarely disposable in the manner of a modern package.

28   Since there was no retail trade, concepts of marketing, advertising, price structures and distribution were irrelevant. Population levels were not large enough to support **mass production**, even in the most limited sense.

4 The Industrial Revolution

29   *Encyclopedia Britannica* describes the Industrial Revolution as "the change that transforms a people with peasant occupations and local markets into an industrial society with world-wide connections". This new type of society makes great use of machinery and manufactures goods on a large scale for general consumption.

30   The Industrial Revolution started in England in about 1700 and spread rapidly through Europe and North America. Some characteristics of this revolution included the following:

• Rural agricultural workers migrated into cities, where they were employed in factories.

• Inexpensive mass-produced goods became available to a large segment of the population; the consumer society was born.

• Factory workers needed commodities and food that were previously produced largely at home.

• Many new shops and stores opened to sell to the newly evolving working class.

• By necessity, some industries were located in nonagricultural areas, requiring that all food be transported into the growing urban settings.

31   These changes increased the demand for barrels, boxes, **kegs**, baskets, and bags to transport the new **consumer commodities** and to bring great quantities of food into the cities. The fledgling packaging industry itself had to mechanize in order to keep up with the growing demand. With large segments of the population living away from food production points, it became necessary to devise ways of preserving food beyond its **natural biological life**.

5 The Evolution Of New Packaging Roles

52 For most of recorded history, people lived in rural communities and were largely self-sufficient. **Bulk packaging** *was the rule, with the barrel being the workhorse of the packaging industry.* Flour, apples, biscuits, molasses, gunpowder, whiskey, nails and whale oil were all transported in barrels. Packaging served primarily to contain and protect. **Individual packaging** *was of little importance until the Industrial Revolution spurred the growth of cities.* The new industrial workers needed to be fed by a separate agricultural system and supported in most of their nonfood needs by the manufacturing skill of others.

53 City dwellers did not have a farm's storage facilities, and so quantities purchased tended to be small and trips to the shop more frequent. This was an open opportunity to create individual packages in the amounts that people preferred to purchase. In practice, it took many years for this to happen, and even today the transformation is not complete.

54 Initially, shops simply adapted the bulk delivery system to consumer selling. The shopkeeper received apples and biscuits in barrels, cheese in large rounds and herbs or medicines in glass jars. He or she would measure and portion these items, often into a container provided by the purchaser. The shopkeeper sold mostly unfinished product.

55 Medicines, cosmetics, teas, liquors and other expensive products were the first prepackaged products, along with awkward items such as tacks or pins. The latter were often wrapped in paper, and the expression "a paper of pins" accurately described the product. In time, many products were sold in a "paper".

56 Products were sold generically. Cheese was cheese, oatmeal was oatmeal, and lye soap was lye soap. Sometimes identifying marks were made with a blackening brush or with a hot **branding iron** on the barrel or **cask** to show origin or manufacturer. Over time, certain **brand marks** became associated with quality products. As individual packaging began to develop, quality producers wished to identify their particular product as a guarantee of quality or composition. The brand mark was carried from the bulk package to **unit packages** or **labels**. It was an early form of product branding, as well as the origin of the term **"brand name"**.

57 The first brand names were inevitably those of the maker. Yardley's (1770), Schweppes (1792), Perrier (1863), Smith Brothers (1866) and Colgate (1873) are a few of the personal names that have survived to this day.

58 Most packages that existed in the mid-1800s were for higher cost goods, and the evolving printing and decorating arts were applied to these early "upscale" packages. Similarly, it was realized that the papers used to wrap a product for sale were easily imprinted with a brand mark, with some message of instruction or a description of the product's virtue. Many early decorations were based on works of art or national symbols or images. Labels were printed with ornate and elaborate scrolls, wreaths, and allegorical figures or impossibly flawless and shapely ladies (some things are difficult to change!). These often combined **typography** in a dozen type styles.

59 Early food can labels had to appeal to simple country folk, so pictures of pastoral life,

barnyards and fruit on a branch were commonly used. Sometimes the label graphics had little to do with the contents, and sometimes the same graphic was used on unrelated products. Another popular practice was to display the gold medals won at one or another of the great national and international fairs held frequently at the time. Many early labels were so attractive that they were saved for decorative use.

Figure 1.4 The Quaker personage as it appeared in 1896 (Such images have to change with time as perception and styles change. The modern Quaker image is not as stem and somewhat slimmer in appearance)

40 A packaging milestone was set in 1877 when the American Cereal Company chose a symbol to represent or trademark their product. The Quaker personage (Figure 1.4) represented purity, wholesomeness, honesty, and integrity—value that by extension also applied to the product. It was perhaps one of the earliest forms of what designers refer to as the "persona", a description of the package or product as if it were a person.

41 After an intense advertising campaign, the company convinced a fair proportion of the population to ask for Quaker Oats rather than just oatmeal. The Quaker figure's success possibly inspired other companies to adopt fictitious personages to represent their products, among them were the Cream of Wheat smiling chef (1893) and the National Biscuit Company's boy in a raincoat (1899).

42 Package decoration follows national art styles and trends. Between 1890 and about 1920, decoration followed the **art nouveau** style popular in that period. This was followed by a period of art deco graphics and designs.

43 The first plastic, based on **cellulose**, was made in 1856, but packaging applications were still a long way off. In 1907, **phenol formaldehyde plastic**, later known as **Bakelite**, was discovered. Bakelite's major packaging application was for closures. A few years later, in 1911, a machine was built to manufacture continuous cellulosic film. DuPont chemists perfected the cellulose **casting process** in 1927 and called their product **cellophane**. Cellulose films dominated the clear film market until the advent of **polyethylene** and **polypropylene**. Bakelite was largely displaced by the newer **thermoplastics** in the 1960s.

44 In earlier days, craftspeople sold their own wares and were able to explain the available choices or how best to use a product. Now the shopkeeper was not there to aid or influence the consumer's purchase. Stores with thousands of products were staffed by persons who had little or no knowledge of the product and their applications. The consumer was face to face with the package, and the package's motivational and informational roles became critical:

- The package had to inform the purchaser.
- The package had to sell the product.

45 Package design and graphics were suddenly much more than a pretty picture, and a whole new profession, package design, was born. The transformation from bulk packaging

6 Packaging In The 20th Century

㊻ *The birth rate after the Second World War and into the* 1950s *was so imposing that it earned its own name*: *the baby boom*. **Demographics**, the study of population structure and trends, was universally realized to be an important factor in designing products and packages.

㊼ Fast-food outlets made their appearance in the 1950s and created a demand for new kinds of packaging. The consumer met disposable **single-service packaging** for the first time, while the fast-food outlets demanded the bulk delivery of ready-to-cook food portions in their own special type of packaging. Later, two other factors joined the fast-food outlets boom to influence packaging: increased levels of public health care and a rapidly growing trend toward eating out rather than at home. Today, this market is large enough to form its own sector, sometimes called the HRI (hospital, restaurant, and institutional) market.

㊽ The 1950s also saw the growth of convenience and **prepared food** packages, such as **cake mixes**, **TV dinners**, **boil-in-bag** foods and **gravy preparations**. A rapidly growing technology added petroleum-derived plastics to the package designer's selection of packaging materials.

㊾ The coming-of-age baby boomers were the largest identifiable population segment in the late 1960s, and this was reflected in a major youth orientation in packaging and products. Sexual morality shifted significantly in the 1960s to allow more suggestive and provocative messages. In the 1960s, this was mostly confined to "cheesecake", images of scantily clad women aimed at selling products to men. Its counterpart, "beefcake", did not become common until the more liberated 1980s. Today both tactics are under increasing criticism as inappropriate methods of promoting goods.

㊿ Consumerism and a concern for the environment became important factors at this time for those who watched for future trends.

51 The 1970s and early 1980s brought numerous changes, many of them legislated. **Child-resistance closures** were mandated for some products. Tamper-evident closures were brought in for others. Labeling laws required listing of ingredients. International agreements were signed to phase out the use of ozone-depleting chlorofluorocarbons (CFCs). Standards for the acceptance of new packaging materials were raised.

52 Microwave ovens became a common household feature, and a significant effort went into devising products and packaging specifically for the microwave. A new health awareness meant not only changes in consuming habits and nutritional labeling but also opportunities for entire new food lines. Yogurt became the "in" food. Bottled water became big business.

53 The last decades of the 20th century witnessed rapid change. The population aged, and many social habits changed. Families became smaller. Single-person households became common. The domestic housewife became a relic of the past as both partners in a marriage sought professional careers or higher income levels. For the modern urban dweller, "convenient"

and "fast" became the operative words. Marketers recognized a whole subclass of people who know only how to boil water or turn on the microwave. If it wasn't ready in five minutes, they didn't want it. If it took more than one dish, their interest wandered.

7 Modern Packaging

7.1 Changing Needs and New Roles

54 Looking back, historical changes are understandable and obvious. That all of them have had an impact on the way products are bought, consumed and packaged is also obvious. What is not so obvious is what tomorrow will bring. Yet, *it is to the needs, markets, and conditions of tomorrow that packaging professionals must always turn their attention.*

55 The forces that drove packaging during the Industry Revolution continue to operate today. The consumer society continues to grow and is possibly best described by a 1980s **bumper sticker**, "Born to Shop". We consume goods today at a rate 4 to 5 times greater than we did as recently as 1935. Most of these goods are not essential to survival; they constitute what we may call "the good life".

56 In the second half of the 20th century, the proliferation of goods was so high that packaging was forced into an entirely new role, that of providing the major purchase motivation rather than presenting the goods itself. On a shelf of 10 competing products, all of them similar in performance and quality, the only method of differentiating became the package itself. Marketers aimed at lifestyles, emotional values, subliminal images, features, and advantages beyond the basic product itself—anything that would make a shopper's hand reach for *their* product rather than the competitor's. In some instances, the package has become the product, and occasionally packaging has become entertainment.

57 Globally, the trend toward urbanization continues. Providing increased tonnages of high-quality food to massive **city complexes** at affordable prices is a problem that continues to challenge packagers. A new concern is the removal of the debris generated by a consumer society and the impact that these consumption rates have on the planet's ecology.

58 The makeup, needs, styles, perceptions and wishes of the consuming public are always changing. The packaging professional must be aware of and keep up with these changes or be lost to history.

7.2 Packaging and the Modern Industrial Society

59 The importance of packaging to a modern industrial society is most evident when we examine the food-packaging sector. Food is organic in nature, having an animal or plant source. One characteristic of such organic matter is that, by and large, it has a limited natural biological life. A cut of meat, left to itself, might be unfit for human consumption by the next day. Some animal protein products, such as seafood, can deteriorate within hours.

60 The natural shelf life of plant-based food depends on the species and plant part involved. **Pulpy fruit** portions tend to have a short life span, while seed parts, which in nature have to survive at least till the next growing season, have a longer life. Stalks and leaves separated from the living plant are usually short-lived.

①　In addition to having a limited natural shelf life, most food is geographically and seasonally specific. Thus, potatoes and apples are grown in a few North American geographical regions and harvested during a short maturation period. In a world without packaging, we would need to live at the point of harvest to enjoy these products, and our enjoyment of them would be restricted to the natural biological life span of each.

②　*It is by proper storage, packaging and transport techniques that we are able to deliver fresh potatoes and apples, or the products derived from them, throughout the year and throughout the country.* Potato-whole, canned, powdered, flaked, chipped, frozen, and instant — is available, anytime, anywhere. This ability gives a society great freedom and mobility. Unlike less-developed societies, we are no longer restricted in our choice of where to live, since we are no longer tied to the food-producing ability of an area. Food production becomes more specialized and efficient with the growth of packaging. Crops and animal husbandry are moved to where their production is most economical, without regard to the proximity of a market. *Most important, we are free of the natural cycles of feast and famine that are typical of societies dependent on natural regional food-producing cycles.*

③　Central processing allows value recovery from what would normally be wasted. By-products of the processed-food industry form the basis of other sub-industries (Figure 1.5). Chicken feathers are high in protein and, properly milled and treated, can be fed back to the next generation of chickens. Vegetable waste is fed to cattle or pigs. Bagasse, the waste cane from sugar pressing, is a source of fiber for papermaking. Fish scales are refined to make additives for paints and nail polish.

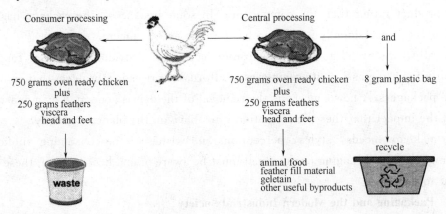

Figure 1.5　By-products collected with central processing can be converted into useful materials

④　The economical manufacture of durable goods also depends on good packaging. A product's cost is directly related to production volume. A facility building 10,000 bicycles per year for local sale could not make bicycles as cheaply as a 3-million-unit-a-year plant intended to capture the national facility. Both would fail in competition against a 100-million-unit world market facility. But for a national or international bicycle producer to succeed, it must be a way of getting the product to a market, which may be half a world away. Again, **sound packaging**, in this case distribution packaging, is a key part of the system.

⑥⑤ Some industries could not exist without an international market. For example, Canada is a manufacturer of irradiation equipment, but the Canadian market (which would account for perhaps one unit every several years) could not possibly support such a manufacturing capability. However, by selling to the world, a manufacturing facility becomes viable. In addition to needing packaging for the irradiation machinery and instrumentation, the sale of irradiation equipment requires the safe packaging and transport of radioactive isotopes, a separate challenge in itself.

7.3 World Packaging

⑥⑥ This discussion has referred to primitive packaging and the evolution of packaging functions. However, humankind's global progress is such that virtually every stage in the development of society and packaging is present somewhere in the world today.

⑥⑦ Thus, a packager in a highly developed country will agonize over choice of package type, hire expensive marketing groups to develop images to entice the targeted buyer and spend lavishly on graphics. In less-developed countries, consumers are happy to have food, regardless of the package. At the extreme, consumers will bring their own packages or will consume food on the spot, just as they did 2,000 years ago.

⑥⑧ *Packagers from the more-developed countries sometimes have difficulty working with less-developed nations, for the simple reason that they fail to understand that their respective packaging priorities are completely different.* Similarly, developing nations trying to sell goods to North American markets cannot understand our preoccupation with package and graphics.

⑥⑨ The significant difference is that packaging plays a different role in a market where rice will sell solely because it is available. In the North American market, the consumer may be confronted by five different companies offering rice in 30 or so variations. If all the rice is good and none is inferior, how does a seller create a preference for his particular rice? How does he differentiate? The package plays a large role in this process.

⑦⓪ The package-intensive developed countries are sometimes criticized for **overpackaging**, and certainly overpackaging does exist. However, North Americans also enjoy the world's cheapest food, requiring only about 11% to 14% of our disposable income. European food costs are about 20% of disposable income, and in the less-developed countries food can take 95% of family income.

⑦① It is simplistic to say that the less-developed countries do not have adequate land to raise enough food, although in some few instances this is true. United Nation's studies have shown that many countries in which hunger exists actually raise enough food for their population. However, without adequate means of preservation, protection and transportation, up to 50% of the food raised never survives for human consumption. Food goes beyond its natural biological life, spoils, is lost, is infested with insects or eaten by rodents, gets wet in the rain, leaks away or goes uneaten for numerous reasons, all of which sound packaging principles can prevent. Furthermore, in a poor economy that can afford no waste, no industries recover secondary value from food by-products.

72　This is a tragic waste. The United Nations maintains staff whose purpose is to increase packaging level and sophistication in less-developed countries. Packaging is perceived to be a weapon against world hunger.

8　Waste Management And Environmental Issues

73　A discussion of packaging today means eventually turning to environmental issues. Packaging is often blamed for a host of ills, and a perception exists in some circles that if only the packaging industry would stop doing something or, conversely, start doing something, all our **landfill** and pollution problems would go away. This, of course, is not true. On the contrary, ample evidence suggests that good packaging reduces waste.

74　Unfortunately, the consumer sees packaging as that part of shopping trip that gets thrown away. Hence, packaging is garbage. No home decorator would dispute the necessity of a paint can or a **caulking tube**. Yet when empty, these, along with other household packages that have fulfilled their function, are suddenly perceived as garbage—unnecessary and a problem.

75　Packaging waste is far less than the typical consumer imagines. In fact, **residential waste** itself is much less than half of what needs to be disposed of. The greater part of what goes into a landfill is construction and **demolition waste** and **industrial waste**. The University of Tennessee provides the following breakdown of total landfill waste:

Residential waste	37.4%	Commercial waste	27.3%
Industrial waste	29.3%	Other sources	6.0%

76　The calculation of what exactly is going into landfills and what is being recycled or otherwise diverted is a significant problem. Individual waste management jurisdictions have various way of measuring the waste stream do not necessarily even agree on what should be measured. For example, Table 1.1 shows, slightly less than one-third of landfill waste is attributable to packaging. Critics point out that these numbers can be deceptive since the Environmental Protection Agency (EPA) does not include construction and demolition waste and some light industrial wastes in its calculation.

Table 1.1　Materials mix by weight in residential solid waste

Material	Packaging	Nonpackaging
Paper	12.7%	19.6%
Wood	4.6%	—
Metal	2.0%	5.7%
Glass	5.7%	0.8%
Plastic	4.1%	5.5%
Other misc.	0.1%	12.1%
Food waste	—	8.1%
Yard waste	—	19.0%
Totals	29.2%	70.8%

77　Most waste-management issues fall under local rather than national jurisdictions. The problem that this poses for industry is that every state or province can pass its own packaging

regulations or mandates, which can take many forms. Examples of past and current regulations in North America include:
- Recycling mandates/laws.
- Material reduction mandates/laws.
- Restrictions on selected materials/package types.
- Material bans or restrictions; for example: heavy metals or poly (vinyl chloride).
- Bans on materials accepted at landfills, such as not accepting as corrugated fiberboard.
- Green labeling requirements/prohibitions.
- Purchasing preference mandates.
- Tax incentives/penalties.
- Deposit laws/advance disposal fees.
- Recycled content requirements.
- Volume-based household garbage removal fees, such as paying by the bag or can.
- Extended producer responsibility/stewardship laws where the producer is responsible for the product and package up to and including proper disposal.

78 Developing packages able to meet dozens of differing waste management requirements would present a formidable challenge.

79 Germany's Packaging Ordinance is generally considered to be the most stringent when discussing collection and recycling of spent packaging. The basic principle of the ordinance is that manufacturers or distributors must take back used packaging for reuse or material recycling. Since this process is difficult for small producers, industry established a comprehensive collection system for qualifying packages that carry a green dot. Because the ordinance allows two ways of collecting used packages, it is described as the "Dual System" (DS or DSD). Most European countries are adopting variations or elements of the DSD system. The successes and problems of the DSD system are monitored globally by both government policymakers and environmentalists.

80 International concern led the International Organization for Standardization (ISO) to the development of the ISO 14000 series of standards. These documents do not set recycling target levels or mandate package types. Rather, ISO 14000 series provides guidance for suitable management policies, auditing methods, environmental labeling practices, corporate environmental performance evaluations, and life cycle analysis (LCA) among other subjects.

81 The guiding principles for developing environmentally acceptable packaging are embodied in the four Rs **hierarchy**:
- Reduce.
- Reuse.
- Recycle.
- Recover.

82 First, all packaging design should use the minimum amount of material consistent with

fulfilling its basic function. This reduction in material use will eliminate any further considerations of reusing, recycling, or recovering other value. Second, where practical, containers or packaging components should be reused. Third, where practical, packaging should be collected and the materials recycled for further use. Finally, before consigning packaging to a landfill, some thought should be given to possibly recovering other value from the waste.

⑧③ Energy is one value that can be recovered from waste, and it is recovered in many parts of the world (Table 1.2). However, *with not-in-my-backyard attitudes and a heavy negative contribution from environmentalists*, **incineration** *has become highly politicized in North America*. It is not likely that this useful technology will be employed in the near future, even though most authorities agree that incineration can make a positive contribution to waste management problems.

Table 1.2 Percent of municipal solid waste incinerated in selected countries

Country	Switzerland	Japan	Sweden	France	United States
Incinerated Waste	74%	66%	50%	35%	15%

⑧④ The incineration of **polyolefin plastics** such as PE and PP (Figure 1.6) is referred to as the cascade model. *PP, for example, is made by joining together into a single structure many molecules of a gas related to propane.* It can be regarded as propane in a solid form. Propane gas can be burned for its energy value directly, or it can be reformed into PP. The plastic can be used as a food container and recycled perhaps to become an office letter tray. When it is finally burned after several useful **reincarnations**, it will still have a large percentage of its original propane energy content. Combustion by-products are water and carbon dioxide.

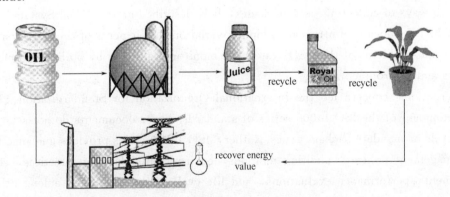

Figure 1.6 The cascade model proposes that polymer monomer gases such as ethylene and propylene can be used to make useful plastics, recycled several times and still have most of their energy value when incinerated.

⑧⑤ Recycling is the R that has caught the imagination and devotion of great parts of the consuming public. Despite its many problems, few public figures would risk anything other than positive commentary. However, many public myths exist about recycling, perhaps the greatest being that placing material in a blue box at your doorstep constitutes recycling. Recycling does not occur until someone uses the material collected.

86 Developing markets for **postconsumer recycled (PCR)** materials is one of the greatest challenges to the packaging industry. Although it is generally expected that the packaging industry will use its own PCR materials, serious impediments remain.

87 For example, the use of PCR materials in immediate contact with food needs to be extensively investigated. In the instance of **pharmaceutical packaging**, such use is simply not allowed. These two markets alone account for significantly more than half of all packaging material use in North American. Another impediment is a guarantee of consistent and reliable supply of the recovered material.

88 A second myth presupposes that since the recycled material is recovered from discards, it should be economical. In most instances, there is no economic advantage to using recycled materials. In many instances, recycled material is more costly, and its use needs to be supported in some way.

89 The cost of landfilling **municipal solid waste** (MSW) is still less than recycling in most areas. The bulk of the recovered material is paper and has low value. Aluminum, the material with the highest value, is only 2% of the collected weight. Revenues generated from the sale of recyclable materials do not always recover collecting and recycling costs. Obviously, someone has to make up the shortfall.

90 The process of recycling cannot ignore market economics. For a short period in the mid-1990s, various material shortages made recycling economically attractive. With excess capacity and lower new material prices in the late 1990s and early 2000s, markets for recycled materials declined and scores of recycling operations closed shop. Many municipalities stopped collecting selected materials because of unfavorable economics. Glass collection, in particular, has been significantly reduced.

91 Environmentalists will maintain that recycling is an issue of the environment, not of economics. This is quite true. However, money expended to recycle a material represents an investment in fuel, water and other resources. When the resource investment to recover a material exceeds the value of the material recovered, then the harm to the environment is greater, not less.

92 The process of collecting and regenerating a packaging material for further use is a complex one for most materials, and it is one that requires significant investment in sophisticated equipment. Identifying, sorting and recycling of metals are comparatively easy. While glass is apparently readily identifiable, individual glass compositions as well as different colors make it difficult to get uncontaminated **feedstock**.

93 Paper products vary considerably in their fiber makeup and quality. However, reasonably efficient sorting systems are in place and some cross-contamination does not present serious problems. On the negative side, paper fiber quality deteriorates with every recycling, and so paper cannot be recycled indefinitely. Paper's low material value requires that collection costs also be low.

94 Plastic materials pose a number of serious recycling problems. Many different plastics

are used in packaging, and many are not mutually compatible. Identifying and sorting the many plastic materials by appearance alone is beyond the abilities of even a conscientious consumer. The plastic industry developed a code for identifying the six most commonly used packaging plastics; it includes an "other" selection as a seventh code (Figure 1.7). The code identifies only the general plastic family: Significant variations can occur within each family.

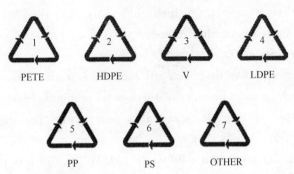

Figure 1.7 A code identifies the main packaging plastic families (PETE is usually abbreviated PET and V is usually abbreviated PVC. Less commonly used plastics and mixed-plastics constructions are classified as "other")

95 Other myths propose that one or another of the many materials used for packaging is more environmentally friendly. There is no magic material. Packagers select appropriate materials to contain, protect/preserve, efficiently transport and sell the product in the most cost-effective manner. **Laminate** constructions, a favorite target, are designed to combine the best characteristics of several different materials, offering properties not available with any single material. Their design is based on the most cost-effective and material-efficient method of achieving the needed result. They are, in fact, environmentally friendly.

96 The dilemma of packaging and waste management is a complex one, not amenable to simplistic solutions. Every packaging professional should be involved in educating the public to the real and vital benefits of packaging. Inflexible environmentalism needs to be replaced with a keen awareness of packaging's legitimate role and the difficult decisions that must be made. In the final analysis, the consumer makes the choices and will direct the course of the industry. The consumer deserves to have the correct information to make those choices.

9 The Modern Packaging Industry

97 Drawing clear-cut boundaries around the packaging industry is difficult. Obviously, those actually manufacturing the physical package (cans, bottles, wraps) are part of the packaging industry. Their function is to take various raw materials and convert them (hence the general classification of this part of the industry as **"converters"**) into useful packaging materials or packages. Viewed from this perspective, packaging becomes a material application science.

98 In most instances, the company forming the physical package will also print or decorate

the package. Thus, part of the printing industry and all its attendant suppliers is also viewed as part of the packaging industry.

99 Many user-sector companies, the firms that package products, are also regarded as part of the packaging industry. Package users can be divided into a number of categories (Figure 1.8) and each of these can be further subdivided. Each subsector has its own unique package design requirements.

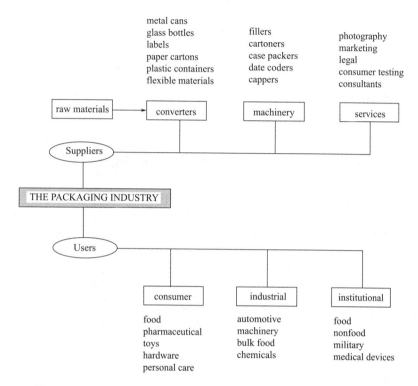

Figure 1.8 The packaging industry can be divided into those that use packaging for their products and those that supply to these users

100 The **supplier** sector, manufacturers of machines for the user sector and the suppliers of ancillary services, such as marketing, consumer testing and **graphic design**, are also important sectors of the packaging industry.

101 Serving these industry sectors is a large number of professional associations. Some are broad-based general-interest associations that cover the entire gamut of packaging concerns. Prominent among these are the Institute of Packaging Professionals (**IoPP**) and the Packaging Association of Canada (PAC). Other associations are more specialized in their packaging focus. Examples are the Packaging Machinery Manufacturers Institute (**PMMI**), the Flexible Packaging Association (FPA), and the Plastic Bottle Institute.

102 International packaging activities are often coordinated by the World Packaging Organization (WPO). The International Organization for Standards (ISO) is an international standards body that issues many international standards affecting packaging.

103 Some associations are not specific to packaging, but their activities are important to the packaging industry. The American Society for Testing and Materials (**ASTM**) and the

Technical Association of the Pulp and Paper Industry (**TAPPI**), for example, supply the bulk of material and package-testing procedures. The International Safe Transit Association (**ISTA**) is concerned with loss prevention and the safe transit of goods. The decisions of various retailing associations can also have a major impact on packaging.

[04] The industry is served by a large number of **trade journals**. Like associations, trade journals can be either broadly based or more specifically focused on a particular package type, material, process, or product category.

[05] Lastly, the modern packaging industry is highly regulated. *Added to the basic complexity of the industry are the jurisdictional complexities of the political world.* Various aspects of packaging are governed by authorities ranging from the federal to the local, and it is not always entirely clear which authority has jurisdiction over a particular issue. Ensuring that all legal requirements are met can be an especially challenging task.

(7283 words)

New Words and Expressions

aseptic [ə'septik] *adj.* 无菌的
sack [sæk] *n.* 麻袋、包 *v.* 掠夺
amphora ['æmfərə] *n.* 双耳细颈椭圆土罐
woodcut ['wudkʌt] *n.* 木刻画
flexography [flek'sɔgrəfi] *n.* 苯胺印刷（术）、柔性印刷
keg [keg] *n.* 小桶
cask [kæsk] *n.* （尤指盛酒精饮料的）桶
label ['leibəl] *n.* 标签
typography [tai'pɔgrəfi] *n.* 印刷样式、排印
cellulose ['seljə,ləus] *n.* 纤维素
cellophane ['selə,fein] *n.* 玻璃纸
polyethylene (PE) [,pɔli'eθili:n] *n.* 聚乙烯
polypropylene (PP) [,pɔli'prəupili:n] *n.* 聚丙烯
thermoplastic [,θə:mə'plæstik] *n.* 热塑性塑料
demographics [,demə'græfiks] *n.* 人口统计学
cheesecake ['tʃi:z,keik] *n.* 乳酪蛋糕、半裸体的女人照片
beefcake ['bi:fkeik] *n.* 牛肉蛋糕、健美男子
overpackaging ['əuvə'pækədʒiŋ] *n.* 过度包装
landfill ['lænd,fil] *n.* 垃圾掩埋法（场）
hierarchy ['haiərɑ:ki] *n.* 层次（级）、等级
incineration [in,sinə'reiʃən] *n.* 焚化、焚烧
reincarnation [,ri:inkɑ:'neiʃən] *n.* 再生
cascade [kæ'skeid] *n.* 小瀑布、瀑布状物
feedstock ['fi:dstɔk] *n.* 原料、给料
laminate ['læmə,nei] *n.* 层压（材料）、叠压
converter [kən'və:tə] *n.* 加工机械

supplier [səˈplaiə] n. 供应商、供货方
filler [ˈfilə] n. 充填机（对干料而言）、灌装机（对液体类物料而言）
cartoner [ˈkɑːtənə] n. 装盒机
capper [ˈkæpə] n. 压盖机、封口机
tamper-evident closure 显窃启盖
foil/fiber composite can 铝箔/纤维复合罐
environmentally acceptable packaging 环境可接受包装
wood pulp 木浆
refillable aluminum can 可再装铝罐
fire-bearer 炉箅托架、火炉子
dry product 干燥物品、固体类物料
glass bead 玻璃珠
glass blowpipe 玻璃器皿吹管
a gob of （玻璃）球坯、滴料
glass beaker 玻璃烧杯
wooden barrel 木桶
arts and crafts 工艺
the Dark Ages 欧洲中世纪、黑暗时代
papyrus reed 纸莎草芦苇
sheet material 片材、板材
custom business 定制业务、定制生意
mass production 大规模生产、批量生产
consumer commodity 消费品
natural biological life 自然生物寿命（周期）
bulk packaging 散装包装、裸装的、大包装
individual packaging 单独包装、小包装
branding iron 烙铁、火印
brand mark 印记、商标
unit package 单元化包装（件）、组合包装
brand name 商标、品牌
art nouveau （流行于19世纪末的）新艺术
phenol formaldehyde plastic (Bakelite) 苯酚甲醛塑料
casting process 流延（平挤）过程（工艺）
single-service packaging 一次性包装
prepared food 预加工食品
cake mix 做蛋糕用的配料
TV dinner （食前加温即可的）冷冻快餐
boil-in-bag 蒸煮袋
gravy preparation 肉汁配制品
child-resistance closure 儿童安全盖

bumper sticker　（汽车上的）保险杠贴
city complex　城市综合体
pulpy fruit　软果
sound packaging　良好包装、完好包装
caulking tube　堵缝管
residential waste　住宅垃圾、生活垃圾
demolition waste　工地废渣料
industrial waste　工业废料
polyolefin plastic　聚烯烃塑料
postconsumer recycled（PCR）　消费后再循环的
pharmaceutical packaging　药品包装
municipal solid waste（MSW）　城市固体垃圾
graphic design　图形设计、平面设计
IoPP　包装专业技术人员协会
PMMI　包装机械制造者协会
case packer　装箱机
date coder　日期打码机
ASTM　美国试验与材料协会
TAPPI　纸浆与造纸工业技术协会
ISTA　国际安全运输协会
trade journal　行业杂志

Notes

1. ...*out of this need came the first "package".* （Para. 11）……出于需要，就出现了第一个"包装"。这是一个倒装句子。

2. *Manufacturing was strictly a **custom business**, and what we have called packages to this point was personally crafted, as were most goods.* （Para. 27）严格说来，制造是一种定制业务。我们称作为包装的东西就像大多数商品一样，此刻只是个人的手工制作。what 引导主语从句，表示"……东西"，as 后边是倒装句。

3. ***Bulk packaging** was the rule, with the barrel being the workhorse of the packaging industry.* （Para. 32）大包装当时是惯用的做法，其中，桶是当时包装工业的主要容器。with 引导了独立主格结构。

4. ***Individual packaging** was of little importance until the Industrial Revolution spurred the growth of cities.* （Para. 32）直到工业革命推动了城市发展，小包装才体现出较大的重要性。until 是连词，用于否定句意思为："直到……才"，"在……以前"。

5. *The birth rate after the Second World War and into the 1950s was so imposing that it earned its own name: the baby boom.* （Para. 46）二战结束后，进入20世纪50年代的人口出生率是如此的令人印象深刻，以至于赢得了一个名词：婴儿潮。"婴儿潮时代"（Babyboomer）是美国家喻户晓的名词，是指二战之后的1946～1964年间，美国共有7900万婴儿出生，约占美国总人口的三分之一。

6. *Yet, it is to the needs, markets, and conditions of tomorrow that packaging profession-

als must always turn their attention. (Para. 54) 然而，包装专业人士必须要把他们自己的注意力转向明天的需求、市场及其环境上。这是一个强调句型，强调 to 后面的东西，"turn…to"意指"转向……"。

7. *It is by proper storage, packaging and transport techniques that we are able to deliver fresh potatoes and apples, or the products derived from them, throughout the year and throughout the country.*（Para. 62）正是通过适当的仓储、包装和运输方法，我们才能一年到头、全国各地输送新鲜的土豆和苹果以及来源于它们的制品。It is…that…是强调句型。

8. *Most important, we are free of the natural cycles of feast and famine that are typical of societies dependent on natural regional food-producing cycles.*（Para. 62）最重要的是，我们摆脱了享受美食和饥荒的自然循环，而这种现象在依赖于自然的、区域性的食品生产周期的社会里是具有代表性的。"feast and famine"意指"盛宴和饥荒"。

9. *Packagers from the more-developed countries sometimes have difficulty working with less-developed nations, for the simple reason that they fail to understand that their respective packaging priorities are completely different.*（Para. 68）来自较发达国家的包装人员有时与欠发达国家的同仁难以合作，其简单的原因是他们不理解各自的包装关注点是完全不同的。difficulty 与 working 之间省略了 in；reason 引导后面的同位语结构。

10. *However, with not-in-my-backyard attitudes and a heavy negative contribution from environmentalists,* **incineration** *has become highly politicized in North America.*（Para. 83）然而，因为"远离我家后院"的态度及来自环境保护主义者大的负面作用，在北美，焚烧处理已经高度政治化。"not-in-my-backyard"曾是美国环保运动的著名口号。

11. *PP, for example, is made by joining together into a single structure many molecules of a gas related to propane.*（Para. 84）例如，PP 的制得是通过把与丙烷相关的气体的多个分子连成一单个结构。"join A into B"意指"连接 A 成为 B"，本句将"连接"的对象即"A"置于后面，遵从了英语语言习惯和语法的要求。

12. *Added to the basic complexity of the industry are the jurisdictional complexities of the political world.*（Para. 105）包装工业基本的复杂性增加了政界许多司法的复杂性。这是一个倒装句。"be added to…"意指"增加到……上"、"被添加到……"。

Overview Questions

1. Although packaging has existed from primitive times, the Industrial Revolution is generally taken as the time when modern packaging was born. What were the changes that lend validity to this statement?
2. Why is food loss so high in less-developed countries?
3. Why is the United Nations so interested in packaging?
4. The four Rs are used as the guiding principles for managing the waste problem. What are the four Rs, in the correct order?
5. What are the two major divisions within the packaging industry? Name two subcategories in each major division.

Lesson 2 Basic Functions of Packaging

1　Introduction

① In lesson 1, "A Historical Perspective on Packaging," the functions of a package were given as
- Containment.
- Protection/Preservation.
- Transport.
- Information/Sales.

② When discussing packaging functions, keep in mind the different packaging levels:

Primary package	The first wrap or containment of the product that directly holds the product for sale.
Secondary package	A wrap or containment of the primary package.
Distribution package (shipper)	A wrap or containment whose prime purpose is to protect the product during distribution and to provide for efficient handling.
Unit load	A number of distribution packages bound together and unitized into a single entity for purposes of mechanical handling, storage, and shipping.

③ Figure 2.1 illustrates some of these levels. In addition, packages are often defined by their intended destination:

Consumer package	A package that will ultimately reach the consumer as a unit of sale from a merchandising outlet.
Industrial package	A package for delivering goods from manufacturer to manufacturer. Industrial packaging usually, but not always, contains goods or materials for further processing.

④ The basic packaging functions have different degrees of importance, depending on the particular packaging level and intended destination. It is common for several packaging levels to contribute to a single function.

⑤ The primary package for a breakfast **cereal** is the inner undecorated bag. Its main function is to contain and preserve the product, and to a lesser extent, to protect it. The secondary package, a paperboard carton, provides physical protection, informs the consumer and motivates the purchase decision. Twelve cartons are packed

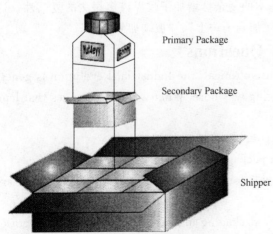

Figure 2.1　Packaging can have many levels. All levels of the system must work together

into a corrugated shipping container to protect the product and to facilitate distribution. The information printed on the corrugated shipper primarily identifies the product for distribution purposes. Finally, corrugated shippers are assembled into a unit load, whose primary purpose is to facilitate transport and distribution.

⑥ In some instances, a package will be required to assume all the functions. The primary package for a **power tool** may be strong enough to protect the product and withstand the rigors of shipping. This single package may feature all the necessary information to inform and motivate the consumer.

⑦ Typically, the inform/sell function plays a less significant role in industrial packaging.

2 The Containment Function

⑧ The first step in preparing a package design is to consider the nature of the product and the kind of packaging needed to contain the product. These considerations include

- The product's physical form:

mobile fluid	viscous fluid	solid/fluid mixture
gas/fluid mixture	**granular material**	paste
free-flowing	non-free-flowing powder	solid unit
discrete items	multicomponent mix	

- The product's nature:

corrosive	corrodible	flammable
volatile	perishable	fragile
aseptic	toxic	abrasive
odorous	subject to odor transfer	easily marked
sticky	**hygroscopic**	under pressure
irregular in shape		

⑨ Throughout this discussion, we will examine the characteristics of various packaging materials and how their qualities influence effective containment packaging design.

3 The Protection/Preservation Function

⑩ In the context of this discussion, "protection" refers to the prevention of physical damage, while "preservation" refers to stopping or inhibiting chemical and biological change.

⑪ To provide physical protection, specifics on what will cause loss of value (damage) must be known. Specifics means knowing not only the general condition, but also a **quantified** measure of the level of that condition at which unacceptable damage starts to occur (Table 2.1).

Table 2.1 Examples of protective packaging problems and concerns

Condition	Quantification or Design Requirement
Vibration	Determine resonant frequencies
Mechanical shock	Determine fragility factor(drop height)
Abrasion	Eliminate or isolate relative movement

续表

Condition	Quantification or Design Requirement
Deformation	Determine safe compressive load
Temperature	Determine **critical values**
Relative humidity	Determine critical values
Water	Design liquid barrier
Tampering	Design appropriate systems

02　The preservation function most often refers to the extension of food **shelf life** beyond the product's natural life or the maintenance of **sterility** in food or medical products. Like the protective function, the preservation function needs to be defined and quantified (Table 2.2).

Table 2.2　Typical preservation packaging problems and concerns

Condition	Quantification or Design Requirement
Oxygen	Determine required barrier level
Carbon dioxide	Determine required barrier level
Other **volatiles**	Determine nature and barrier level
Light	Design **opaque** package
Spoilage	Determine nature/chemistry
Incompatibility	Determine material incompatibilities
Loss of sterility	Determine mechanism
Biological deterioration	Determine nature
Deterioration over time	Determine required shelf life

4　Food Preservation

4.1　The Nature of Food

03　Food is derived from animal or vegetable sources. Its organic nature makes it an unstable commodity in its natural form. Left on their own, foodstuffs can deteriorate rapidly. sometimes becoming unfit for human consumption within hours. Various means can increase the natural shelf life of foods, thus reducing dependence on season and location. To understand how the natural life of foodstuffs is prolonged, it is necessary to understand how food products **deteriorate.** Food spoilage can occur by three means.

04　"*Internal biological deterioration*" describes biological functions that continue even though the food has been harvested. Fruits continue to ripen and vegetables continue to respire. Fresh meat exhibits many of the processes associated with living tissue. For example, **myoglobin**, which gives meat its red color, continues to interact with atmospheric oxygen.

05　In some instances, internal biological factors are used to advantage. Fruit, for example, is often picked green or in a firm state; final ripening is a controlled process allowed to take place on the way to the market. Beyond a certain point, however, all biological activity will lead to spoilage and loss of product.

06　"*External biological deterioration*" refers to the action of **microorganisms.** What is food to us is also food to a host of other organisms. Molds, bacteria and yeasts are present in most foods. Often they are harmless or even beneficial. In other instances, they can be deadly.

17 **Abiotic deterioration** describes those changes that are chemical or physical in nature and that are not dependent on a biological agent. For example, atmospheric oxygen will chemically react with (oxidize) many substances. Vitamin C is no longer a useful nutrient once oxidized. Oxidized oils and fats have a **rancid taste.**

18 What is generally described as "taste" more correctly refers only to the sweet, sour, salty, and bitter sensations that we detect with the taste sensors located on our tongue. Other mouth sensations are texture or "mouthfeel", temperature and chemical burning such as the effect of pepper.

19 We are also capable of detecting complex volatile substances, variously known as **essential oils** or **"sensory active agents"**. We detect essential oils when minute quantities, in gaseous form, pass over sensors located in our nasal passages. Our sense of smell is highly developed, and We are capable of differentiating thousands of smells or aromas compared to the four tastes we detect in our mouths.

20 What we perceive as a food product's flavor is a combination of what we detect with our sense of taste combined with what we detect with our sense of smell. Because essential oils are volatile, they are easily lost through evaporation or oxidation. Great care must be taken to ensure that essential oils characteristic of a product are not lost. Preservation of essential oils retains the food's full flavor at retail. **Off flavors** can also permeate into foods. Some products are virtual blotters for any **stray volatiles** in the atmosphere. Absorption of even part-per-million quantities of undesirable volatiles can impact off-flavors to some food products. Contamination from improperly dried or cured inks and adhesives is a common example of this problem. Volatiles can permeate packaging materials and making the problem of contamination or isolation even more difficult.

21 The above discussion addresses the importance of controlling the gain or loss of essential oils in food products. Similarly, it is equally important to retain essential oils in the many nonfood products whose appeal lies partly or entirely on the smells associated with that product. Perfumes, **colognes** and room fresheners are essentially blends of pure essential oils. Most health and beauty aids such as cosmetics, soaps, shampoos and toothpastes also contain essential oils in their formulations.

22 Water vapor is similar to an essential oil in that it readily permeates many packaging materials. Moisture loss or gain can be a deteriorating factor, depending on the nature of the food. A snack food loses quality as it gains moisture while a cake loses quality as it loses moisture.

23 The creation of **high-barrier packaging** systems is partly in response to the need for packaging that will either hold desirable gases and volatiles in the package or prevent undesirable volatiles from entering the package.

24 Temperature can promote undesirable changes that are abiotic in nature. The most common of these are the **irreversible** changes encountered when some fruits are frozen. The for-

mation of ice crystals punctures the fruit's fragile cell walls, and the fruit loses its desirable character.

㉕ *Specific food categories have their own characteristics with meats, fish, and produce being important categories.*

4.2 Meat products

㉖ Meats are an ideal medium for microorganisms because they contain all the necessary nutrients to sustain growth. In addition to biological action, fatty tissue is susceptible to oxidation, and the entire mass can lose water.

㉗ Reduced temperature **retards** microorganism activity, slows evaporation and slows chemical reactions such as those associated with oxidation. At 0 ℃ (32 ℉) and 85% relative humidity (R. H.), beef **carcasses** keep for about 21 days. Pork and lamb keep about 4 days. Beef retail cuts on open display at 5 ℃ (41 ℉) keep for 1 or 2 days. Proper packaging and storage of retail cuts can increase this to 10 days.

㉘ An important marketing factor with red meat is the bright red color associated with freshness. This color results from different oxidation states of myoglobin. Fresh-cut beef tends toward a purplish red color that comes from a slightly **oxygen-deficient** state. Exposure to controlled amounts of oxygen produces the bright cherry red of **oxymyoglobin** so desired by the consumer in North America (Consumers in most other countries do not have an aversion to beef that is not bright red). In fact, neither state is wrong. Since North American consumers prefer the bright red appearance, though, Packagers use plastic films that allow the correct amount of oxygen into the package to maintain the bright red appearance.

4.3 Fish

㉙ The preservation of fish is a difficult challenge because of three main factors:
- **Psychrophilic** bacteria may be present.
- Many fish oils are unsaturated and are easily oxidized.
- Typical fish proteins are not as stable as red meat proteins.

㉚ Chilling does not affect the activity of psychrophilic bacteria to the extent it does **mesophyllic** types, so the "keeping quality" of fresh fish is limited. Frozen fish is typically kept at much lower temperatures (−30 ℃/−22 ℉) than other frozen foods in order to ensure the control of psychrophilic bacteria.

4.4 Produce

㉛ Harvested fruits and vegetables continue to respire and mature. Furthermore, they contain large amounts of water and will wither if water loss is excessive. No two fruits or vegetables are alike, and the rate at which biological and abiotic changes occur varies with the species. Peas, green beans, and leafy vegetables have high respiration rates compared with those of apples oranges, and pears. Potatoes, turnips and pumpkins respire slowly and are easy to store. Moisture loss is more rapid with lettuce than with a turnip because of the large available surface area.

㉜ Few fruits will ripen below 5 ℃ (41 ℉). **As a rule of thumb**, a 10 ℃ (18 ℉) temperature drop will typically increase shelf life by a factor of three (providing freezing is avoided).

Freezing of many produce items will damage cell structure, and breakdown is very rapid after **thawing**.

53 The growth, maturation and ripening of a fruit or vegetable is controlled by various **hormones** and gases. Increasing the amount of carbon dioxide while reducing the amount of oxygen slows the respiration rates. However, some oxygen must always be present to keep the fruit alive. These techniques are used in **modified atmosphere packaging**.

54 **Ethylene** gas, produced by plant tissues, is associated with the ripening of many fruits, and its control is effectively used to retard or accelerate the ripening process.

55 Bananas are particularly sensitive to the atmosphere around them. However, they can remain in a mature but green state for up to six months in atmospheres of 92% nitrogen, 5% oxygen, 3% carbon dioxide and no ethylene. The bananas will ripen normally when transferred to "ripening rooms" containing a few parts per million of ethylene.

56 As the above brief discussion illustrates, atmosphere and temperature control are key requirements for extending the shelf life of fresh produce. Packaging for these products must be tailored to the individual product's needs, and **trade-offs** are necessary. The ideal humidity for most produce is about 90%. At these levels bacterial and fungal growth is greatly encouraged. Furthermore, sealed plastic bags are subject to condensation and wetting, which will only **aggravate** the problem. The compromise seen for many produce items is a **perforated** or **vented** plastic wrap. This allows respiration while providing for some containment and restraint to the loss of moisture. Another option is to select packaging films with high **gas-transmission rates**.

57 Some recently developed plastic films have excellent moisture barrier and very low oxygen barrier. One application of this film is for precut salad bags. *The high moisture barrier keeps bagged produce from drying out, while the good supply of oxygen allows the produce to respire naturally.* The shelf life of precut salads packaged with this material is about ten days.

4.5 Barrier Packaging

58 We have noted that the movement of gases into or out of a package can lead to undesirable changes in the product. An important factor in the preservation of products that contain gaseous or volatile components or that are susceptible to change through the action of such components is the ability to control the movement of these gases and volatiles (Figure 2.2).

59 Stopping the movement of a gas requires barrier packaging. This packaging construction either retains desirable gases and volatiles inside the package or prevents undesirable gases and volatiles from entering the package. Often, barrier packaging must address both capabilities. Of the materials a packager can choose from, only glass and metal provide absolute barriers to all gases and volatiles. While glass and metal are superior in this property, they have associated disadvantages, and packagers frequently seek alternatives with plastic materials. However, all plastics have a measurable **permeability**. The actual permeability varies widely depending on the plastic selected and the nature of the permeant gas or volatile. It is important to understand both the nature of the permeant and the properties of the candidate

plastics. The term "high barrier" plastic is a relative, nonspecific term and should not be taken to mean "absolute" barrier.

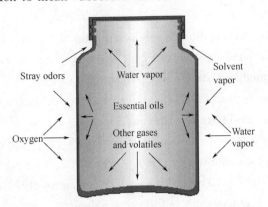

Figure 2.2 A barrier packaging material is one that slows down or stops the movement of selected gaseous substances into or out of a package

④⓪ Barrier packaging can harm some products. Fresh produce, for example, continues to respire after harvesting and would shortly consume all the oxygen in an oxygen-barrier package. This would lead to reduced shelf life. Plastic bags for produce commonly have vent holes punched in them to allow for a free exchange of atmospheric gases.

4.6 Microorganisms

④① A large part of food preservation depends on the control of microorganisms. These can be present in various forms. Bacteria or microbes are **unicellular microscopic** organisms that reproduce by splitting into two identical cells (**binary fission**). Bacteria grown **exponentially** and can divide as fast as every 20 minutes. Certain bacterial species can form **spores** that are highly resistant to killing.

④② Molds or fungi are **multicellular** and unicellular plantlike organisms. Neither is capable of producing **chlorophyll** or **carbohydrates**. Instead, they depend on outside sources for nutrients. *Molds form filamentous branching growths called mycelia and reproduce by spores. Yeasts are similar organisms that reproduce by budding. The propagation and spread of molds and yeasts is typically slower than for bacteria because of the reproduction method.*

④③ Typical of any living entity, each microorganism type has a preferred environment in which to exist and **propagate** and other environments under which it will not. By manipulating the four principal environmental factors that regulate microorganism growth-temperature, moisture, acidity (pH), and nutrient source-microorganisms can be controlled or eliminated. Microorganisms are often classified by their preferred reproduction environment, the most important being the following:

Mesophyllic	Prefer ambient conditions, 20～45 ℃ (68～113 ℉)
Psychrophilic	Prefer cool conditions, 10～25 ℃ (60～77 ℉)
Thermophilic	tolerate heat; will propagate at 30 to 75 ℃ (86～167 ℉)
Aerobic	need oxygen to propagate
Anaerobic	propagate only in the absence of oxygen

④④ Some microorganisms act only on the food. They or their by-products may change the nature of the food to either its benefit or its detriment, but They do little harm when ingested. **Pathogenic** organisms, on the other hand, cause sickness or death. **Pathogens** fall into basic classes:

- Those that produce harmful toxins as by-products in the food they infest.
- Those that infest the food and then grow in the human body to produce illnesses.

45 Six basic methods can extend the normal biological shelf life of food. The methods are used alone or in combination. They are:
- Reduced temperatures.
- Thermal processing.
- Water reduction.
- Chemical preservation.
- Modified atmospheres.
- Irradiation.

46 Each method can slow the natural biological maturation and spoilage of a food product, reduce biological activity or inhibit the chemical activity that leads to abiotic spoilage. Each method requires its own unique blend of packaging materials and technology.

① **Reduced Temperature and Freezing**

47 Reducing temperatures below the ambient temperature has many beneficial effects that will lead to a longer shelf life. Doing so:
- Slows chemical activity.
- Slows loss of volatiles.
- Reduces or stops biological activity.

48 While chilling a food product will increase shelf life, actual freezing provides the greatest benefits. Bacteria and molds stop developing at about $-8℃$ ($-18℉$), and by $-18℃$ ($0℉$) chemical and microorganism activity stops for most practical purposes. Freezing kills some microorganisms, but not to the extent of commercial usefulness.

49 Ice crystal formation is greatest between 0 and $-5℃$ (32 to 23 ℉). Ice crystals can pierce cell walls, destroying the texture of many fruits and vegetables. Rapid freezing reduces this damage.

50 Freezer conditions will cause ice to **sublimate**, and serious food **dehydration**, commonly referred to as **freezer burn**, will occur. Snug, good **moisture-barrier packaging** with a minimum of free air space will reduce freezer dehydration. Complete filling of the package is desirable because ice will sublimate inside the package, dehydrating the product and leaving ice pieces in the voids.

51 Frozen food packages materials must remain flexible at freezer temperatures, provide a good moisture barrier and conform closely to the product. When paperboard is used as part of the package, it should be heavily waxed or coated with polyethylene to give protection against the inevitable moisture present in the freezing process.

52 Poultry packaging in high-barrier **poly（vinylidene chloride）(saran) bags** is an excellent example of an ideal freezer pack. Prepared birds, placed into bags, pass through a vacuum machine that draws the bag around the bird like a second skin. The tight barrier prevents water loss and freezer burn for extended periods, as well as preventing passage of oxygen that would oxidize fats and oils.

② **Thermal Processing**

53　Heat can destroy microorganisms. The degree of treatment depends on the:
- Nature of the microorganism to be destroyed.
- **Acidity** (pH) of the food.
- Physical nature of the food.
- Heat tolerance of the food.
- Container type and dimensions.

54　In many instances, it is not necessary to kill all microorganisms. **Pasteurization**, a mild heat treatment of 60 to 70℃ (140 to 150 ℉), kill most, but not all, microorganisms present. Pasteurization is used when:
- More severe heating would harm the product.
- Dangerous organisms are not very heat resistant (such as some yeasts).
- Surviving organisms can be controlled by other means.
- Surviving organisms do not pose a health threat.

55　"Hot filling" refers to product filling at elevated temperatures up to 100℃ (212 ℉). Hot filling is used to maintain sterility in products such as jams, **syrups** and juices.

56　Some products can tolerate high temperatures for short time periods. Ultra-high temperature (UHT) processing of milk and fruit juices uses temperatures in the range of 135 to 150℃ (275 to 302 ℉), but for a few seconds or less. The high temperature is enough to kill most **pathogens**. UHT is the basis of most flexible aseptic drink packaging. The term "aseptic" as applied to packaging refers to any system wherein the product and container are sterilized separately and then combined and sealed under aseptic conditions. Metal cans were sterilized and filled with puddings, sauces, and soups in the 1940s. In the 1970s aseptic packaging was adapted to institutional **bag-in-box** systems.

57　**Sterilizing** package and product separately eliminates the need for the elevated temperatures and pressures used in conventional canning methods. Eliminating the need for extreme sterilizing conditions allows aseptic packaging materials to have lower physical strengths and lower temperature tolerance. *Commercial systems, such as Tetra Pak, Combibloc, and Bosch, use hydrogen peroxide to sterilize simple paper, foil and polyethylene laminates, and then fill the formed package with UHT-treated product.*

58　Several aseptic systems use the heat of forming plastic as a "free" sterilant. **Thermoformed** plastic containers can be kept sterile until after they are filled and sealed. Sterile solutions are filled into blow-molded plastic bottles at the molding machine as a guarantee of sterility.

59　Unlike aseptic packaging, normal canning maintains only **nominal cleanliness** in the food and the container. After the food has been sealed in the container, it is subjected to temperatures high enough to kill pathogens and achieve **commercial sterility**. Temperatures of 110 to 130℃ (230 to 265 ℉) are typical. The actual cook time depends on many factors, calculated in advance to ensure commercial sterility. One of the most important factors is the rate of heat penetration to the farthest and most insulated portion of the product, usually the container's geometric center.

⑩ Sealed cans are an oxygenless environment. At pH levels above 4.5, conditions are **conducive** to the growth of **clostridium botulinum**, a particularly dangerous anaerobic bacterium that produces heat-resistant toxins. Generally, the less acid the food, the longer the cook times needed to ensure destruction of clostridium botulinum. Foods with acidities high enough to prevent harmful pathogens from propagating can be heat-processed by immersion in boiling water.

㉑ Most canned-food cooking takes place in large pressure cookers, or **retorts**, that raise temperatures considerably above the atmospheric boiling point. Keeping food at these temperatures for a long time results in overcooking and gives some foods their "canned" taste or texture. Canning is not successful for many foods because the cooking cycle would produce objectionable changes in taste or texture.

㉒ Cans are subjected to alternating positive and negative pressure effects as they are heated and cooed. They suffer mechanical abuse, too. The rigorous temperature and pressure conditions needed to achieve commercial sterility restricted retorting to rigid metal cans and glass jars for many years. As flexible materials became more heat resistant and stronger, other constructions became feasible alternatives to rigid containers.

㉓ The **retortable pouch** is a laminate of polyester (for toughness), foil (for an oxygen barrier) and a heat-sealable **polyolefin**. Retortable pouch material is shipped in roll form, creating significant transport and storage savings. Since the pouch can be as little as 12 mm (1/2 inch) thick, thermal processing time can be reduced, thus improving food texture and nutritional qualities. There are also attractive implications for waste disposal. Despite these advantages, North Americans have not readily accepted retort pouching. The largest customer is the military.

③ Water Reduction

㉔ Drying is an old and well-established method of preserving food. The essential feature of drying is that moisture content is reduced below that required for the support of microorganisms. An added advantage is reduced bulk and reduction of other chemical activity. Available moisture can be reduced by simple heat drying or, less obviously, by the addition of salt or sugar. Concentrated salt and sugar solutions tie up free water and make it unavailable to microorganisms. Jams and **marmalades** having high sugar contents do not require refrigeration for this reason.

㉕ In addition to managing moisture content as a method for controlling microorganisms, moisture control is important for keeping many foods at their most desirable moisture levels.

㉖ Most foods are hygroscopic and exist in a state of **equilibrium** with the relative humidity in the immediate atmosphere. If a food is sealed in a closed container, the food will either gain or lose moisture until an equilibrium moisture content in the air space is reached.

㉗ **Equilibrium relative humidity** (E. R. H.) is the atmospheric humidity condition under which a food will neither gain nor lose moisture to the air. This value is often expressed as A_w, the water activity. A food with an A_w of 0.5 is at an equilibrium relative humidity of

50%. Table 2.3 lists the moisture content and the desired E. R. H. for some common foods.

Table 2.3 Typical moisture content and E. R. H ranges

Product	Typical Moisture/%	E. R. H.
Potato chips, instant coffee	3% or less	10% to 20%
Crackers, breakfast cereals	3% to 7%	20% to 30%
Cereal grains, nuts, dried fruit	7% to 20%	30% to 60%
Salt		75%
Sugar		85%

68 Ambient R. H. ranges from very low to very high during the course of a year, and a food's moisture content will be changing continuously as it adjusts to the current R. H.. However, the best mouthfeel for many foods is at a specific moisture content.

69 The A_w for sugar is 0.85, which explains why we rarely have problems with sugar caking. Salt is somewhat lower at 0.75 and does take up moisture on the most humid days. Both would present problems at conditions of 90% humidity.

70 Knowledge of a food's E. R. H. or A_w provides a good indication of the package required by the food. Very low-E. R. H. foods are hygroscopic and will draw available moisture from the air. These foods require a barrier package that will not permit the entry of atmospheric moisture.

71 Dried foods such as **potato chips** and **instant coffee** have a moisture content of 3% of lower and an E. R. H. of 10% to 20%. Since ambient relative humidity is rarely this low, there is a great tendency for these products to take up water. They require packaging materials with high moisture-barrier properties. Potato chips are also rich in oil (about 30%), so that they also need a high oxygen barrier. In-package **desiccants** and **oxygen scavengers** are sometimes used to increase the shelf life of very sensitive products.

72 Dried foods with E. R. H. values of 20% to 30% have less stringent moisture-barrier requirements and are easier to package. Depending on the food, oxygen or other barriers may still be needed. Many crackers, biscuits, and breakfast cereals fall into this category.

73 Foods with an E. R. H. of 30% to 60% can often be stored for long periods with little or no barrier packaging since their E. R. H. corresponds to typical atmospheric conditions. Cereal grains, nuts and dried fruits are in this group. Again, if the food has a high oil content, oxygen barriers may be needed. Bacteriological activity is rarely a problem with low-or reduced-moisture foods since one of the essentials of bacterial growth has been removed.

74 High E. R. H. foods lose moisture under typical atmospheric conditions. **At first thought**, it may seem that effective packaging would include a good barrier to stop the loss of moisture; however, a cake with an E. R. H. of 90% would soon establish a relative humidity of 90% inside a sealed package, creating ideal conditions for mold growth. The packaging challenge is to control moisture loss, retarding it as much as possible, but not to the extent that a high humidity is established within the package.

④ Chemical Preservatives

75 Various natural and synthetic chemicals and **antioxidants** are used to help extend the keeping quality of foods. Generally insufficient by themselves, they are used in conjunction with other preservation methods. The use of most of them is strictly controlled by law, although what is allowed varies from country to country.

76 Chemical preservatives work in various ways. Some, such as **lactic, acetic, propionic, sorbic** and **benzoic acids**, produce acid environments. Others, such as alcohol, are specific **bacteriostats**. Carbon dioxide, found in beers and carbonated beverages creates an acid environment and is also a bacteriostat. Smoking and curing of meat and fish is partly a drying process and partly chemical preservation. **Aliphatic and aromatic wood distillation** products (many related to **creosote**) are acidic and have variable bacteriostatic effects. Varying amounts of salt pretreatment accompanies most smoking.

77 Antioxidants and oxygen absorbers can reduce oxidation. Some oxygen absorbers have been used indirectly, contained in separate pouches within the sealed package. The absorber, usually a fine iron powder, scavenges any available oxygen still in the package. Package material may also incorporate antioxidants.

⑤ Modified Atmosphere Packaging

78 Modified atmosphere packaging (MAP) recognizes that many food **degradation** processes have a relationship with the surrounding atmosphere. One mode of degradation is removed if a product prone to oxidation is packaged in an atmosphere free of oxygen. MAP involves the introduction of a gas mixture other than air into a package, that mixture is then left to equilibrate or change according to the nature of the system. A related process, **controlled atmosphere packaging** (CAP), is used in storage and warehousing where the atmosphere can be monitored and adjusted.

79 **Vacuum packaging** is one type of MAP. It has the effect of eliminating some or all oxygen that might contribute to degradation. However, the method is not universally useful, since products such as fruits and vegetables have respiratory functions that must be continued. Another difficulty is that red meat will turn brown or purple without oxygen. Pressures created by the external atmosphere surrounding a vacuum-packaged product can physically crush delicate products or squeeze water out of moist products. Other types of MAP solve these problems.

80 Ambient air is about 20% oxygen and 80% nitrogen, with a trace of carbon dioxide. Altering these proportions alters product response. This forms the basis of MAP shelf life extension. Table 2.4 lists atmospheric combinations for some common food products.

Table 2.4 Typical modified atmospheres for selected food products

Product	Oxygen	Carbon Dioxide	Nitrogen
Red meat	40%	20%	40%
White meats/pasta	—	50%	50%
Fish	20%	80%	—
Produce	5%	—	95%
Baked goods	1%	60%	39%

81　Oxygen is biologically active, and for most products, is associated with respiration and oxidation. Oxygen is normally reduced to slow down the respiration rate of produce and reduce oxidation activity. Red meat is the single exception, where high oxygen levels are used to keep the bright red **"bloom"** associated with freshness.

82　With most other meats, baked goods, **pastas** and dairy products, oxygen is reduced to the absolute minimum consistent with not creating an oxygenless atmosphere that would encourage the growth of anaerobic bacteria. Produce needs at least some oxygen to continue natural respiration.

83　Carbon dioxide in high concentrations is a natural bacteriostat. Levels of 20% and higher are used to create conditions unfavorable to most microorganisms. Carbon dioxide is highly soluble in water, creating a mild acid, and moist products can dissolve enough carbon dioxide to create a partial vacuum. In some instances, the resulting external pressure is undesirable.

84　Nitrogen, unlike the previous two gases, is biologically inert. Its **solubility** in water is negligible, and it is tasteless. Nitrogen is used as a "filler" gas or to displace oxygen.

85　Most packaging materials used in MAP for everything other than produce must have good gas-barrier properties to all three gases. This is true even if the package does not contain the gas. A package containing only carbon dioxide and nitrogen is a system where atmospheric oxygen is trying to penetrate the package and establish an equilibrium partial pressure. The **integrity** of all seals is of paramount importance.

86　The natural respiration of a fruit or vegetable consumes oxygen and produces carbon dioxide and moisture. **Ventilated** or low-barrier packaging is needed to ensure a supply of oxygen and to rid the package of excess moisture.

87　MAP has increased natural shelf life by 2 to 10 times. Cooked pasta, for example, will keep for up to 21 days in an atmosphere of 50% carbon dioxide and 50% nitrogen. The atmosphere used must be tailored specifically to the food item and the type of package being used.

⑥ Irradiation

88　Radiation is energy categorized by wavelength and includes radio waves, microwaves, **infrared radiation**, visible light, ultraviolet light and X rays. These types of radiation increase in energy from radio to X rays; the shorter the wavelength, the greater the energy. Given sufficient energy, waves can **penetrate** substances. With more energy still, they will interact with the molecules of the penetrated substance.

89　Short-wavelength radiations have enough energy to cause energy to **ionization** of molecules, mainly water. Ionization can disrupt complex molecules and leads to the death of living organisms. **Enzymes**, vitamins and other similar complex molecules can also be destroyed. Excessive exposure will produce enough chemical changes that the taste of food or the chemistry of an enclosing container will be altered.

90　Irradiation has been used to increase the keeping quality of various foods. **Cobalt 60**, a radioactive isotope, is the principal source of ionizing radiation (gamma rays). Since the cobalt source is radioactive, it must be shielded with about 1.8 meters of concrete and lowered into a pool of water when not in use. All safety precautions pertaining to radioactive hazards must be observed. It should be noted that while the energy source is radioactive, gamma rays cannot make other substances radioactive. Irradiation is a unique process in that it is carried out at **ambient temperatures** and can penetrate packaging material or products.

91　Low irradiation doses have been used to reduce microbial or insect populations. In addition, irradiation has been found to inhibit **sprouting** in onions and potatoes and delay opening of mushroom caps. A common use is to reduce microbial loadings on such heat-sensitive items as herbs and spices. In addition to treating food, gamma rays can sterilize packaging materials. A significant proportion of hospital supplies are sterilized with gamma irradiation. In another application, irradiation induces cross-linking in some polymers to yield tougher films, some of them with attractive **heat-shrink** properties.

92　Irradiation of consumable food is an issue that is not fully resolved, and the process is carefully controlled in most countries. Critics argue that irradiation causes chemical change and that we are unsure of the prolonged effects of consuming irradiated food. Proponents reply that normal heat processing also causes significant chemical changes in a food whose details we are also unsure of, but that this has never been viewed as a problem. On this basis, food irradiation is prohibited in some countries and highly regulated in most. However, the use of irradiation to achieve sterility for medical devices, packaging materials and **personal care products** does not present a problem and is a useful technology.

93　Labeling is another **contentious** issue. The irradiation symbol (Figure2.3) must be accompanied by a statement such as "treated by irradiation" or "irradiated". The term **"ionizing energy"** is being **touted** to replace the more dangerous sounding "irradiation". Some claim that the design of the irradiation symbol is misleading.

94　Canada has given **product clearances** for potatoes, onions, wheat flour, and spices. The United States has given product clearance for potatoes, wheat flour, spices, fresh fruit and vegetables, pork, chicken, **ground meats** and dehydrated vegetables. Israel and the Netherlands have the broadest range of food irradiation clearance.

Figure 2.3　The international food irradiation symbol

5　The Transport Function

95　The transport function entails the effective movement of goods from the point of production to the point of final consumption. This involves various transport modes, handling techniques and storage conditions. In addition to the general physical rigors of distribution, there are a number of **carrier rules** that will influence package design. Examples of some of the information required to design successful distribution packaging appear in Table 2.5.

Table 2.5 Typical transport handling and storage information

truck	rail	aircraft
cargo ship	storage duration	storage conditions
handling methods	unitizing methods	specific shipping unit
weight considerations	**stock-picking**	dimension limits
carrier rules	environmentally controlled storage	

⑯ Transportation and distribution is generally regarded as an activity that is hazardous to the product being moved. In many instances, the stresses that the product will experience are greater than the durability of the unprotected product. In such instances, it will be necessary to design additional packaging to isolate or cushion the product from the external forces.

⑰ Packaging contributes to the safe, economical, and efficient storage of a product. Good package design takes into account the implications of transport and warehousing, not just for the distribution package and unitized load, but for every level of packaging.

6 The Information/Sales Function

⑱ The communication role of packaging is perhaps the most complex of the packaging functions to understand, measure and implement because of the many levels at which this communication must work. Law or customs **dictate** certain messages without much **leeway** in their presentation. Examples of such message are:

- Specific name of the product (what is this?).
- Quantity contained.
- Address of the responsible body.

⑲ However, to promote the contained product effectively, a package must appeal to the potential customer at all levels. A good package is said to have a **"persona"**, or personality. If the designer has done an effective job, that persona will appeal to the targeted audience.

⑳ The targeted audience itself needs to be identified and studied. This is the **realm** of demographics and **psychographics.**

The package itself communicates by many channels such as:

- Selected material.
- Shape and size.
- Color.
- Predominant typography.
- Recognizable symbols or icons.
- Illustrations.

㉑ A brand of peanut butter aimed at family consumption might come in a **plastic tub** with a **snap-on lid.** The text may simply state that it is an economy peanut butter. The tub would have minimal or no illustration. A gourmet peanut butter, on the other hand, would more likely come in a glass jar with an old-fashioned-looking **screw-on closure.** The label would have an upscale name in a carefully selected old-fashioned font. *Features such as embossing or gold* *stamp printing would further promote the gourmet persona.* The whole package might be offered in a wooden box or placed in a **wicker basket**. Similar products, two totally different packaging treatments and resulting personas.

02　All of the communication channels must be balanced and supportive of one another to produce a persona with appeal and instant recognition. All supporting material, such as promotions and advertisements, must agree with the image projected by the package.

03　Producing a well-balanced package persona requires an intimate familiarity with not just the structural qualities of packaging materials, but also the emotional qualities that they project. *A thorough understanding of the various printing processes and the specialized decorating techniques used to create particular effects or decorate unusual surfaces is essential.*

(6545 words)

New Words and Expressions

cereal ['siriəl] *n.* 谷类、谷物
hygroscopic [,haigrə'skɔpik] *adj.* 易潮湿的
quantify ['kwɔntifai] *v.* 量化、定量
abrasion [ə'breiʒən] *n.* 磨损、磨耗
deformation [,di:fɔ:'meiʃən] *n.* 变形
tamper ['tæmpə] *n.* 偷换、窜改
sterility [stə'riləti] *n.* 无菌性
volatile ['vɔlətl] *n.* 挥发物
opaque [əu'peik] *adj.* 不透明的、迟钝的
spoilage ['spɔilidʒ] *n.* 损坏、变质
deteriorate [di'tiəriəreit] *v.* 恶化、变坏
myoglobin ['maiəu,ɡləubin] *n.* 肌红蛋白
microorganism [,maikrəu'ɔ:ɡənizəm] *n.* 微生物
abiotic [,eibai'ɔtik] *adj.* 非生物的、无生命的
cologne [kə'ləun] *n.* 古龙香水
irreversible [,iri'və:səbl] *adj.* 不可逆的
retard ['rita:d] *v.* 延迟、使减速
carcase ['ka:kəs] *n.* 肉牛酮体
oxymyoglobin [,ɔ:ksimaiə'ɡləubin] *n.* 氧合肌红蛋白
psychrophilic [,psaikrəu'filik] *adj.* 好寒性的
mesophyll ['mesəu,fil] *n.* 叶肉
thawing [θɔ:iŋ] *n.* 解冻、融化
hormone ['hɔ:məun] *n.* 激素、荷尔蒙
ethylene ['eθili:n] *n.* 乙烯
aggravate ['æɡrəvet] *v.* 加重
perforate ['pə:fəreit] *v.* 打孔
vent [vent] *v.* 排放、发泄
permeability [,pə:miə'biliti] *n.* 渗透性

unicellular [ˌjuːniˈseljələ] adj. 单细胞的
microscopic [ˌmaikrəˈskɔpik] adj. 微观的
exponentially [ˌekspəuˈnenʃəli] adv. 以指数方式
spore [spɔː] n. 孢子
multicellular [ˌmʌltiˈseljulə] adj. 多细胞的
chlorophyll [ˈklɔːrəfil] n. 叶绿素
carbohydrate [ˌkɔːbəˈhaidreit] n. 碳水化合物、糖类
propagate [ˈprɔpəgeit] v. 传播、繁殖
thermophilic [ˌθəːməuˈfilik] adj. 适温的、喜温的
aerobic [eəˈrəubik] adj. 需氧的
anaerobic [ˌæneiəˈrəubik] adj. 厌氧的
pathogenic [ˌpæθəˈdʒenik] adj. 致病的、病原的；
pathogen [ˈpæθədʒən] n. 病原菌、致病菌
sublimate [ˈsʌblimeit] v. 升华
dehydration [ˌdiːhaiˈdreiʃən] n. 脱水
saran [səˈræn] n. 聚偏二氯乙烯，同 PVDC
acidity [əˈsidəti] n. 酸度、pH 值
pasteurization [ˌpæstəraiˈzeiʃən] n. 加热杀菌法、巴斯德杀菌法
syrup [ˈsirəp] n. 糖浆、果汁
sterilize [ˈsterəlaiz] v. 消毒、杀菌
thermoform [ˈθəːməuˌfɔːm] v. 加热成型
conducive [kənˈdjuːsiv] adj. 有益的、有助于……的
retort [riˈtɔːt] n. 蒸煮锅
polyolefin [ˌpɔliˈəuləfin] n. 聚烯烃
marmalade [ˈmɔːməled] n. （橘子或柠檬等水果制成的）果酱
equilibrium [ˌiːkwiˈlibriəm] n. 均衡、平衡
desiccant [ˈdesikənt] n. 干燥剂
antioxidant [ˌæntiˈɔksidənt] n. 抗氧化剂、防老化剂
lactic [ˈlæktik] adj. 乳化的
acetic [əˈsitik] adj. 醋的、乙酸的
propionic [prəuˈpiɔnik] adj. 丙酸的
sorbic [ˈsɔːbik] adj. 山梨酸的
bacteriostat [bækˈtiriəstæt] n. 抑菌剂
aliphatic [ˌæliˈfætik] adj. 脂肪质的
aromatic [ˌærəˈmætik] adj. 芳香族的
creosote [ˈkriət] n. 木馏油
degradation [ˌdegrəˈdeiʃən] n. 降解
pasta [ˈpɑːstə] n. 意大利面
solubility [ˌsɔljuˈbiləti] n. 溶解度

integrity [in'tegrəti] *n.* 完整性
ventilated [,ventl'et] *adj.* 通风的
penetrate ['penitreit] *v.* 渗透、穿透
ionization [,aiəni'zeʃən] *n.* 离子化
enzyme ['enzaim] *n.* 酶
sprout [spraut] *n.* 芽、萌芽、苗芽
contentious [kən'tenʃəs] *adj.* 有异议的、诉讼的
tout [taut] *v.* 兜售、招揽顾客
dictate [dik'teit] *v.* 命令、口述
leeway ['li:wei] *n.* 灵活性、回旋余地
persona [pə:'səunə] *n.* 人、角色
realm [relm] *n.* 领域
psychographics [,saikə'græfiks] *n.* 消费心理学
primary package　一次包装（件）、内包装
secondary package　二次包装（件）、中包装
distribution package（shipper）　运输包装（件）(同 transport packaging)
unit load　单元化装载（集装）
consumer package　销售包装（件）
industrial package　工业包装（件）
power tool　电动工具
granular material　颗粒料
critical value　临界值
relative humidity　相对湿度
shelf life　货架寿命、保存期限
rancid taste　（油脂变质后的）哈喇味、腐臭味
essential oil　香精油
sensory active agents　感官活性剂
off flavor　异味、败味
stray volatile　游离挥发物
high-barrier packaging　高阻隔性包装
oxygen-deficient　缺氧的
as a rule of thumb　根据经验
modified atmosphere packaging（MAP）　气调包装
trade-off　折中、权衡
gas-transmission rate　气体透过率
binary fission　（细胞的）二分体
freezer burn　冻斑、冷冻食品表面干燥变硬
moisture-barrier packaging　防潮包装
poly（vinylidene chloride）（PVDC）　聚偏二氯乙烯
bag-in-box　盒中袋

nominal cleanliness 名义洁净度
commercial sterility 商业无菌
clostridium botulinum 肉毒杆菌
retortable pouch 蒸煮袋、软罐头
equilibrium relative humidity（E. R. H.） 相对平衡湿度
potato chips 油炸薯片
instant coffee 速溶咖啡
oxygen scavenger 去氧剂
at first thought 乍一想
benzoic acid 苯甲酸
wood distillation 木材蒸馏法
controlled atmosphere packaging（CAP） 可控气氛包装
vacuum packaging 真空包装
infrared radiation 红外线辐射
cobalt 60 钴 60
radioactive isotope 放射性同位素
ambient temperature 室温
heat-shrink 热收缩性
personal care products 个人护理用品
ionizing energy 游离能、电离能量
product clearance 产品许可证
ground meat 碎肉
carrier rule 运送规则
stock-picking 选货
plastic tub 塑料管
snap-on lid 按扣盖、搭锁盖（可咯嗒一声盖住的）
screw-on closure 旋盖
wicker basket 柳条篮

Notes

1. *Specific food categories have their own characteristics with meats, fish, and produce being important categories.*（Para. 25）特定的食品类别有着它们自己的特性，其中肉、鱼以及农产品最重要。这里 with 引导分词 being 的独立主格。

2. *The high moisture barrier keeps bagged produce from drying out, while the good supply of oxygen allows the produce to respire naturally.*（Para. 37）而氧气的良好供给使得农产品能自然地呼吸，高阻湿性保持袋中农产品不至于干燥。while 引导的是并列语句。

3. *Molds form filamentous branching growths called mycelia and reproduce by spores. Yeasts are similar organisms that reproduce by budding. The propagation and spread of molds and yeasts is typically slower than for bacteria because of the reproduction method.*（Para. 42）霉菌助长了丝状分枝的生长称为菌丝，并通过孢子繁殖。酵母

菌是类似的生物体，通过发芽繁殖。由于繁殖方式不同，霉菌和酵母的繁殖和传播，通常比细菌的繁殖慢得多。

4. *Commercial systems, such as Tetra Pak, Combibloc, and Bosch, use hydrogen peroxide to sterilize simple paper, foil and polyethylene laminates, and then fill the formed package with UHT-treated product.*（Para. 57）利乐、康美、博世这些商业体系用双氧水对单独的纸、铝箔消毒，再复合聚乙烯，然后将经过超高温处理的产品充填到成型的包装中。

5. *Features such as embossing or gold stamp printing would further promote the gourmet persona.*（Para. 101）如饰以凹凸印或烫金等特色印刷将会进一步提升美食家的形象。句中 embossing or gold stamp printing 是两种用于产生装饰效果的专门的装潢技术，是指"凹凸印或烫金"。

6. *A thorough understanding of the various printing processes and the specialized decorating techniques used to create particular effects or decorate unusual surfaces is essential.*（Para. 103）对各种印刷工艺和用于产生特殊效果或装潢独特表面的专门装潢技术的深入理解是必不可少的。used to create … or decorate … 为过去分词短语，作定语，修饰 the specialized …。understanding 为主语，is essential 为系表结构，作谓语。

Overview Questions

1. Name the four principal functions of a package.
2. Containment is a primary packaging function. List 12 product characteristics that will affect your choice of material and package design.
3. In providing for the protection/preservation function, it is essential to know and quantify what factors?
4. Name the six ways of extending the natural shelf life of foods. For each of the methods, briefly note the mechanism by which the keeping quality of a food product is increased.
5. List five means by which a package communicates its persona to the observer.

UNIT 2

Packaging Materials And Containers

Lesson 3 Paper and Paperboard

1 Introduction

① Paper is defined as a **matted** or **felted** sheet usually composed of plant fiber. Paper has been commercially made from such fiber sources as **rags** (linen), **bagasse** (sugar cane), cotton, and straw. Modern paper is made almost exclusively from **cellulose fiber** derived from wood.

② Although the word "paper" is derived from the Egyptian term, **"papyrus"** was not a true paper in the modern sense. The invention of paper by blending cellulose fibers didn't occur until the beginning of the second century A. D. Ts'ai Lun, a member of the court of the later Han Dynasty, is generally **credited with** developing the first real papermaking process in 105 A. D.. It wasn't until the beginning of the 18th century that workable machines were developed to mass produce paper and certain forms of paperboard. *The* **"Fourdrinier machine"** *was the first on the market and produced a* **homogenous** *(single-ply) sheet of boxboard in various thicknesses. It was soon joined by the* **"Cylinder machine"** *which formed a multi-layered (multi-ply) type of paperboard.* These machines were first installed in the United States around 1830. Highly refined version of the two are still the primary machines used to produce paper and paperboard throughout the world.

③ The paper industry has few definitive terms. For example, **paperboard**, **boxboard**, **cardboard**, and **cartonboard** are all terms used to describe heavier paper stock. "Paper" and "paperboard" are nonspecific terms that can be related to either material **caliper** (thickness) or **grammage** (weight). The International Organization for Standardization (ISO) states that material weighing more than 250 grams per square metre (51 lbs per 1000 ft^2) shall be known as paperboard. U. S. practice calls material that is more than 300μm (0.012 in) thick paperboard.

2 Representative Papermaking Machines

④ There are many variations on papermaking machines, each imparting its own character to the resulting paper. Only three representative classes will be discussed here: Fourdrinier, cylinder, and **twin-wire machines.**

2.1 Fourdrinier machines

⑤ Fourdrinier machines (Figure 3.1) pump **furnish** from a **headbox** directly onto a moving wire screen through which the water is continuously drained. Fourdrinier machines may have a second headbox situated downstream of the first headbox to add further quantities of furnish onto the partially dewatered initial lay-down.

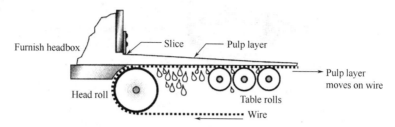

Figure 3.1 Furnish pours out of the headbox of a fourdrinier machine and onto an endless wire or screen where excess water can be drained. The fibers remain trapped on the screen

⑥ However, it is impractical to add more headboxes to produce thick paperboard on a fourdrinier machine, since the water from each successive addition must drain through fibers that have already been laid down. This limits the thickness of paper produced on a conventional fourdrinier machine. Heavier caliper boards can be made by bringing together the wet pulp layers laid out by two or more completely separate Fourdrinier machines and pressing these together before the sheet is sent to the **dryer**.

⑦ Most free water is removed at the "wet end" of a **papermaking machine** (Figure 3.2). *At the* **couch roll**, *the wet paper has enough strength to be removed from the wire and passed around a series of heated drying drums where moisture content is brought down to finished-product specifications.* The dried paper may go through some further treatments at the "dry end" of the papermaking process before being wound up into a mill roll.

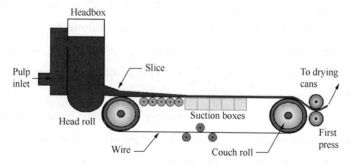

Figure 3.2 Paper is dewatered at the wet end of a Fourdrinier machine

2.2 Cylinder Machines

⑧ A cylinder machine (Figure 3.3) rotates a screen drum in a **vat** of furnish. (The paper is sometimes called vat paper.) As the water pours through the screen, fiber accumulates on the outside of the screen. This thin layer of matted fiber is transferred onto a moving felt belt that passes sequentially over further rotating cylinders, each of which deposits another fiber layer.

⑨ *Cylinder machines* **dewater** *furnish at the cylinder and paste a thin layer of fiber*

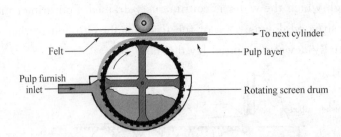

Figure 3.3　A single cylinder station on a cylinder-type machine

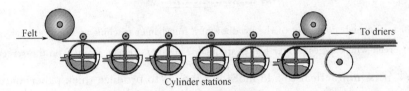

Figure 3.4　A cylinder machine with six cylinders at which a paper layer can be formed

against the felt (Figure 3.4). *The fibers of subsequent layers do not* **intermingle**, *and therefore the bond between the layers is weak*. The dry end is similar to that of the Fourdrinier machine.

⑩　Cylinder machines do not have the Fourdrinier machine's limitation on the number of stations, and six-or seven-station machines are common. Higher-caliper boards for **folding and setup cartons** are usually cylinder boards. An advantage of cylinder machines is that low-quality fiber can be used to fill or bulk the middle of a board, while higher quality bleached fibers can be used on one or both liners.

⑪　Cylinder board has definite layers, or plies, and individual plies can often be easily separated. Poor interply bonding can produce a variety of packaging problems related to ply **delamination**. Generally, papers are made on Fourdrinier or twin-wire formers, whereas heavier paperboard products are made on cylinder-type machines. Extremely heavy boards are made by laminating several thinner sheets.

⑫　A typical cylinder board construction (Figure 3.5) may have a top **liner** composed of good-quality bleached pulp with some short fibers, possibly **sized** and **clay coated** to produce a smooth, attractive printing surface. The underliner may also be composed of a good-quality stock, possibly bleached to provide a smooth, opaque base for the top liner.

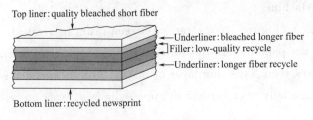

Figure 3.5　Cylinder boards are multiply boards. An advantage is that the plies can all be different

③ Filler plies use the most economical recycled pulps, since they have little impact on properties such as stiffness. The bottom liner is a better quality pulp to add stiffness. If appearance is not a factor, the liner may be good-quality **recycle pulp**. If appearance is critical or if the paperboard will be printed on both sides, the bottom liner will also be **bleached stock**.

2.3 Twin-Wire Machines

④ **Verti-formers** and twin-wire formers (Figure 3.6) inject the furnish between two moving wire screens. The advantage is that dewatering takes place on both sides of the paper and is therefore

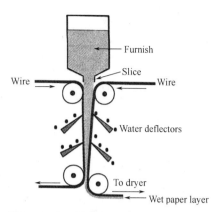

Figure 3.6 Water can be simultaneously removed from both sides of the paper on a twin-wire paper former

fast. These machines can produce single and multi-ply sheets with identical formation at both faces.

3 Machine Direction And Cross Direction

⑤ Depositing a fiber-and-water **slurry** onto a moving wire belt tends to align fibers in the direction of travel, known as the **machine direction** (MD). The direction across the papermaking machine and across the fiber alignment is the **cross direction** (CD)(Figure 3.7). Because of this fiber alignment, paper is an **anisotropic material**; measured properties differ depending on the direction in which the property is measured. Figure 3.8 shows the relationship of tear, stiffness, and **fold endurance** to machine direction. Paper specification sheets normally show physical values measured in both directions. Package designers need to be aware of paper's directionality. For example, it is much easier to create a tear strip along the machine direction rather than trying to force it in the cross direction.

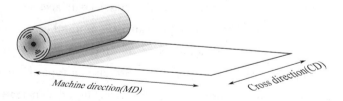

Figure 3.7 Fibers in a manufactured paper sheet tend to align themselves in the machine direction

⑥ Cylinder machines tend to align fibers more than Fourdrinier machines. Tensile strength ratios in MD and CD for a typical Fourdrinier board are about 2∶1, whereas for a cylinder board the ratio might be 4∶1 or higher, meaning that the MD **tensile strength** is four times greater than the CD tensile strength. *The greater the degree of fiber alignment, the greater the difference in a given property when measured in MD and CD. The ratio of a property*

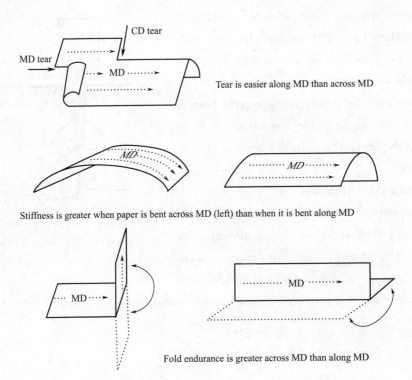

Figure 3.8 The relationship between MD and tear, stiffness, and fold endurance properties *in the two directions is often used as a **gauge** of fiber alignment.*

4 Surface Or Dry-End, Treatments And Coatings

17 After the paper is formed and dried, it is usually passed between multiple sets of heavy rolls (Figure 3.9). This **"calendaring"** operation has many variations, but the prime objective is to iron and smooth out the surface of the paper stock to make it more suitable for printing. Calendering also compresses the paper sheet, giving a denser product and a **glossier** surface.

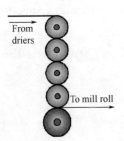

Figure 3.9 Calendering consists of passing the formed dried paper between sets of heavy rolls. The paper surface may be dampened to help in smoothing it

18 Finished paper can have a number of surface sizings and/or coatings applied at the dry end in order to further improve surface characteristics. Starch is a typical surface sizing used to fill surface voids and reduce liquid penetration rate. To meet the highest opacity, gloss, brightness, and printing-detail requirements, papers are coated with **pigments** such as clay, **calcium carbonate**, and **titanium dioxide**. These are bound together and to the paper surface with adhesives. Coated papers are usually called "clay-coated" regardless of the actual formulation. Coated papers are calendered to maintain a high-quality, smooth surface.

19 If clay-coated paper is passed under highly polished chrome drums that are rotating

counter to the paper or faster than the paper speed, the clay coatings will be polished. These papers are referred to as machine-glazed or chrome-coated papers.

⑳ Poorly bonded clay coatings can cause printing problems. In a typical situation, pigment particles come free or are picked from the paper surface. These contaminate the ink and adhere to the printing plate. In addition, highly sized and clay-coated boards can be difficult to bond with water-based adhesive because of poor liquid penetration and the inability of the adhesive to bond to the underlying fibers. Adhesives need to be appropriate to the paper's properties, and, where necessary, coated boards should have perforations in the adhesive-bond areas so that adhesive can penetrate to the body of the paper.

㉑ With such a large number of principal variables (and many lesser ones), it is easy to understand that virtually no two paper mills in the world will produce an identical paper, and that even a single mill might have difficulties producing within a tight tolerance level from year to year.

5 Paper Chracterization

5.1 Caliper and Weight

㉒ The prime specifying values for paper or paperboard are its caliper, or thickness, and a value expressing a density relationship. Both inch/pound and metric units are used, and different units may be used for different mill products. Take care to know exactly what units are being quoted.

㉓ In inch/pound units:

• Caliper is expressed in thousandths of an inch or in "points". One thousandth of an inch is 1 point. (For example, a 0.020 in board would be 20 points.)

• **Containerboard** for the **corrugated board** industry and most paperboards are specified by the weight in pounds per 1,000 sq. ft, the **"basis weight"**.

• Fine papers can be specified by the weight in pounds per **ream**. A ream is 500 sheets, but the actual sheet size can vary depending on the product. In most instances a ream is taken to be 3,000 sq. ft.

㉔ In metric units:

• Caliper is expressed in micrometres properly abbreviated as "μm". (The symbol is difficult to find on a keyboard, and it is commonly written as "μm".) Millimetres (mm) are sometimes used, but micrometres have the advantage of eliminating the decimal point.

• Paper mass/unit area relationship is reported as "grammage", defined as being the mass (weight) of paper in 1 square metre (m^2).

The metric conversion factors are

$$\text{lbs}/1{,}000 \text{ sq. ft} \times 4.88 = \text{grams}/m^2$$

$$0.001 \text{ inch} \times 25.4 = 1\mu m \text{ (usually rounded to } 25\mu m\text{)}$$

$$1 \text{ mm} = 1{,}000 \mu m$$

5.2 Brightness

㉕ Brightness is a measure of the total **reflectance** of white light. Values are expressed on a scale of 1 to 100, with 100 being the brightness of pure **magnesium oxide.** (Paper cannot be made to a brightness value of 100.) Brightness should not be confused with "whiteness", which is a color description much as red or yellow is. Most quality grades have reflectance values in the mid-80s. The brighter the board, the more brilliant the graphic possibilities. Paper stock brightness should be specified for quality printing.

5.3 Paper and Moisture Content

㉖ Paper is hygroscopic and absorbs and loses moisture according to the ambient relative humidity (R.H.) and temperature. Paper at 20% R.H. will contain about 4% moisture, while at 80% R.H., it will contain about 15% moisture. (Equilibrium paper moisture content varies slightly depending on whether equilibrium is reached from a lower humidity or a higher humidity.) This effect is known as **hysteresis. By convention**, accurate moisture analysis is done by bringing paper to equilibrium from a lower humidity.

㉗ The physical properties of paper vary dramatically with **moisture content**, and in some applications the moisture content of the paper during processing must be controlled. Where strength is an important design factor, paperboard should be selected to perform at the highest anticipated R.H.. Because physical characterization values depend on moisture content, all paper testing must be done at a precisely controlled temperature and humidity. Internationally, the standard conditions are specified as 23°C and 50% R.H..

㉘ Paper is **hygroexpansive**: when it absorbs moisture, it expands; when it dries out, it shrinks. Between 0 and 90% R.H., the dimensions can change 0.8% in the MD and 1.6% in the CD. A 1% shrinkage over a 1m (3 ft) **carton blank** is 10 mm (about 0.4 in). Such a difference can play **havoc** with printing and **die-cutting** register.

㉙ Whenever a paper sheet is laminated to or coated with a material that is not affected by moisture (for example, plastic film, aluminum foil, or heavy print or varnish), there is the potential for **curling** when the humidity changes. *If the paper gains moisture and expands while the surfacing laminate or coating remains the same, the paper will curl toward the surfacing material. When the paper loses moisture, it will shrink and curl away from the surfacing material* (Figure 3.10).

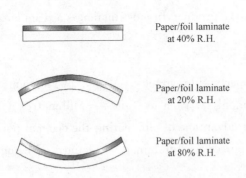

Figure 3.10 Paper's hygroexpansive nature can cause unwanted curling when paper is bonded to an environmentally stable surface

5.4 Viscoelasticity

⑤⓪ Paper is more or less **viscoelastic**, depending on the rate at which load is applied. Simply put, the faster a load is applied, the greater the **apparent strength**. Over long loading periods, paper fibers move and distort or "**creep**". Long-term **static compressive strength** is much less than **dynamic compressive strength**. Boxes tested at 500kilogram-load compressive strength in the laboratory (a fast loading rate is dynamic) could be expected to fail in about a year when a 250kilogram static load is applied in the warehouse.

⑤① The tendency for paper fibers to move or acquire a set also means that the longer a **knocked-down** folding carton blank is kept flat, the greater the permanent set developed at the **creasing scores**. The break-open force required to open the carton on a cartoning machine increases, and at some point may reach a level that will cause erecting problems.

6 Paper Types

⑤② Most papers are made on Fourdrinier or twin-wire machines. Unlike multi-ply boards, papers are usually composed of a single pulp type or blend. In a few instances, additional layers are laid on from additional headboxes.

⑤③ Paper manufacturers have **proprietary names** for paper, and there is little uniformity in the names of the different types and grades. Many papers are designed for particular printing processes, and it is important for the paper supplier to know which process will be used. Some mills produce **specialty papers**, such as colored stock or stock with textured surface finishes. The following list of generic paper types gives a general idea of **paper-mill** products.

6.1 Newsprint and Related Grades

⑤④ **Newsprint** is composed of up to 95% economical mechanical or **groundwood pulps**. Runability is a prime requisite, and higher quality pulps are sometimes added to improve strength qualities. Standard newsprint is about 49 grams/m^2 (10 lbs/1,000 sq. ft). Newsprint has relatively low brightness, typically 55 to 65, and has low physical strength properties. Newspaper inks are primarily oil and carbon, and high oil absorbency is a desirable characteristic in newsprint.

6.2 Book Papers

⑤⑤ Like newsprint, book and catalog papers are also mostly based on **mechanical pulp**, but are sized and clay coated to varying degrees to improve surface appearance and printing qualities. Inexpensive pocket books are printed on heavier calipers of similar paper. The papers used in popular magazines are about 50% mechanical pulp and are coated and calendered to provide good white printing surfaces.

6.3 Commercial Papers

⑤⑥ Commercial papers cover a broad range of products, many of them specifically designed for particular applications. They are generally based partly on groundwood, and they are invariably coated. Such papers are used for higher quality journals and books, and for general office purposes. Brightness usually ranges from 73 to 85. Papers used for photocopying are designed to withstand the heat of the copying process with a minimum of paper curl and distortion. They also must have controlled electrostatic surface resistance to provide uniform image transfer.

6.4 Greaseproof Papers

㊲ **Greaseproof papers** are made from **chemical pulps** that have been highly refined to break up fiber bundles. The fine fibers pack densely, providing for a structure that does not readily absorb fluids. **Glassine papers** are supercalendered, semitransparent greaseproof papers. **Sulfuric acid** treatment further reduces the fibrous quality of **parchment papers**-thus, increasing their water and grease barrier properties and transparency. Greaseproof papers are used for snack foods, cookies, butter and **lard** wraps, and other oily or greasy products. Occasionally, they will be used for their semitransparency properties. Greaseproof paper may be treated with high density polyethylene or poly (vinylidene chloride) to further increase water and grease holdout. Plastic films have replaced these grades for many applications.

6.5 Natural Kraft Paper

㊳ Natural **kraft** is the strongest of the common packaging papers and is used wherever maximum strength is needed. The light brown papers are used in industrial bags, in carryout grocery bags and as inner plies in multiwall bags. **Impact strength** is imparted to some kraft paper used for industrial bags by **creping** the paper. The slight extensibility gives the paper greater shock resistance (i.e., tensile energy absorption). Paper for carryout bags can be machine finished (calendered) to provide a pleasing surface gloss. The fibrous nature and color of natural kraft paper does not **lend itself to** fine printing.

6.6 Bleached Krafts and Sulfites

㊴ Where strength and appearance are important, kraft paper can be bleached. For applications where the material will be printed, the kraft is coated to smooth the surface. Flour and sugar bags are typical applications. Low-density bleached kraft and sulfite papers are used to produce "dry waxed" paper. Highly calendered bleached kraft and sulfites are used for "wet waxing", where the wax coats the surface and produces a high surface gloss.

6.7 Tissue Paper

㊵ "Tissue" is a generic term for any light paper. In packaging, tissue is used for protective wrapping and as a laminating component.

6.8 Label Paper

㊶ Label papers are similar to book papers. Generally, the printing quality required will determine the grade to be used. For example, uncoated papers are used only for line copy and simple text. Machine-finished papers provide good results up to about a 110 line screen, and supercalendered paper is needed for 120 and 133 line screens.

㊷ Most papers used for label printing are coated on one or both sides. Coated papers can be made in **matte**, dull, gloss, and high-gloss finishes. Cast-coated paper is made by drying a clay coating in contact with a heated, highly polished chromed drum. The process results in a very smooth, extremely glossy surface.

6.9 Pouch Papers

㊸ **Pouch papers** are supercalendered virgin kraft papers that have been treated with a **plasticizer** to make them more pliable. They are high density, very strong, and have a smooth surface finish. Pouch papers are used for soap wrappers.

6.10 Containerboards (linerboard and medium)

㊹ Containerboard specifically refers to linerboard and medium produced for use in the manufacture of corrugated board. Linerboard is a solid kraft board made specifically for the liners (facings) of corrugated fiberboard. Performance characteristics are governed by carrier specifications. Most linerboard is natural kraft run on a machine with two headboxes. The **inner face** is somewhat rougher to promote good gluing to the medium. The **outer liner** of a linerboard is sometimes made of bleached kraft stock. The bleached liner is not heavy enough to totally hide the natural kraft underlayer, resulting in a somewhat **mottled** or **off-white** appearance.

㊺ **Medium** is the material used to produce the fluted core of corrugated fiberboard. The desired properties are glueability, stiffness, and runability. Most medium is made to an industry specification of 127 grams/m^2 and 229 μm caliper (26 lbs/1,000 sq. ft and 0.009 in). Medium is made from semichemical hardwood pulps and from recycled kraft containers. Medium board is the one paper product where **lignin** content is an asset; it imparts **thermoformability** properties to the board.

7 Paperboard Grades

7.1 Chipboard, Cardboard, Newsboard

㊻ **Chipboard** is made from 100% recycled fiber and is the lowest cost paperboard. Colors range from light gray to brown. Chipboard is unsuitable for printing and has poor folding qualities due to its short fiber length. Chipboards often contain **blemishes** and impurities from the original paper application. They are used for **setup boxes, partitions,** backings, and other applications where appearance and foldability are not critical.

㊼ The term "cardboard" is not commonly used among paper professionals but is occasionally applied to describe chipboard-type products. **Newsboard** is a low-grade board made mostly from recycled newspapers.

7.2 Bending Chipboard

㊽ **Bending chipboard** is a slightly better grade of chipboard, still primarily composed of recycled fiber, but with enough higher quality fiber to allow **scoring** and folding. Usually a gray or light **tan**, it is the lowest cost folding-carton board.

7.3 Lined Chipboard

㊾ **Lined chipboard** has had a white face-liner applied to improve appearance and print quality. Bending characteristics and filler and back liner colors vary depending on ply composition. Lined chipboard may or may not be clay coated. **Single white-lined, double white-lined,** and **clay-coated newsback** are varieties of lined chipboard.

7.4 Single White-Lined (SWL) Paperboard

⑤⓪ The top liner of single white-lined paperboard is 100% new pulp or high-quality recycled pulp. The back is usually gray or light **tan**. It is a smooth board with a typical brightness of 60 to 70. SWL is used for folding boxes, where the appearance of the back is not critical. The board may or may not be clay coated.

7.5 Clay-Coated Newsback (CCNB)

⑤① This is a single white-lined paperboard in which the back and sometimes filler plies are made from recycled newsprint. The back liner has a neutral pale gray color that is judged to be more aesthetically acceptable for some applications than the mixed browns of other SWL boards.

7.6 Double White-Lined (DWL) Paperboard

⑤② Double white-lined paperboard is the same as SWL except that both front and back have been lined with a white pulp. Usually, the white-lined back is somewhat less finished than the front. DWL is used where the internal appearance of the box is important or where both sides will be printed. The primary face is usually clay coated.

7.7 Solid Bleached Sulfate (SBS)

⑤③ SBS is a strong premium paperboard composed of 100% bleached sulfate pulps. SBS boards are white throughout. They are used in applications where total board appearance is of primary importance and where maximum physical properties are required for the given weight of board. Because SBS boards are made from virgin pulps, properties vary less during a run than for a similar lined chipboard. Many high-speed packaging operations prefer SBS for its consistent line performance even though for other considerations a lined chipboard may have been adequate.

7.8 Food Board

⑤④ Highly sized SBS paperboards are often called food board. These are used for wet foods, freezer boxes, and other applications where good performance under wet conditions is important.

7.9 Solid Unbleached Sulfate (SUS)

⑤⑤ SUS is a maximum strength unbleached kraft paperboard. Where surface appearance is important, the board will be heavily clay coated, sometimes with a double coat to provide the opacity needed to hide the brown kraft body. Beverage baskets (i. e., six-packs) are a major application of clay-coated solid unbleached sulfate.

(4047 words)

New Words and Expressions

matted ['mætid] *adj.* 无光泽的
felted ['feltid] *adj.* 黏制的
rag [ræg] *n.* 破布
bagasse [bə'gæs] *n.* 甘蔗渣
papyrus [pə'paiərəs] *n.* 纸莎草、草纸
homogenous [hə'mɔdʒinəs] *adj.* 同质的
paperboard ['peipəbɔːd] *n.* 纸板、卡纸

boxboard ['bɔksbɔːd] *n.* （纸盒、纸箱用）硬纸板（或木板）
cardboard ['kɑːdbɔːd] *n.* 纸板（总称）、尤指中厚度纸板
caliper ['kælipə] *n.* （材料）厚度
grammage ['græmidʒ] *n.* 克重
furnish ['fəːniʃ] *n.* 纸浆配料、浆料
headbox ['hed,bɔks] *n.* 流浆箱
dryer ['draiə] *n.* 干燥机
vat [væt] *n.* 大桶、制剂桶
dewater [diːwɔːtə] *v.* 脱水
intermingle [intəmiŋg(ə)l] *v.* 混合、混杂
delamination [diːlæmiˈneiʃən] *n.* 层离、分层
liner ['lainə] *n.* 内衬
size [saiz] *v.* 涂胶
deflector [diˈflektə] *n.* 导流板
slurry ['slʌri] *n.* 混合液、浆料
gauge [geidʒ] *n.* （厚度）计量单位
calendaring ['kæləndəriŋ] *n.* 压光、压延成型
glossy ['glɔsi] *adj.* 光滑的、有光泽的
pigment ['pigm(ə)nt] *n.* 色素、颜料
containerboard [kənˈteinəbɔːd] *n.* 箱板纸
ream [riːm] *n.* 令（纸张计数单位）
reflectance [riˈflekt(ə)ns] *n.* 反射率
hysteresis [,histəˈriːsis] *n.* 滞后
blank [blæŋk] *n.* 盒坯、箱坯
havoc ['hævək] *n.* 破坏、损坏
curling ['kəːliŋ] *n.* 卷曲
viscoelastic [,viskəuiˈlæstik] *adj.* 黏弹性的
creep [kriːp] *v.* 蠕变
newsprint ['njuːzprint] *n.* 新闻纸
lard [lɑːd] *n.* 猪油
kraft [krɑːft] *n.* 牛皮纸
crepe [kreip] *v.* 起绉、绉纸
matte [mæt] *adj.* 无光泽的
plasticizer ['plæstisaizə] *n.* 可塑剂
linerboard ['lainəbɔːd] *n.* 瓦楞纸、板面纸
mottled ['mɔtld] *adj.* 杂色的
medium ['miːdiəm] *n.* 瓦楞芯纸、瓦楞原纸
lignin ['lignin] *n.* 木质素
chipboard ['tʃipbɔːd] *n.* 粗纸板、灰纸板
newsboard ['njuːzbɔːd] *n.* 旧报纸做的纸板

thermoformability 热成型
blemish ['blemiʃ] n. 瑕疵、污点
partition [pɑːtiʃən] n. 隔离物、隔板（同 divider）
score [skɔː] v. 压痕
tan [tæn] n. 棕色、褐色
cellulose fiber 纤维素纤维、木质素纤维
be credited with 被认为
Fourdrinier machine 长网纸机
cylinder machine 圆网纸机
cartonboard （折叠纸盒用）纸板
twin-wire machine 双网（夹网）纸机
papermaking machine 造纸机
couch roll 伏辊
folding carton 折叠纸盒
setup carton 固定（自立）纸盒
clay coated 瓷土涂布
recycle pulp 回收浆
bleached stock 漂白浆
verti-former 竖式造纸机
machine direction 纸张纵向、机器方向
cross direction 纸张横向
anisotropic material 各向异性材料
fold endurance 耐折度
tensile strength 抗张强度
calcium carbonate 碳酸钙
titanium dioxide 二氧化钛
corrugated board 瓦楞纸板
basis weight 基重、基本重量
magnesium oxide 二氧化镁
by convention 按照惯例
moisture content 湿量、水分（含量）
hygroexpensive 湿润膨胀
carton blank 盒坯
die-cutting 模切
apparent strength 表观（视）强度
static compressive strength 静态抗压强度
dynamic compressive strength 动态抗压强度
knocked-down 拆散压扁（纸盒纸箱和其他容器压扁后储存和运输）
creasing score 压痕线
proprietary name 专利商品名

specialty paper　特种纸
paper-mill　造纸厂
groundwood pulp　磨木浆
mechanical pulp　机械浆
greaseproof paper　防油纸
chemical pulp　化学浆
glassine paper　玻璃纸
sulfuric acid　硫酸
parchment paper　羊皮纸
natural kraft paper　天然牛皮纸
impact strength　冲击强度
lend itself to　有助于
tissue paper　棉纸、薄纸
label paper　标签纸
pouch paper　纸袋纸
liner facing　（瓦楞纸外层的）面纸
inner face　里面纸
outer face　外面纸
off-white　灰白色
setup box　自立纸箱
bending chipboard　耐折纸板
lined chipboard　贴面灰纸板
single white-lined（SWL）paperboard　单面贴有白色面纸的纸板
clay-coated newsback（CCNB）　瓷土涂布新闻纸
double white-lined（DWL）paperboard　双面白色贴面纸板
solid bleached sulfate（SBS）　漂白硫酸盐硬纸板
food board　包装食品用纸板
solid unbleached sulfate（SUS）　非漂白硫酸盐硬纸板

Notes

1. *The "Fourdrinier machine" was the first on the market and produced a homogenous (single-ply) sheet of boxboard in various thicknesses. It was soon joined by the "Cylinder machine" which formed a multi-layered (multi-ply) type of paperboard.* (Para. 2) 长网造纸机是市场上第一个出现的造纸机，它可以生产不同厚度的（单层）纸板。不久就出现了可以制造出多层纸板的圆网造纸机。Fourdrinier machine 指"长网纸机"，造纸的第一步即纤维和水的混合浆料涌到运动着的网带上；Cylinder machine 指"圆网纸机"，靠圆形网的转动带出制浆料的一种纸机；还有 twin-wire machine 即"夹网式纸机"，即浆料通过立式布置的一对运动着的网袋，优点是脱水发生在正在形成的纸页两侧。

2. *At the couch roll, the wet paper has enough strength to be removed from the wire and passed around a series of heated drying drums where moisture content is brought down to*

finished-product specifications. (Para. 7) 在伏辊上，湿纸有足够的强度从周围的带子上通过一系列的热干燥鼓除去水分，将含水量降低到成品的规格。

3. *Cylinder machines **dewater** furnish at the cylinder and paste a thin layer of fiber against the felt. The fibers of subsequent layers do not intermingle, and therefore the bond between the layers is weak.* (Para. 9) 圆网纸机在滚筒上进行纸浆脱水并且把一薄层的纤维黏结紧贴在毡布上，之后的纤维层由于没有混合在一起，因此层与层之间的连接不牢固。

4. *The greater the degree of fiber alignment, the greater the difference in a given property when measured in MD and CD. The ratio of a property in the two directions is often used as a **gauge** of fiber alignment.* (Para. 16) 纤维排列程度越大，对纸张纵向和横向的给定性能的差别就越大。在这两个方向上性能的比率经常被用作衡量纤维排列的尺度。

5. *If the paper gains moisture and expands while the surfacing laminate or coating remains the same, the paper will curl toward the surfacing material. When the paper loses moisture, it will shrink and curl away from the surfacing material.* (Para. 29) 当纸张表面的复合材料或涂层保持不变，如果纸张吸收了水分并且延展，纸张就会朝着表层材料卷曲。而当纸张失去水分时，它就会收缩并背向表层材料卷曲。

Overview Questions

1. Paper is made with different machines, each machine imparting certain characteristics. In most discussions, paper is referred to as coming from two types of papermaking machines. What are these machines, and what is the principal characteristic that indicates what kind of machine a paper came from?
2. Is there a difference between "whiteness" and "brightness" when describing paperboard? Define these terms.
3. Paper is hygroexpansive. What does this term mean, and where might this property affect your packaging?
4. Which of the following characteristics is highest in MD: tear, fold endurance, or stiffness?
5. What is the moisture content of paper at standard test conditions? Paper is hygroscopic. What does this term mean?

Lesson 4 Corrugated Fiberboard Boxes

1 Historical Perspective

① The most common type of shipping container being used commercially today is the **corrugated box.** *The first patents for making corrugated paper were recorded in England in 1856. In the United States the first patents were granted to A. L. Jones in 1871 for an* **unlined** *corrugated sheet for packing lamp chimneys and similar fragile objects.*

② The first user of a box made of double-lined corrugated board was a cereal manufacturer, which obtained acceptance in the official **freight classification** for this type of shipping container in 1903. By the end of World War I, about 20 percent of the boxes were corrugated or **solid fibreboard,** and 80 percent were of wood construction. By the end of World War II, these figures had reversed, and 80 percent of all shipments were being made in fibreboard boxes.

③ Now more than $17.3 billion worth of corrugated containers are being produced in about 770 plants in the United States. Many of these are sheet plants, which buy **combined board** from other plants and do only the printing and cutting.

④ The proper name for a fibre shipping container is "box", rather than "carton" or "case", although all three terms are commonly used to describe the same type of container.

⑤ The railroads were the first continental mass movers of goods. Since **common carriers** are liable for loss or damage of goods in their care, they had an early interest in the quality of shipping containers. The first rules for constructing corrugated containers were established in the United States by **the railroad's Freight Classification Committee** in 1906. These rules, updated many times, continue in use as rule 41 of **the Uniform Freight Classification** (UFC). A similar set of rulings was later adopted by the trucking industry as **National Motor Freight Classification** (**NMFC**) item 222.

⑥ Broadly speaking, the classification systems require that boxes shipped by rail or truck meet certain construction requirements. Briefly, the rules provided that specified board grades were to be used to construct corrugated boxes, depending on the weight and dimensions of the intended box. The box construction was to be described on a box maker's class stamp on the bottom of the box.

⑦ A key aspect of earlier rules was the sole use of the **Mullen burst test** for designating corrugated board grades. This led to the establishment of essentially five linerboard basis weights from which corrugated board could be constructed. Until recently, these rules guided corrugated container design.

⑧ Over time, however, methods of transport and materials handling changed. **Box compression strength** became an overriding consideration. The burst test is a measure of a board's resistance to **rupture** and is somewhat related to the board's **tensile properties,** but it has no direct correlation to a container's ability to hold a load stacked upon it. Advances in paper manufacturing made it possible to make stiffer papers, but these would not necessarily meet the existing burst-test standards.

⑨ *The **edge crush test** (ECT) was proposed as a more suitable test for grading corrugated board, the advantage being that an ECT value could be used to calculate the anticipated compression strength of a container.* After much debate, a dual grading system was put into effect in 1991, by which a corrugated box could be designed to meet the original criteria based on the Mullen burst test or to meet new criteria based on ECT values.

2 Corrugated Board

2.1 Construction

⑩ Corrugated fiberboard, or combined board has two main components: the linerboard and the medium. Both are made of a special kind of heavy paper called containerboard. Linerboard is the flat facing that adheres to the medium. The medium is the **wavy, fluted** paper in between the liners (Figure 4.1).

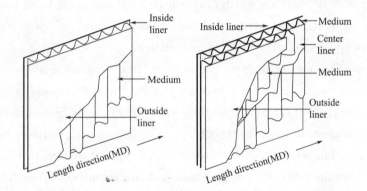

Figure 4.1 Corrugated fiberboard

⑪ The following illustrations demonstrate the four types of combined board:

Single Face One medium is glued to one flat sheet of liner board (Figure 4.2).

Single Wall The medium is between two sheets of liner board (Figure 4.3). Also known as **Double Face.**

Figure 4.2 Single face corrugated   Figure 4.3 Single wall corrugated board

Double Wall Three sheets of linerboard with two mediums in between (Figure 4.4).

Triple Wall Four sheets of linerboard with three mediums in between (Figure 4.5).

Figure 4.4 Double wall corrugated board

Figure 4.5 Triple wall corrugated board

12 The facings are typically kraft linerboard manufactured on a Fourdrinier paper machine with two headboxes. Corrugating medium is a one-ply sheet and frequently contains hardwood and recycled fiber. The corrugating machine forms the medium into a fluted pattern and bonds it to the linerboard facings, usually with a starch-based adhesive. A material with only one facing sheet (single face) is flexible in one axis and is sometimes used for a protective wrapping.

2.2 Flutes

13 *Architects have known for thousands of years that an arch with the proper curve is the strongest way to span a given space. The inventors of corrugated fiberboard applied this same principle to paper when they put arches in the corrugated medium. These arches are known as flutes and when anchored to the linerboard with a starch-based adhesive, they resist bending and pressure from all directions.*

14 When a piece of combined board is placed on its end, the arches form rigid columns, capable of supporting a great deal of weight. When pressure is applied to the side of the board, the space in between the flutes acts as a cushion to protect container's contents.

15 The flutes also serve as an insulator, providing some product protection from sudden temperature changes. At the same time, the vertical linerboard provides more strength and protects the flutes from damage.

16 Flutes come in several standard shapes or **flute profiles** (Figure 4.6, A, B, C, E, F, etc.). A-flute was the first to be developed and is the largest common flute profile. B-flute was next and is much smaller. C-flute followed and is between A and B in size. E-flute is smaller than B and F-flute is smaller yet.

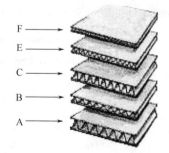

Figure 4.6 Common flute profiles (A, B, C, E, F, etc.)

17 In addition to these five most common profiles, new flute profiles-both larger and smaller than those listed here-are being created for more specialized boards. Generally, larger flute profiles deliver greater vertical compression strength and cushioning. Smaller flute profiles provide enhanced structural and graphics capabilities for primary (retail) packaging.

18 Different flute profiles can be combined in one piece of combined board. For instance, in a triple wall board, one layer of medium might be A-flute while the other two layers may be C-flute. Mixing flute profiles in this way allows designers to manipulate the compression

strength, cushioning strength and total thickness of the combined board.

⑲ Table 4.1 designates the four standard flute (i.e., A-, B-, C-, and E-flute) sizes. A still finer flute, called F-flute or **microflute**, is being produced by a few board manufacturers. "**Take-up factor**" is the length of medium per length of finished corrugated board; for example, it takes 1.54 m of medium to make 1 m of A-flute.

Table 4.1 Standard flute configurations

Flute	Flutes/Metre	Flutes/Foot	Thickness[①]	Factor
A	100~120	30~36	4.67 mm(0.184 in)	1.54
B	145~165	44~50	2.46 mm(0.097 in)	1.32
C	120~140	36~42	3.63 mm(0.142 in)	1.42
E	280~310	86~94	1.19 mm(0.047 in)	1.27

① Not including facings.

⑳ Finished board is described by component grammage or basis weight, starting from the outside. The outside of a corrugated sheet has the smoother finish. Embossed lines from the corrugating rolls are usually visible on the inside. Corrugated board described as 205/127C/161 would have the following components:

Outside liner=205 grams

Medium=127 grams, formed to a C-flute

Inside liner=161 grams

2.3 Fiberboard Grades

㉑ Facings and corrugating medium material can be made to virtually any weight or thickness; however, historical North American freight rule requirements have led to the standardization of certain traditional grades (Table 4.2). Linerboard and medium are specified in metric by their grammage: the mass in grams per square metre. "Basis weight" refers to the weight in pounds per 1,000 square feet. This is sometimes abbreviated to lb/MSF. The most common medium weights are shown in Table 4.3.

Table 4.2 The most commonly used linerboard grades, based on Mullen test grading. These are being phased out in favor of edge crush test (ECT) grades.

North American Grades		European Grades
Grammage	Basis Weight	Grammage
127g	26 lb	125 g
161g	33 lb	150 g
186 g	38 lb	—
205 g	42 lb	200 g
—	—	225 g
—	—	250 g
337 g	69 lb	300 g
Other grades	—	400 g
—	—	440 g

㉒ A generation of newer linerboards, referred to as high-performance boards, is being made to meet ECT rather than Mullen burst test and basis weight requirements. *Since the alternate carrier rules call for these boards to meet stiffness values measured by an edge crush test, there are no standard basis weights of the kind found in the older Mullen and basis weight system.* Users have found-to their advantage-that in many situations they can get satisfactory performance using lighter grades of the high-performance boards.

Table 4.3 The most commonly used corrugating medium weights

Grammage	*Basis Weight*	*Grammage*	*Basis Weight*
127g	26lb	161g	33lb
147g	30lb	195g	40lb

㉓ Linerboard is made in several material constructions, natural kraft being the most common. A linerboard with a whiter surface to provide better graphics is made by using bleached fiber in the second headbox of the paper-making machine. Because the bleached fiber layer is thin, some of the background kraft shows through, giving it a somewhat mottled appearance. "Mottled white" and "**oysterboard**" are common commercial names used to describe this board. Solid bleached white kraft is used where top-quality graphics are needed, but at a significant cost markup. It is typically used in preprint applications.

㉔ Varying amounts of recycled or secondary fiber are used for producing both linerboard and medium, a practice that can be expected to grow as recycling increases. Recycled board is made to the same specifications as **virgin** containerboard, so that stiffness and burst values are similar. Recycled board is sometimes slightly thicker to compensate for the somewhat weaker recycled fiber. Other properties will depend on the source and quality of the fiber.

㉕ Recycled board typically has a smoother surface finish and lower **coefficient of friction** (**CoF**) than virgin kraft. A good recycled board has an excellent printing surface. Some recycled board is not as stiff as virgin kraft and performs better on equipment such as wraparound casers-machines that wrap a sheet of corrugated board around the product rather than putting it in a previously manufactured box.

2.4 Corrugating Adhesives

㉖ Standard corrugated board is made with a **starch-based adhesive** applied at about 10 to 14 grams per square metre. Starch is not tolerant of high moisture levels and loses strength quickly. Since corrugated board itself loses about 50% of its compression strength between 50% R.H. and 90% R.H., this does not pose a limitation. Where higher resistance is needed, starches can be modified or supplemented by the addition of various polymeric materials. A **weather-resistant** adhesive would maintain box properties at a somewhat higher level for a longer period.

㉗ Water-resistant adhesive would be required for those applications where the finished container will be in actual contact with water for periods of time. *These are more expensive than weather-resistant adhesives and would be used only with corrugated board that is waxed or otherwise treated.*

2.5 Board Manufacture

28 A corrugating machine is made up of a number of operating stations that take the appropriate linerboards and mediums, shape the flutes, and join the fluted medium to the linerboards.

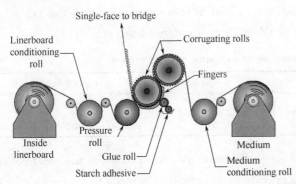

Figure 4.7 The single-facer of a corrugating machine is where the flutes are formed and bonded to the inside liner

29 At the **single-facer** station (Figure 4.7), medium is **preconditioned** with heat and steam to make it **pliable** and capable of being formed into a fluted configuration.

30 Linerboard is also pretreated to bring it to the same temperature and moisture conditions as the medium. The medium is fluted by being passed between large rolls with a geared surface pattern matching the desired flute geometry.

31 Older machines have brass **"fingers"** to hold the fluted medium in the forming rolls until it can be joined to the linerboard. These fingers, spaced a few inches apart, remove some adhesive and lightly score the medium. Newer "fingerless" machines use a vacuum system to hold the medium to the forming rolls. Board from a fingerless machine is said to have better compression properties.

32 Adhesive is applied to the flute tips, and preconditioned linerboard is pressed against the medium, where heat and pressure **gel** the starch adhesive. This liner frequently has visible lines embossed into the surface, and boxes are usually constructed with this liner to the inside; hence the term **"inside liner"**.

33 The single-faced material is flexible in one axis and is sent to the bridge, where it is **draped** in an **overlapping** wave pattern as it travels to the **double-backer** station. The purpose of the bridge is to isolate the two ends of the corrugating machine.

34 At the double backer section, adhesive is applied to the flute tips on the other side of the medium, and the outer linerboard is matched to the single-faced board. The assembled corrugated board is no longer flexible in any axis, and the final heat setting and cooling is done between two long, flat belts. The edges are trimmed, and the finished board is slit to widths and cut to lengths corresponding to the required order. The finished sheets are then stacked ready for subsequent operations (Figure 4.8).

35 Most corrugated board is manufactured in a balanced construction: the outer and inner liners have identical grammage. For some purposes, an unbalanced construction may be specified. For example, going from a board with 205g (42 lb) liners to one with 337g (69lb) liners, the next available balanced construction, is a substantial step up. Someone looking for only a small gain in performance may elect to upgrade one liner only.

36 Unbalanced constructions tend to have more problems with board **warpage**, but this

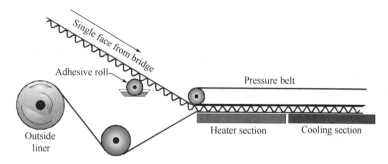

Figure 4.8 The double-backer section of corrugating machine where a second linerboard is applied to the single-faced material coming from the single-facer unit

should not preclude their use where the economics justify it. For better printing, the heavier liner is placed on the outside of the finished box. For better compression strength, the heavier liner is put on the inside.

3 Fiberboard Characterization Tests

㊲ *A number of paper characterization tests are used to describe certain basic properties in the corrugating plant and, to varying degrees, in subsequent operations for purposes of design input and quality control.* Only the most common are noted here. Most board tests are described in methods provided by the Technical Association of the Pulp and Paper Industry (TAPPI).

3.1 Mullen Burst Test (TAPPI T 810)

㊳ Briefly, the Mullen burst test involves forcing a **rubber diaphragm** against a facing of the fiberboard [Figure 4.9(a)] until the facing bursts. The burst test or bursting strength values are reported as pounds per square inch (psi) or the kilopascals (kPa) (1psi = 6.89476kPa). Burst values are not significantly affected by the medium.

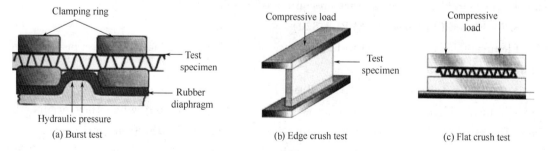

Figure 4.9 Standard corrugated board burst and crush tests

㊴ The burst test has limited use as a design value, since it is not related to the most critical properties of corrugated board from a box user's point of view. Burst tests are related to paper tensile strength, while the box user is often interested in stiffness and compression. Cloth will have a high Mullen burst value but has no compression strength.

3.2 Edgewise Compression Test (TAPPI T 811)

㊵ In an **edgewise compression test** (**ECT**), a small specimen (2 in×2 in) is placed between the **platens** of a compression tester and loaded until failure occurs. [Figure 4.9(b)] Edgewise compression values are a function of the stiffness contributed by the facings and the medium. ECT values have a direct relationship to the projected **stacking strength**.

㊶ One objection to the Mullen burst test is that it has no relationship to a box's compression strength, while a number of formulas have been devised that relate ECT values to expected box compression strength.

㊷ Carrier rule changes allow either Mullen values or ECT values to specify the boards used to construct a corrugated container. The majority of shippers choose to use ECT values.

3.3 Flat Crush Test (TAPPI T 808).

㊸ The **flat crush test** is similar to the edge compression test except that the specimen is compressed in the flat [Figure 4.9(c)]. The test provides a measure of flute **rigidity**.

3.4 Combined Weight of Facings

㊹ The combined weight of facings is required on the box maker's class stamp when the box is constructed to Mullen test rules. This measurement describes the combined linerboard weight (i.e., the weight of the inner and outer liners on a singlewall board) per 1,000 square feet of corrugated board. This information is not required on the class stamp when using the ECT system.

3.5 Thickness of Corrugated Board (TAPPI T 411)

㊺ Reduced board thickness (caliper) is an excellent indicator of reduced compression strength. Caliper can be reduced by improper manufacture, excessive printing pressure, and improper handling and storage.

3.6 Gurley Porosity (TAPPI T 460 and T 536)

㊻ **Gurley porosity** measures the time it takes for a given volume of air to pass through a paper. The lower the number, the more porous the paper. Porosity can vary from 2 seconds to 200 seconds but averages between 10 and 20 seconds. The porosity of paper is sometimes the **culprit** when problems occur at **vacuum-cup** transfer points.

3.7 Flexural Stiffness (TAPPI T 820)

㊼ Flexural stiffness is related to box compression strength. Reduced stiffness is a good indicator of damage during fabrication.

3.8 Water Take-up Tests (TAPPI T 441)

㊽ Gluing problems are not common with corrugated board, but when they occur, the **Cobb size test**, which measures the amount of water absorbed by the facing in a given time, is frequently asked for. This "how-much" value is not a good test for gluing problems. What is really needed is a "how-fast" value, and the time to absorb a single drop of water or ink would be a better comparative test.

㊾ The Cobb test is used to measure water absorption for materials specified to be used for hazardous product containers.

3.9 Puncture Test (TAPPI T 803)

㊿ The **puncture test** measures the energy required to puncture a board with a triangular **pyramidal** point affixed to a **pendulum arm**. "Beach puncture" is used mostly to quantify the puncture resistance and stiffness of triple wall corrugated. The **box maker's stamp** on triple wall containers calls for a puncture test rather than a bursting test.

51 A triangular pyramidal point with 25mm (1inch) sides, mounted on a weighted swinging pendulum, is forced through the board under test. The energy absorbed in forcing the tip completely through the board is reported in "puncture test units".

3.10 Pin Adhesion (TAPPI T 821)

52 **Pin adhesion** quantifies the strength of the bond between the medium's flute tips and the linerboard facings.

3.11 Ply Separation (TAPPI T 812)

53 **Ply separation** evaluates the board's resistance to ply separation when exposed to water. It is used mostly to differentiate boards made with conventional adhesives from those made with weather-resistant adhesives.

3.12 Coefficient of Friction (TAPPI T 815 and ASTM 04521)

54 **Coefficient of friction** (CoF) can affect machinability and load stability. One method of determining CoF consists of placing a weighted **sled** of the test material onto a plane surface faced with the material. The angle is gently increased, and the plane angle at which **slippage** is first observed is noted. The tangent of the average slide angle is reported as the static CoF. A **stress/strain machine** can be used to directly measure the force required to pull a sled along a flat surface. A stress/strain machine method will give both static and dynamic CoF values.

55 A CoF of less than 0.30 is considered to be unacceptable, and values between 0.30 and 0.40 are **marginal.** Boxes with a CoF of between 0.40 and 0.50 are fairly stable; normal untreated boxes fall within this range. Boxes with values below 0.30 usually need to be treated in some way. This may be as simple as brushing them with water after sealing and stacking or as complex as special **antiskid** surface treatments and coatings.

4 Corrugated Boxes

4.1 Selecting the Correct Flute

56 Consult closely with your corrugated supplier to select the correct flute and board weight. Although there are apparently many choices, in practice a corrugated supplier stocks few. **Barring** any established technical reason to do otherwise, use a carrier classification and C-flute as good starting points.

57 E-and F-flutes are not associated with shipping containers but rather are replacements for thicker grades of solid paperboard. The general **rule of thumb** says that E-or F-flute should be considered if a folding carton design calls for boards thicker than 750μm (30 point). E-and F-flutes are mostly used to replace paperboard for heavier or special protective primary packs. They are an excellent choice where the primary container may become a distribution container for some part of its travels. Small tools, hardware, small appliances, and housewares often fall into this category.

58 A-flute, one of the originally specified flutes, is not in common use today. A-flute's

almost 5 mm (1/4 in) thickness occupies more space than C-flute and has significantly greater **deflection** before bearing a load when compressed.

(59) In theory, A-flute's thicker section should give it the highest top-to-bottom compression strength of the three flutes. This is true under laboratory conditions. However, A-flute has the lowest flat crush resistance (Table 4.4). This makes A-flute with 127g (26-pound) medium almost impossible to machine and transport without destroying the flute structure. Engineering studies suggest that A-flute is most efficient when constructed with a 195g medium. Some authorities regard 127gram medium C-flute to have the minimum acceptable flat crush for shipping applications. Heavy mediums cannot be made into the small B-flute.

Table 4.4 Relative flute flat crush values

Medium Grammage	*A-Flute*	*C-Flute*	*B-Flute*
127g	0.70	1.00	1.15
161g	0.90	1.25	1.45
195g	1.10	1.50	N.A.

(60) A-flute is useful for **cushion pads** and the construction of triple wall board grades, where the added thickness is an advantage. With heavy mediums, A-flute makes a strong, rigid box. B-flute is used for canned goods or other products where box stacking strength is not required. B-flute's high flat crush strength is an advantage when supporting heavy goods such as bottles or cans. It can also be used to advantage for lighter load applications where a high stack strength is not needed, or where the distribution environment is very short.

(61) C-flute has about 10% better stacking strength than the same board weights in B-flute. It is best for applications where the corrugated container must bear some or all of the warehouse load. C-flute is sometimes chosen over B-flute for boxes that will hold glass bottles, despite C-flute's lower flat crush strength. It is felt that the thicker flute will provide more protection for the glass. Table 4.5 summarizes the differences among the various types of flutes.

Table 4.5 Comparison of corrugated board characteristics

Characteristic	*A-Flute*①	*B-Flute*	*C-Flute*	*E-Flute*
Stack strength	best①	fair	good	poor
Printing	poor	good	fair	best
Die cutting	poor	good	fair	best
Puncture	good	fair	best	poor
Storage space	most	good	fair	least
Score/bend	poor	good	fair	best
Cushioning	best	fair	good	poor
Flat crush	poor	good	fair	fair

① A-flute is subject to the limitations of flat crush when light mediums are used.

4.2 Box Styles

⑥② The primary use for corrugated combined board is boxes. Over 90 percent of all goods in most developed countries are shipped in corrugated boxes. These boxes can be used for everything from apples to washing machines. By changing the design of corrugated boxes, combining layers of corrugated or adding interior packaging, a corrugated box can be manufactured to efficiently ship and store almost any product.

⑥③ Many standard box styles can be identified in three ways: by a descriptive name, by an **acronym** based on that name, or by an international code number. For example, a **Regular Slotted Container** could also be referred to as an **RSC** or as ♯0201.

⑥④ The numerical code system, known as the **International Fibreboard Case Code**, was developed by the **European Federation of Manufacturers of Corrugated Board (FEFCO)** and the **European Solid Fiberboard Case Manufacturer's Association (ASSCO)** to avoid confusion when communicating in different languages. This code has been adopted by the **International Corrugated Case Association (ICCA)**.

⑥⑤ There are many standard corrugated box styles (**Slotted Boxes, Telescope Boxes, Folders, Rigid Boxes (Bliss Boxes), Self-Erecting Boxes** and **Interior Forms**) -so many, in fact, that it is impossible to describe them all here. In addition, corrugated boxes can be custom-designed to meet the specific needs of any box user. A manufacturer's representative will have more information about additional box style options.

⑥⑥ Slotted box styles (International Fibreboard Case Code: 02 Series) are generally made from one piece of corrugated or solid fiberboard. The blank is scored and **slotted** to permit folding. The box manufacturer forms a joint at the point where one **side panel** and one **end panel** are brought together. Boxes are then shipped flat to the user. When the box is needed, the box user **squares up** the box, inserts product and closes the **flaps**. The International Fibreboard Case Code refers to these styles as Slotted-Type Boxes, while the carrier classifications call them Conventional Slotted Boxes.

⑥⑦ "Regular Slotted Container" (RSC or ♯0201) is the **workhorse** corrugated box style (Figure 4.10). All his flaps have the same length, and the two outer flaps (normally the lengthwise flaps) are one-half the container's width, so that they meet at the center of the box when folded. If the product requires a flat, even bottom surface, or the protection of two full layers, a **fill-in pad** can be placed between the two inner flaps.

⑥⑧ This is a highly efficient design for many applications. There is very little manufacturing

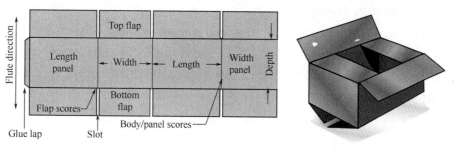

Figure 4.10 Parts of a regular slotted container (RSC) blank

waste. The RSC can be used for most products and is the most common box style.

4.3 Manufacturer's joint

⟨69⟩ *A flat piece of corrugated fiberboard, which has been cut, slotted and scored, is called box blank. For some box styles, in order to make a box, the two ends of the box blank must be fastened together with tape, staples or glue. The place where these two ends meet is known as the* **manufacturer's joint**.

⟨70⟩ Liquid adhesives are most often used to join the two surfaces. Often there is a **glue tab**, extending along one end of the box blank. This tab is scored and folded to form one corner of the box when joined. The tab can be joined to either the inside or the outside of the box. If there is no tab, the box must be joined using tape. Not all boxes have manufacturers joints; for example, the bliss box does not. However, most widely used box styles have a manufacturer's joint.

4.4 Dimensioning

⟨71⟩ Dimensions are given in the sequence of length, width and depth. Internationally, the words *length*, *breadth* and *height* may be used to express these dimensions. The dimensions of a box are described based on the opening of an **assembled box**, which can be located on the top or the side, depending on how it is to be filled. The opening of a box is a rectangle; that is, it has two sets of parallel sides. The longer of the two sides is considered its length, the shorter of the two sides is considered its width. The side perpendicular to length and width is considered the depth of the box.

⟨72⟩ Dimensions can be specified for either the inside or the outside of the box. Accurate **inside dimensions** must be determined to ensure the proper fit for the product being shipped or stored. At the same time, **palletizing** and distributing the boxes depends on the **outside dimensions**. The box manufacturer should be informed as to which dimension is most important to the customer.

5 Carrier Rules

5.1 Application

⟨73⟩ The Uniform Freight Classification (UFC) and National Motor Freight Classification (NMFC) were established to categorize articles for shipment via common carrier with respect to value, density, **fragility**, and potential for damage to other freight. Rail and motor carriers publish the classifications. The classifications specify the conditions under which specific articles can be shipped and at what rates. *When shipping by rail, refer to UFC. When shipping by truck, refer to NMFC. UFC rule 41 and NMFC item 222 are the most frequently used in describing corrugated packaging.*

⟨74⟩ There are four basic steps for determining **authorized** packaging:

• Fully identify the product. The product must be identified by its transportation description, not by trade or popular names. The **bill of lading** description is often the best reference.

• Select the proper governing classification. If you know that the product will be shipped exclusively by rail or truck, use that classification. If there is some question, consult both and follow the more demanding requirement.

• Use the "Index to Articles" to find the applicable item number.

• Consult the proper article to find the required packaging.

75 Failure to comply with regulations can subject the shipper to **penalties** such as higher freight rates, refusal of acceptance by the carrier, or nonpayment of damage claims.

5.2 Summary of Rules for Corrugated Box Construction

Carrier rules for corrugated box construction can be summarized as follows:

76 Specified boards (using either Mullen burst test or ECT values) shall be used for a given product weight, providing the box does not exceed a specified dimensional limit. The dimensional size limit for a box is determined by adding an outside length, width, and depth.

77 Table 4.6 summarizes the construction requirements for corrugated boxes. The values given for ECT may change as more information becomes available. The rules also require that a **box manufacturer's certificate** (BMC) on the bottom of the container (Figure 4.11).

Figure 4.11 Box manufacturer's certificates using burst test and ECT values

78 If the user elects to use Mullen values to specify box construction, the board must meet the minimum burst requirements set out in Table 4.6, Part A. If the user elects to use ECT values to specify box construction, the board must meet the minimum ECT values set out in Table 4.6, Part B. Note that although the two parts are side by side, Mullen values and ECT values have no correlation; the boards that have the values listed in Parts A and B of the Table 4.6 are not equivalent.

79 Using the freight classifications as a design specification has a number of disadvantages. For example:

• No reference is made to duration of warehouse storage.

• No reference is made to flute type.

• No reference is made to distribution hazards.

• The tables essentially apply only to regular slotted container (RSC) designs.

• The contribution of different medium weights is not taken into account.

Freight classifications therefore do not necessarily form the most effective basis for a design.

Table 4.6 Summary of carrier rules for corrugated boxes

PART A[①]				PART B[①]
Maximum weight of Box and Contents (lbs)	Maximum Outside Dimension, Length, Width and Depth Added(in)	Minimum Burst Test, Single Wall, Double Wall or Solid Fiberboard (lbs per sq. in) or Minimum Puncture Test, Triple Wall Board (in oz per in of tear)	Minimum Combined Weight of Facings, including Center Facing(s) of Double Wall and Triple Wall Board or Minimum Combined Weight of Pliers, Solid Fiberboard, Excluding Adhesives (lbs per 1,000 sq. ft)	Minimum Edge Crush Test (ECT) (lbs per in width)
Single Wall Corrugated Fiberboard Boxes				
20	40	125	52	23
35	50	150	66	26
50	60	175	75	29
65	75	200	84	32
80	85	250	111	40
95	95	275	138	44
120	105	350	180	55
Double Wall Corrugated Fiberboard Boxes				
80	85	200	92	42
100	95	275	110	48
120	105	350	126	51
140	110	400	180	61
160	115	500	222	71
180	120	600	270	82
Triple Wall Corrugated Fiberboard Boxes				
240	110	700	168	67
260	115	900	222	80
280	120	1100	264	90
300	125	1300	360	112
Solid Fiberboard Boxes				
20	40	125	114	
40	60	175	149	
65	75	200	190	
90	90	275	237	
120	100	350	283	

① Mullen(Part A) and ECT(Part B) are presented side-by-side, but there is no correlation between the values.

(5516 words)

New Words and Expressions

unlined [ʌnˈlaind] *adj.* 无贴面纸的
rupture [ˈrʌptʃə] *v.* 破裂
wavy [ˈweivi] *adj.* 波浪状的
fluted [ˈfluːtid] *adj.* 有凹槽的
virgin [ˈvəːdʒin] *adj.* 没有处理过的

precondition [priːkənˈdiʃ(ə)n] v. 预处理
pliable [ˈplaiəbl] adj. 易曲折的、柔软的
finger [ˈfiŋgə] n. 导向板
gel [dʒel] v. 胶化
drape [dreip] v. 用布帘覆盖、使呈褶皱状
overlap [əuvəˈlæp] n. 重叠
warpage [ˈwɔːpeidʒ] n. 翘曲、弯曲
platen [ˈplæt(ə)n] n. 压盘、压板
rigidity [riˈdʒidəti] n. 硬度、刚性
culprit [ˈkʌlprit] n. 问题、麻烦
pyramidal [piˈræmid(ə)l] adj. 椎体的
sled [sled] n. 翘板
slippage [ˈslipidʒ] n. 移动、滑动
marginal [ˈmɑːdʒin(ə)l] adj. 少量的
antiskid [ˌæntiˈskid] adj. 防滑的
barring [bɑːriŋ] prep. 除……之外
deflection [diˈflekʃən] n. 偏差、挠曲
acronym [ˈækrənim] n. 首字母缩略词
slot [slɔt] v. 开槽
flap [flæp] n. 摇翼、摇盖
workhorse [ˈwəːkhɔːs] adj. 负重的
palletize [ˈpælətaiz] v. 托盘化
fragility [frəˈdʒiliti] n. 易碎性、脆值
authorized [ˈɔːθəraizd] adj. 经授权的、经认可的
penalty [ˈpen(ə)lti] n. 罚款、处罚
corrugated box   瓦楞纸箱
freight classification   货物分类
solid fibreboard   硬纸板
common carrier   公共承运商
railroad's Freight Classification Committee   铁路货物分类委员会
Uniform Freight Classification（UFC）   统一货物分类
National Motor Freight Classification（NMFC）   国家汽车货物分类
Mullen burst test   纸箱耐破度测试
box compression strength   纸箱抗压强度
tensile property   拉伸性能
edge crush test   边压强度测试
combined board   （瓦楞纸）复合板、瓦楞纸板
single face   单面瓦楞纸板（两层）
single wall   单瓦楞纸板（三层）
double face   双面瓦楞纸板（三层）

double wall　双瓦楞纸板（五层）
triple wall　三瓦楞纸板（七层）
flute profile　楞型
microflute　微瓦
take-up factor　收缩率
oysterboard　灰纸板
coefficient of friction（CoF）　摩擦系数
starch-based adhesive　淀粉黏结剂
weather-resistant　耐候的
single-facer　单面机
inside liner　里纸
double-backer　双面机
rubber diaphragm　橡皮膜
edgewise compression test（ECT）　边压测试
stacking strength　堆码强度
flat crush test　平压强度
Gurley porosity　格利孔隙度
vacuum-cup　真空吸盘
cobb size test　科布施胶度试验
puncture test　戳穿试验
pendulum arm　摇臂
Beach puncture　戳穿试验（即 puncture test）
box maker's stamp　纸箱制造商证章
pin adhesion　点黏合性测试
ply separation　分离层测试
stress/strain machine　应力/应变机
rule of thumb　经验法则
cushion pad　缓冲垫
regular slotted container（RSC）　普通开槽箱（即 0201 箱）
International Fibreboard Case Code　国际纤维板箱代码
European Federation of Manufacturers of Corrugated Board（FEFCO）　欧洲瓦楞纸板制造商联合会
European Solid Fiberboard Case Manufacturer's Association（ASSCO）　欧洲硬纸板箱制造商协会
International Corrugated Case Association（ICCA）　国际瓦楞纸箱协会
slotted box　开槽型纸箱
telescope box　套合型纸箱
folder box　折叠型纸箱
rigid box（Bliss box）　固定型纸箱
self-erecting box　能自动装配的纸箱

interior form　内附件
side panel　（纸箱）侧板
end panel　端板
square up　将纸箱撑开成直角
fill-in pad　填充衬垫
manufacturer's joint　制造接头
glue tab　糊头
assembled box　组装盒
inside dimension　内尺寸
outside dimension　外尺寸
bill of lading　提货单
box manufacturer's certificate（BMC）　纸箱制造证章

Notes

1. *The first patents for making corrugated paper were recorded in England in 1856. In the United States the first patents were granted to A. L. Jones in 1871 for an unlined corrugated sheet for packing lamp chimneys and similar fragile objects.*（Para. 1）有记录表明英国在1856年最先获得了瓦楞纸生产的发明专利。在美国，无贴面瓦楞纸板生产的第一个专利于1871年被授予了A. L. Jones，当时主要是为包装灯罩和类似的易碎物品。were granted to 意指"被授权"。

2. *The edge crush test（ECT）was proposed as a more suitable test for grading corrugated board, the advantage being that an ECT value could be used to calculate the anticipated compression strength of a container.*（Para. 9）提议边压测试以更适于度量瓦楞纸板的等级，其优点是边压强度值可以用来计算瓦楞纸箱容器的预期抗压强度。was proposed as 意指"被提议作为……"。

3. *Architects have known for thousands of years that an arch with the proper curve is the strongest way to span a given space. The inventors of corrugated fiberboard applied this same principle to paper when they put arches in the corrugated medium. These arches are known as flutes and when anchored to the linerboard with a starch-based adhesive, they resist bending and pressure from all directions.*（Para. 13）几千年以来，建筑师就因为利用合适的拱形曲度建造强度高的跨度桥而闻名。瓦楞纸板的发明者们把芯纸制成拱形也是利用相同的原理。这些拱形被称为瓦楞，当用淀粉黏结剂将它黏结在面纸上时，它们就能抵抗来自各个方向的弯曲和压力。known for 意指"因……闻名"；are known as 意指"被作为"。

4. *Since the alternate carrier rules call for these boards to meet stiffness values measured by an edge crush test, there are no standard basis weights of the kind found in the older Mullen and basis weight system.*（Para. 22）由于新的货运法规要求包装纸板要满足由边压测试所确定的刚度值，所以在以前的耐破度测试和基本重量体系里未发现存在此类标准的规定。call for 意指"要求"。

5. *Water-resistant adhesive would be required for those applications where the finished con-*

tainer will be in actual contact with water for periods of time. These are more expensive than weather-resistant adhesives and would be used only with corrugated board that is waxed or otherwise treated.（Para. 27）当容器在和水实际接触一段时间的应用场合，就需要应用防水性黏合剂。这些要比使用耐候性黏合剂更贵，并且只能应用于经过涂蜡或者经处理的瓦楞纸板。

6. A number of paper characterization tests are used to describe certain basic properties in the corrugating plant and, to varying degrees, in subsequent operations for purposes of design input and quality control.（Para. 37）若干纸张的性能测试值在瓦楞纸生产工厂里被用来描述纸的基本性能，在不同程度上，用于以设计输入和质量控制为目的的后续操作。to varying degrees 为插入语，意指"在不同程度"。

7. A flat piece of corrugated fiberboard, which has been cut, slotted and scored, is called box blank. For some box styles, in order to make a box, the two ends of the box blank must be fastened together with tape, staples or glue. The place where these two ends meet is known as the manufacturer's joint.（Para. 69）平张瓦楞纸板经模切、开槽、压痕后成为箱坯。对有些箱型，为了制成一个纸箱，箱坯的末端必须用胶带、钉子或者胶水接合在一起。箱坯两端的接合处就作为制造接头。

8. When shipping by rail, refer to UFC. When shipping by truck, refer to NMFC. UFC rule 41 and NMFC item 222 are the most frequently used in describing corrugated packaging.（Para. 73）当用铁路运输货物时，参照 UFC。当用卡车运输货物时，参照 NMFC。UFC 的第 41 条款和 NMFC 的第 222 条款常用于描述瓦楞包装。UFC rule 41 是美国铁路货物运输规定（UFC）中的一项条款，NMFC item 222 是美国国家公路货物运输规定（NMFC）中的一项条款，均包括了对瓦楞纸箱和硬纸板箱的要求。

Overview Questions

1. What are the five corrugated board flutes called, in order from largest to smallest? Give examples of the typical applications for each flute.
2. What does "combined weight of facings" describe? Where is this value used?
3. What is an edgewise compression test? What are the advantages and disadvantages of its use?
4. Why regular slotted container is one of the most popular box styles?

Lesson 5　Metal Containers

1　Background

① Steel is one of the older packaging materials and was originally used for round, square, and rectangular boxes and **canisters**. Tea and tobacco were two of the first products packaged in **tin-plated**, mechanically **seamed** or **soldered** steel containers with friction or **hinged lids**. Today such labor-intensive metal boxes are limited to custom and upscale applications. The old-fashioned appearance of a fabricated metal box is effectively used by package designers to create nostalgia for specialty and gift-type containers.

② Of all the metal packaging forms, none has had as much impact on society as the **sanitary** food can. Thermal processing of food packed into hand-soldered cylindrical metal cans started in the early 1800s, and soon developed into a major industry. Metal cans exhibit advantages of being relatively inexpensive, capable of being thermally processed, rigid, easy to process on high-speed lines, and readily recyclable. Metal offers a total barrier to gas and light. Despite market changes resulting from freezing and plastic-based packaging, metal cans remain an important means of delivering a shelf-stable product.

③ *Originally, all steel containers were fabricated from flat sheets that were cut to size, bent to shape, and mechanically* **clinched** *or soldered to hold the final shape.* Food cans were three-piece construction, a formed sidewall and a top and bottom end (Figure 5.1).

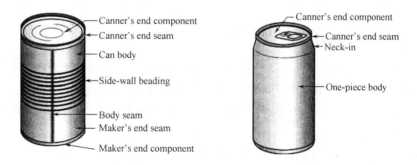

Figure 5.1　Three-piece (left) and two-piece (right) can construction

④ With time, ways of drawing metal (shaping metal by pushing it through a die) were developed. Shallow drawn containers with friction or **slip covers** were used for pastes, salves, greases, and other semisolid products. Later, two-piece **shallow drawn cans** with **double-seamed** (folded) ends were used for sardines.

⑤ **Two-piece cans** have a body and bottom in a single piece with a separate attached end (Figure 5.1). Immediate advantages are reduced metal usage, improved appearance, and the elimination of a possible leakage location. However, while **three-piece cans** can easily be changed in length and diameter, two-piece cans require more elaborate tooling that is dedicated to one can form.

⑥ *Improvements in metallurgy and processing allowed **deeper draws** and **multiple draws** and eventually led to a **draw-and-iron** process in which the walls of a drawn container were made thinner by an ironing step.* Aluminum joined steel as a can material.

⑦ **Ductile** metals, such as tin, lead, and aluminum, can be formed into tubular shapes by impact extrusion. Originally, only tin and lead were used to make **collapsible tubes-a** tube that can be collapsed, or squeezed to expel the contents. Today, impact-extruded collapsible tubes are made from aluminum, except for a small number of special applications requiring the chemical properties of tin or lead.

⑧ **Impact-extrusion** technology has advanced to the point where heavier gauge aluminum extrusions can be used for pressurized aerosol containers.

2 Common Metal Container Shapes

⑨ Stock metal cans come in a great variety of sizes and shapes. A quick list of the most common would contain the following:

• Three-piece steel sanitary food cans.

• Aerosol cans, made by two methods:

① Three-piece steel cans with a welded body and two ends.

② One-piece, impact-extruded aluminum cans necked-in to accept the valve cup.

• Steel or aluminum two-piece drawn-and-ironed beverage cans.

• Two-piece steel or aluminum cans made by drawing or by draw and redraw. Full-opening, **ring pull-top cans** are used for fish products, potted meats, and dips. Double-seamed, conventional-top cans can be used for many canned food products.

• Cans with hinged lids, usually steel, used for medications, confections, small parts and novelties.

• Flat round cans of drawn steel or aluminum with slipcovers. Used for **ointments**, salves, confections, shoe polish, and novelties (Figure 5.2).

• Three-piece steel or aluminum ovals, typically fitted with a **dispensing spout** and used for oils (Figure 5.2).

• Traditional pear-shaped, three-piece steel ham cans.

• **Oblong** steel three-piece F-style cans used mostly to contain aggressive solvents (Fig-

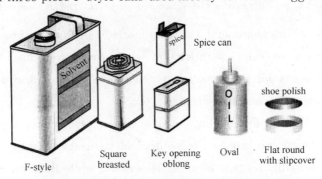

Figure 5.2 Examples of specialized can shapes

ure 5.2). The "F" name comes from **Flit insecticide**, an early, large-volume user. There are no A-, B-, C-, D-, or E-style cans.

• Oblong **key-opening cans**, three-piece steel, used for luncheon meat products.

• Multiple friction cans of three-piece steel, used primarily by the paint industry. Also referred to as **double-and triple-tight cans**.

• Three-piece, square-breasted steel cans. Larger designs are specific to the **talcum, bath, and baby powder** markets. Spice cans are smaller three-piece cans with a perforated metal or plastic top, used for spices and dry condiments.

• Industrial pails and drums. The most common are 20litre pails (5gal U.S.) and 210litre drums (55 gal U.S.).

• Two-piece, low-profile steel or aluminum ovals, with full-opening ring pull-tops for seafood products.

• Impact extrusions, which are used for one-piece aluminum aerosol cans and collapsible metal tubes.

3 Three-Piece Steel Cans

⑩ Steel three-piece can bodies can be mechanically seamed, bonded with adhesive, welded, or soldered (Figure 5.3). Aluminum cannot be soldered and cannot be welded economically. Welded sanitary three-piece can bodies are therefore made exclusively of steel. Mechanical seaming or clinching is used only for containers intended for dry product, where a **hermetic seal** is not important.

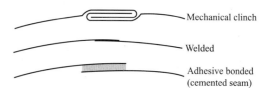

Figure 5.3 Mechanical, welded, and adhesive-bonded side seams for three-piece cans

⑪ Adhesive bonding, or cementing, uses a thermoplastic (or other) adhesive extruded onto a hot can blank. The blank is shaped into a cylinder on a body former. The thermoplastic adhesive is heated, and the seam is "bumped" and quickly chilled to set the bond.

⑫ *Adhesive bonding is an attractive body-assembly method for those applications where the can will not be subjected to thermal processing. Unlike welded cans, adhesive-bonded constructions can have full wraparound lithography.* At one time, three-piece beverage containers were adhesive bonded. Some frozen juice concentrate and paint cans are adhesive bonded.

⑬ *To solder a can, engaging hooks are bent into the can blank similar to the approach for a mechanical seam, the body is formed, and the engaging hooks are flattened to hold the cylindrical shape.* The seam is treated with a flux and is passed over a roll rotating in a bath of molten solder. Solders are typically composed of 97.5% lead and 2.5% tin. Lead extraction by food products was always a potential problem with soldered seams, and the industry quickly adopted welding technology when it became available. Soldered cans are no longer permitted for food in North America. Some soldering is still done for industrial and nonfood applications. The lead content of many of the solders used has been reduced or eliminated.

④ *Welded cans are strong and eliminate potential lead hazards. Most three-piece steel food cans are welded by a process initiated in Europe by Soudronic.* The body sheet is formed into a tube with a slight overlap along the joint. In the most common process, the joint is passed between two continuous copper wire electrodes; an electrical current passing through the joint heats and fuses the metal (Figure 5.4). **Lithographed can blanks** require about 6 mm (0.25in) of undecorated strip along the weld edges to ensure good electrical contact for welding. The welded seam line is about 30% thicker than the two base metal sheets. Cans shorter than 75 mm (3in) are too short to be welded individually and are made by welding a body twice the required length and then cutting it into two cans.

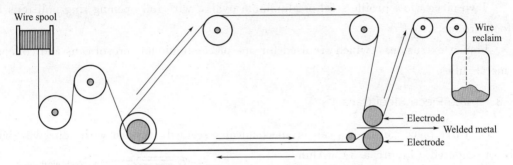

Figure 5.4 The can-welding process

⑤ All three-piece can bodies are pressure tested and have the ends **flanged** to receive the can top and bottom ends. The can maker applies one can end and sends the other end to the user for double seaming after the can is filled.

⑥ Sanitary food cans that may be thermally processed have bead patterns embossed into the can sidewalls to improve resistance to collapse because of external pressure. This prevents collapse (paneling) during pressure differentials encountered during retorting and enables the can to withstand an internal vacuum. Sidewall beading requires more material, reduces top-to-bottom compression strength, and complicates labeling. Many sidewall bead geometries are designed to maximize **hoop** strength while minimizing the accompanying problems.

⑦ Can ends intended for thermal processing are stamped with a series of circular expansion panels (Figure 5.5). This allows for movement of the end panels so that the contents are able to expand and contract without bulging or otherwise distorting the can. The **chuck** panel is designed to give the proper clearance to the double-seaming chuck used to seal the can end

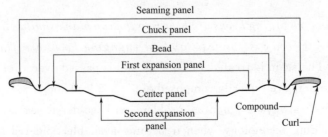

Figure 5.5 Typical can-end embossing pattern

to the body. A vital can-end component is the compound applied around the perimeter curl. This compound acts as a caulk or **sealant** when the end is mated and double-seamed to the can body (Figure 5.6 and Figure 5.7).

Figure 5.6 Double-seaming is the attachment of the can end to the body. It involves two curling steps

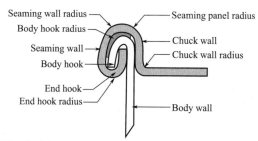

Figure 5.7 The double seam is a critical can component. Every angle, radius, and dimension must be correct to ensure a hermetic seal

4 Two-Piece Cans

⑱ Three methods of making steel or aluminum two-piece cans exist:
- Draw.
- Draw and redraw (DRD).
- Draw and iron (D & I).

4.1 Draw Processes

⑲ **Shallow-profile cans**—one whose height is less than their diameter—can be drawn directly from a circular metal blank. The metal blank is stamped or drawn through a die and re-formed into a new shape. The thickness of the finished can sidewall and bottom remain essentially the same as in the original blank. The process is sometimes referred to as "shallow draw".

Figure 5.8 Straight lines become distorted in different directions during drawing

⑳ Blanks for drawn cans may be decorated prior to drawing. Art must be distorted so that when the metal is re-formed, a correct image will develop (Figure 5.8). Cans that have continuous decoration across the sidewalls and bottom have been printed prior to drawing.

4.2 Draw-and-Redraw Process

㉑ A single-draw operation is limited in how far the metal can be reshaped. Cans having a height equal to or greater than the can diameter will usually require a second draw in what is called the **"draw-and-redraw"** process. The first draw produces a shallow cup. The second reduces the diameter as the can is deepened. Cans having a height significantly greater than the can diameter would require a third draw. If the container is to be thermally processed, sidewall beads are rolled into the walls in a separate step. Body flanges for engaging the can end are rolled on in a manner similar to that used in three-piece can manufacture.

4.3 Draw-and-Iron Process

22 Carbonated beverage cans are made by the draw-and-iron (D & I) process. A blank disk is first drawn into a wide cup (Figure 5.9, step 2). In a separate operation (Figure 5.9, step 3), the cup is redrawn to the finished can diameter and pushed through a series of ironing rings, each minutely smaller in diameter than the previous one (Figure 5.10). The rings "iron", or spread, the metal into a thinner sheet than the original disk.

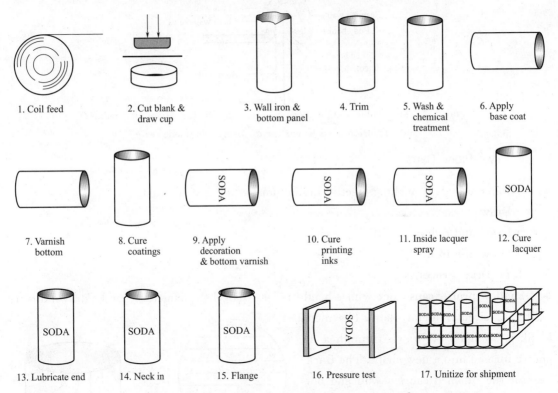

Figure 5.9 The manufacturing sequence for a necked D & I can

23 The bottom of a D & I can has the same thickness as the starting disk; however, the sidewalls are considerably reduced in thickness, and the metal area of the final can is greater than that of the initial disk. Necking operations reduce the diameter of the can top, thereby reducing the end-piece diameter. This results in significant metal savings, since the end piece is much thicker than the sidewalls.

24 The thin walls of the D & I container restrict its use to systems that will not undergo severe thermal processing and that will lend support to the walls. Carbonated beverage cans, where the internal pressure of the carbon dioxide keeps the walls from **denting**, is the prima-

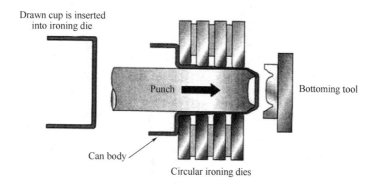

Figure 5.10 The second draw and the ironing stages are all accomplished in one continuous movement. The punch and the ironing rings are shown in this exaggerated illustration. The punch finishes its stroke against the bottoming tool

ry application. Noncarbonated juices packed in D & I cans rely on internal pressure created by inert nitrogen gas introduced into the container.

㉕ Both steel and aluminum are used to produce D & I beverage cans. Aluminum alloys such as 3004 are used for can bodies, whereas the softer 5082 and 5182 alloys are used for can ends. Soft drink producers can use either steel or aluminum equally well. Beer, however, is particularly sensitive to traces of dissolved iron while being relatively insensitive to aluminum.

㉖ Draw-and-iron manufacturing has been used to produce beverage cans as large as 2 litres.

5 Impact Extrusion

㉗ Impact extrusion forms ductile metals such as tin, lead, and aluminum into seamless tubes. Tin and lead were the first metals to be formed by this method, and until the 1960s most collapsible tubes were either lead or tin. Tin's high cost prohibits its use except for collapsible tubes for certain pharmaceuticals. Lead, once the mainstay of the toothpaste market, is now used only for applications where its chemical inertness is an asset. Most impact extrusions are made from aluminum.

㉘ In impact extrusion, a metal **slug** is located on a shaped striking surface, or **anvil**. A **punch** strikes the slug with great force. Under the enormous impact pressure, the metal flows like a liquid straight up along the outside of the striking punch, forming a round, cylindrical shape (Figure 5.11). Tube height can be up to seven times its diameter.

㉙ The tube's shoulders and tip are formed as part of the process. Tubes with a dispensing hole in the tip will have a hole in the slug, while tubes that need a dispenser with a thin web of metal over the opening (a blind end) will start with a solid slug. **Embossed** shoulders are another option.

㉚ The force of the impact work hardens the aluminum and makes it stiff. Collapsible tubes are **annealed** to remove the stiffness. The tubes are trimmed to length, threads are turned into the tube neck, and the tubes are sent for finishing.

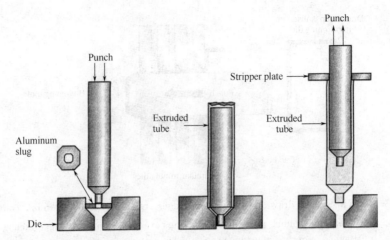

Figure 5.11 Impact-extrusion sequence

31　Neck designs are negotiated with the supplier and are specified by a number indicating the opening size in 64ths of an inch. Thus, a No. 12 neck has an opening of 3/16 inch. Figure 5.12 shows some commonly used tips.

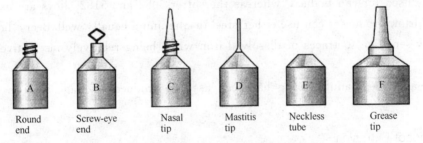

Figure 5.12 Typical impact-extruded tube tips

32　The round(A) end is the most common tip, and blind-end tubes have a metal **membrane** across the end when a positive seal is required. Screw-eye openings(B) are used mostly for adhesives that would bind a normal screw-cap closure. **Nasal tips**(C) are used for nasal ointments and for products that require local point application. Eye tips are similar and are used for **ophthalmic** medications and fluids that are dispensed by the drop. **Mastitis tips**(D) are similar to nasal and eye tips but have a fine, needlelike tip. Neckless tubes(E) can be used to dispense powders and for single entire-content applications. Grease tips(F) are used for dispensing greases when an applicator tip is required.

33　Metal tubes have a number of distinctive characteristics compared with laminate and plastic collapsible tubes:

- They form the best barrier to all gases and flavors.
- They have the best **dead-fold** characteristics (ability to be flattened or rolled up). This feature is particularly important for some pharmaceutical applications, where air **suckback** into the partly empty tube could contaminate the contents or expose the product to oxygen.
- They can be decorated in a manner that takes advantage of their metallic character.

- They have a wide range of lining options because of the metal's ability to withstand high curing temperatures.

54 Tubes are normally coated with a white **enamel** base and then cured. Tubes are printed by dry offset (**offset letterpress**), similar to any round container. Most manufacturers offer six colors.

55 By starting with heavier slugs, strong cylinders can be made by impact extrusion. These have been used to hold special greases and caulks and as **humidor** tubes for expensive cigars. A major application is for aerosol products, where the **sleek**, seamless appearance of these cans is an asset. When a cylinder is used as an aerosol can, a stiff sidewall is desirable, and those cylinders are not annealed. The sidewall is trimmed to length, turned down, and curled over to accept the spray nozzle base (Figure 5.13).

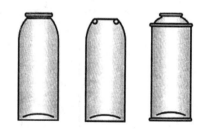

Figure 5.13 Two impact-extruded aerosol can designs (left, center) and a three-piece welded-steel aerosol can (right). All aerosol cans have bottoms that are domed upward against internal pressure

6 Aerosols

56 **Aerosol packaging** refers to products packaged in a pressurized container having a valve that permits controlled product release as required. Depending on the formulation, the valve system and the means of pressurizing, aerosols can be designed to release product in forms ranging from fine mists to heavy pastes.

57 Personal care products such as perfumes, shaving creams, **deodorants**, and hair sprays make up the largest segment of the aerosol market, followed closely by household products such as polishes, cleaners, and room fresheners. Paints, automotive products, and insect sprays have smaller market portions. Food applications are limited.

58 Although the principles of expelling fluids from a container by internal pressure were known earlier, the first practical application came in about 1942, when the U.S. military adopted a heavy metal tank pressurized with a **fluorocarbon** to disperse a fine insecticide mist. The advantage of aerosols was their ability to disperse product into much finer particles that stayed suspended in the air for a much longer time than was available from hand pumps and other systems. This "bug bomb" was used extensively in the Pacific during World War II to reduce the incidence of insect-borne diseases among the troops. By 1946 the first civilian aerosol insecticides made from modified beer cans entered the market, followed quickly by room fresheners and window cleaners.

6.1 Aerosol Propellants

59 A typical aerosol product has a **liquid phase** and a **vapor phase** (Figure 5.14). The liquid phase contains the product to be expelled. The vapor phase is at an increased pressure and will force the product up the dip tube and expel it through the **nozzle** whenever the valve is opened. The product typically occupies about 75%, but never more than 92.5%, of the

available space. Well-designed aerosol containers will deliver 95% or better of the contained product.

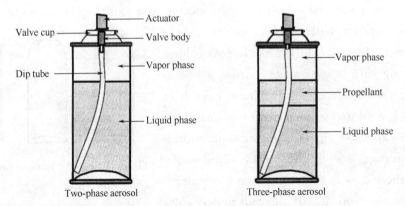

Figure 5.14   In a two-phase aerosol, the propellant is dissolved in the product.
In a three-phase system, the propellant forms a separate layer

⓰ A large part of aerosol design concerns the selection of a suitable propellant to develop the driving pressure. In principle, product can be expelled simply by charging the container with compressed air. The problem with this method is that as product is expelled and the head-space volume increases, the container pressure drops proportionately. Carbon dioxide and **nitrous oxide** gases have some solubility in water, and to that extent, the vapor-phase pressure can be maintained by gas coming out of the solution to replace lost head-space gas. These gases are used in some aerosols.

⓱ The ideal propellant is a gas that can be easily compressed and liquified at the desired operating pressures of an aerosol system. **Chlorofluorocarbons** (CFCs), **hydrocarbons**, **vinyl chlorides**, and **dimethyl** ether exhibit this property.

⓲ CFCs, characterized by high density and nonflammability, were among the first propellants. In the early 1970s, CFCs were implicated in **depletion** of the atmospheric **ozone** layer, and by 1979, they were banned from utility aerosols, and were **phased out** of commercial production. A new group of **halogenated hydrocarbons**, based on **chlorodifluoromethanes** and **halogenated dimethyl ethers**, have a reduced impact on the ozone depletion, and are marketed by Du Pont under the Dymel trade name. Some of these products are referred to as **HCFCs**.

6.2   Other Pressurized Dispensers

⓳ Many products cannot be delivered using standard aerosol technology. For example, mixing most propellants with a food product would be objectionable. In other instance, even small amount of propellant dissolved in the product could cause unwanted foaming. These and other situations are best served by a design variation wherein product and propellant are in separate chambers.

⓴ The two common systems address this issue. One uses a collapsible inner bag to hold the product. The other applies the pressure through a **piston** (Figure 5.15). The propellant charge in both instances is introduced through a port in the can bottom that is subsequently sealed with an elastomeric plug.

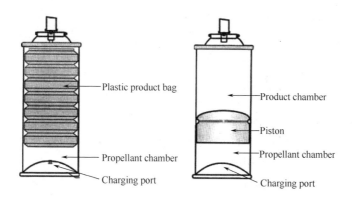

Figure 5.15　Aerosols with propellants in chambers separate from the product

(3396 words)

New Words and Expressions

canister ['kænistə] *n.* （茶叶、烟等）罐
seamed [si:md] *v.* 接口、卷边
soldered ['səuldəd] *v.* 焊接
clinch [klintʃ] *v.* 钉
beading ['bi:diŋ] *n.* 珠状凸缘
ductile ['dʌktail] *adj.* 柔软的、易延展的
sanitary ['sæni,təri] *adj.* 清洁的、卫生的
ointment ['ɔintm(ə)nt] *n.* 药膏、油膏
oblong ['ɔblɔŋ] *adj.* 椭圆形的、长方形的
talcum ['tælkəm] *n.* 滑石
electrode [i'lektrəud] *n.* 电极
flanged [flændʒd] *adj.* 带凸缘的，法兰的
hoop [hu:p] *n.* 加强环、箍
chuck [tʃʌk] *n.* 卡盘
sealant ['si:lənt] *n.* 密封剂
trim [trim] *v.* 修剪
lacquer ['lækə] *v.* 涂漆、使表面光泽
dent [dent] *v.* 削弱、使产生凹痕
slug [slʌg] *n.* 金属块
anvil ['ænvil] *n.* 铁砧
punch [pʌn(t)ʃ] *n.* 冲头
embossed [im'bɔsd] *adj.* 有浮雕图案的
anneal [ə'ni:l] *v.* 退火
membrane ['membrein] *n.* 薄膜、隔膜
ophthalmic [ɔf'θælmik] *adj.* 眼的、眼科的
enamel [i'næm(ə)l] *n.* 瓷釉
humidor ['hju:midɔ:] *n.* 雪茄盒

sleek [sli:k] adj. 圆滑的、井然有序
deodorant [di'əud(ə)r(ə)nt] n. 除臭剂
fluorocarbon [ˌfluərə(u)'ka:b(ə)n] n. 碳氟化合物
propellant [prə'pelənts] n. 推进剂
nozzle ['nɔzl] n. 管口
chlorofluorocarbon (CFC) [ˌklɔro'flurokarbən] n. 含氯氟烃
hydrocarbon [ˌhaidrə(u)'ka:b(ə)n] n. 碳氢化合物
dimethyl [dai'meθil] n. 乙烷、二甲基
depletion [di'pli:ʃən] n. 消耗、损耗
ozone ['əuzəun] n. 臭氧
chlorodifluoromethane [ˌklɔurəˌdifluərə'meθein] n. 氯二氟甲烷
piston ['pist(ə)n] n. 活塞
tin-plated 马口铁、镀锡钢板
hinged lid 铰链盖
slip cover 滑盖
shallow drawn can 浅冲拔罐
double-seamed 二重卷边的
two-piece can 两片罐
three-piece can 三片罐
deeper draw 深冲罐
multiple draw 多级拉深罐
draw-and-iron 变薄拉深罐
collapsible tube （金属）软管
impact-extrusion 冲挤
ring pull-top can 易拉罐
slipcover 套盖
dispensing spout 喷洒口
Flit insecticide 杀虫剂
key-opening can 卷开罐（带有开罐钥匙的金属罐）
double-tight can 双重密封罐
triple-tight can 三重密封罐
hermetic seal （真空）密封
lithographed can blank 彩印罐坯
body hook 身钩
end hook 盖（底）钩
shallow-profile can 浅罐
draw-and-redraw 深冲拉拔、二次拉深
nasal tip 鼻型口
mastitis tip 针型口
dead-fold 折叠充分、残余褶皱

suck-back　倒吸
offset letterpress　凸版印刷
aerosol packaging　喷雾包装
liquid phase　液相
vapor phase　气相
nitrous oxide　一氧化氮
vinyl chloride　氯乙烯
phase out　淘汰、停止
halogenated hydrocarbon　卤代烃
halogenated dimethyl ether　卤代烷二甲醚
HCFC（hydrochlorofluorocarbon）　氢氯氟碳化合物

Notes

1. *Originally, all steel containers were fabricated from flat sheets that were cut to size, bent to shape, and mechanically clinched or soldered to hold the final shape.*（Para. 3）最初，所有的钢制容器都是将片材切割成一定尺寸，弯曲成一定形状，然后再用机械勾连或焊接的方法，成为最后的形状。

2. *Improvements in metallurgy and processing allowed deeper draws and multiple draws and eventually led to a draw-and-iron process in which the walls of a drawn container were made thinner by an ironing step.*（Para. 6）冶金加工技术的改进允许深拉和多次拉深加工，最终实现了变薄拉深的加工过程，拉深罐的罐壁通过多级拉深变薄。

3. *Adhesive bonding is an attractive body-assembly method for those applications where the can will not be subjected to thermal processing. Unlike welded cans, adhesive-bonded constructions can have full wraparound lithography.*（Para. 12）对于那些不需要热处理的罐，黏结法对于罐身装配是一个更具有吸引力的方法，不像焊接罐，黏结罐的罐体结构可以周身印刷。be subjected to 表示"经受……"。

4. *To solder a can, engaging hooks are bent into the can blank similar to the approach for a mechanical seam, the body is formed, and the engaging hooks are flattened to hold the cylindrical shape.*（Para. 13）为实现罐的焊接，咬合钩状物弯曲成罐坯（类似于机械卷边方法），罐身成形，然后啮合钩状物压平以保持罐身为圆柱形。

5. *Welded cans are strong and eliminate potential lead hazards. Most three-piece steel food cans are welded by a process initiated in Europe by Soudronic.*（Para. 14）焊接罐牢固且消除了潜在的铅危害，多数钢制三片食物罐是通过由起始于欧洲的Soudronic的技术焊接而成的。Soudronic公司即是在电阻焊、激光焊接以及在金属包装业关键生产系统上处于全球领先水平的瑞典"苏德罗尼克公司"。

Overview Questions

1. List the methods of creating a body seam on a three-piece steel can. What are the advantages and limitations of each?

2. What are the advantages and disadvantages of a two-piece can?
3. Collapsible tubes can be made of metal, plastic, or laminates. What advantages does a metal tube have over the other possible choices?
4. How can you tell the difference between a can that was drawn and one that was drawn and ironed?
5. What are the principal requirements of an aerosol propellant? About what proportion of municipal waste is classed as packaging?

Lesson 6 Glass Containers

1 Glass Types And General Properties

① "Glass" refers to an **inorganic** substance fused at high temperatures and cooled quickly so that it solidifies in a **vitreous** or noncrystalline condition. That is, the molecular structure of solid glass is practically the same as liquid glass, but the cooled glass is so viscous that the mass has become rigid. Glass has no distinct melting or solidifying temperatures. There is gradual softening with heat and gradual solidifying with cooling.

② Many metal oxide materials can be formed into a "glassy" condition; however, all commercial glass is based on **silica** (quartz), the principal component of sand. Common beach sand is unsuitable for making commercial glass, since it contains impurities and varies widely in composition. Large deposits of high-purity silica sands are available in various parts of the world.

③ Glass production relies on many formulations. *Silica sand fused with about* 10% **sodium compounds** (*usually carbonates*) *produces sodium silicate, or "water glass", a water-soluble glasslike form.* Insolubility is imparted by adding calcium compounds. **Soda**-lime-silica glass, or more simply soda-lime glass, is the type most commonly used for most commercial bottles and jars. Table 6.1 lists the ingredients commonly used to make soda-lime glass. The percentages of the different ingredients vary slightly depending on the manufacturer and the exact composition of the raw materials available.

Table 6.1 Typical soda-lime glass-making ingredients

Ingredient	Percent by Weight	Ingredient	Percent by Weight
Silica sand(silicone oxide)	68 to 73	Soda ash(sodium carbonate)	12 to 15
Limestone(calcium carbonate)	10 to 13	Alumina(aluminum oxide)	1.5 to 2

④ Other mineral compounds may be used to achieve improved properties. **Decolorizers** added to clear glass overcome the slight color imparted by mineral impurities. Other additives aid in processing. **Colorants** and **opacifying agents** change the finished appearance. Standard glass colorants are:

- **Chrome oxides** for **emerald** (green) glass.
- Iron and sulfur for amber (brown) glass.
- **Cobalt oxides** for blue glass

⑤ Besides soda-lime glasses, there are many other glass types used for special applications. They are rarely-if ever-used for packaging purposes. For example, lead compounds provide a soft glass (crystal glass) with exceptional optical properties that may be used for upscale perfume bottles. Boron compounds (borax, boron oxide) give low thermal expansion and high heat-shock resistance. Borosilicate glasses also have exceptionally low extract-ables and are used to contain the most critical parenteral drugs, those administered by injection.

⑥ Glasses other than soda-lime can cause problems if they are included with regular container glass for recycling. For example, borosilicate glasses-of which **Pyrex** bakeware is probably the most visible example-have a significantly higher melt temperature than soda-lime glass. *Along with Pyrex items, window glass, laboratory glassware, china and household glassware should not be included in glass collected for recycling.*

⑦ Glass has many advantages as a packaging material:
- It is inert to most chemicals.
- Foods do not attack glass, nor do they leach out materials that might alter taste.
- Its impermeability is important for long-term storage of products sensitive to volatiles loss or oxidation by atmospheric oxygen.
- Clarity allows product visibility.
- It is generally perceived as having an upscale image.
- The rigidity of glass means that container shapes and volumes do not change under vacuum, under pressure, or when the container is picked up or handled.
- It is stable at high temperatures, making it suitable for hot-fill and **retortable** products.

⑧ Despite these advantages, many traditional glass markets have been eroded or displaced by plastic materials. The disadvantages of glass are its breakability (2.5g/cc) and its high weight. Glass manufacture is heavily dependent on energy supplies, and high energy costs affect the cost of glass.

2 Bottle Manufacture

2.1 Blowing the Bottle or Jar

⑨ Depending on their geometry, glass containers are made by two slightly different processes, **"blow-and-blow"** and **"press-and-blow"**. As just discussed, both processes require two molds: a **blank mold** that forms an initial shape, or **parison**, and a **blow mold**, in which the final shape is produced. The blank or parison mold forms the neck, the finish (the part that receives the closure) and a partially formed body known as a parison. A blank mold comes in a number of sections:
- The finish section.
- The cavity section (made in two halves to allow parison removal).
- A guide or funnel for inserting the gob.
- A seal for the gob opening once the gob is settled in the mold.
- Blowing tubes through the gob and neck openings.

⑩ Molten glass flows by gravity through draw-off orifices with openings ranging from 12 to 50 mm (1/2 to 2 in), depending on the bottle size. Mechanical shears, which are 25 mm (1 in) below the orifice, synchronized with the draw-off flow rate and bottle-forming machine speed, snip off "gobs" of molten glass (Figure 6.1). Each gob makes one container. The falling gob is caught by a spout and directed to one of the blank molds.

01 A mass-production bottle-making machine is typically made up of 6, 8, or 10 individual sections, hence the term I. S. machine. Each section is an independent unit holding a set of bottle-making molds. For large bottles, a set consists of a blank mold and a blow mold. Higher production speeds are achieved for smaller bottles by the use of double or triple gobs on one machine. A mold set then consists of a block of two or three blank molds and a similar block of blow molds. Each blow mold has a number that is imprinted on the bottles made by that mold.

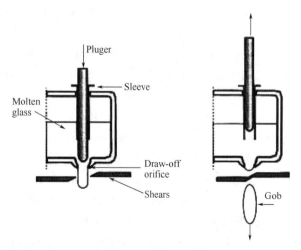

Figure 6.1 Furnace draw-off orifice and gob shears

02 Glass containers produced by the two processes differ only in the way that the parison is produced. In the blow-and-blow process, the bottle is blown in the following sequence (Figure 6.2):

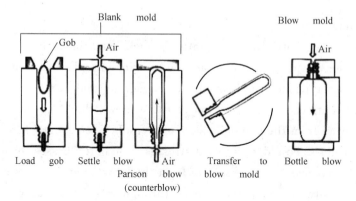

Figure 6.2 Blow-and-blow bottle manufacture

• The gob is dropped into the blank mold through a funnel-shaped guide. Note that the blank mold is upside down. Gob temperature at this point is about 985°C (1800°F).

• The guide is replaced by a parison bottomer, and air is blown into the mold (called the settle blow) to force the glass into the finish section. At this point the bottle finish is complete.

• The parison bottomer is replaced by a solid bottom plate, and air is forced through the bottle finish (called the counterblow) to expand the glass upward and form the parison.

• The parison is removed from the blank mold, using the neck ring (transfer bead) as a gripping fixture, and is rotated to a right-side-up orientation for placement into the blow mold. The parison is supported in the blow mold by the neck ring.

• Air forces the glass to conform to the shape of the blow mold. The bottle is cooled so

⑬ In the press-and-blow process, the gob delivery and settle-blow steps are similar to those in blow-and-blow forming. However, the parison is then pressed into shape with a metal plunger rather than blown into shape (Figure 6.3). The final blowing step in a separate blow mold is identical to that in the blow-and-blow process.

⑭ The blow-and-blow process is used for narrow-necked bottles, while press-and-blow is used to make wide-mouthed jars. Recent advances have allowed press-and-blow to be used for increasingly smaller necked containers. The advantage of press-and-blow is better control of glass distribution.

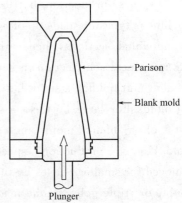

Figure 6.3 Press and blow forms the parison by mechanical action

Typical production rates range from 60 to 300 bottles per minute, depending on the number of sections in a machine, the number of gobs being extruded, and the size of the container.

⑮ The blown bottle is removed from the blow mold with takeout **tongs** and placed on a dead plate to air cool for a few moments before transfer to a conveyor that transports it to the annealing oven.

2.2 Annealing

⑯ The walls of a glass bottle are comparatively thick, and the cooling of such a cross-section will not be even. In theory the inner and outer skins of a glass section will become rigid long before the internal temperature has reduced enough to produce the same degree of rigidity there. The still-contracting inner portion of the wall will build up internal stresses as it tries to contract away from the immobile skin surfaces. Uneven cooling can develop substantial stresses in the glass. To reduce internal stresses, the bottle passes through an annealing oven, or **"lehr"**, immediately after removal from the blow mold.

⑰ The lehr is a controlled-temperature oven through which the glassware is carried on a moving belt at a rate of about 200 to 300 mm (9 to 12 in) per minute. The glass temperature is raised to about 565°C (1,050°F), and then gradually cooled until the containers exit at close to room temperature with all internal stresses reduced to safe levels. This process typically takes about an hour. Improperly annealed bottles are fragile and tend to have high breakage rates in normal transport and filling. Hot-filling will also produce unacceptable breakage levels.

2.3 Surface Coatings

⑱ The inner and outer surfaces of glass containers have slightly different characteristics coming from the mold. The outer surface comes in contact with the mold and takes the grain of the mold surface. However, both surfaces are **pristine**, **monolithic**, sterile, and chemically inert.

⑲ Pristine glass has a comparatively high coefficient of friction, and surface scratching, or "bruising" can occur when bottles rub together on high-speed filling lines. Scratched glass has significantly lower breakage resistance, and glass is typically coated to reduce the coefficient of friction. Two coatings are usually used. The hot-end coating applied at the entrance to the annealing lehr is usually tin or **titanium tetrachloride**. It strengthens the glass surface and acts as a primer or bonding-agent coat for the cold-end friction-reducing coat applied at the lehr exit.

⑳ Many different cold-end coatings are available, depending on the filling process and end use. **Oleic acid, monostearates,** waxes, silicones, and polyethylenes are typical cold-end coatings. The label adhesive will need to be compatible with the cold-end coating.

2.4 Inspection and Packing

㉑ Visual inspection has been largely replaced by mechanical and electronic means. Squeeze testers pass the containers between two rollers that subject the container walls to a compressive force. **Plug gauges** check height, perpendicularity, and inside and outside finish diameters. Optical devices inspect for stones, blisters, checks, **bird swings**, and other blemishes and irregularities by rotating the container past a bank of photocells (Figure 6.4).

㉒ Faulty containers (offware) are ejected from the line and sent for crushing into **cullet**. Glass containers can be transported in reusable corrugated shippers, which are reloaded with the filled bottles. Others are shipped in tiers on pallets, a method best suited for high-speed production lines where automatic equipment can be used to clear tiers off the pallet and feed them into the filling machine.

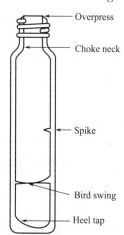

Figure 6.4 Flat bottles sometimes develop "bird swing" and "spike" defects. Spikes are glass projections inside the bottle, and a bird swing is a glass thread joining the two walls

3 Bottle Design Features

3.1 Bottle Parts and Shapes

㉓ Figure 6.5 illustrates the terms used to describe the various parts of a bottle. Viscous glass flows most easily into molds with smooth, round shapes. Round bottles are easiest to manufacture since they are an expansion of the circular parison, eliminating complex material-distribution problems. Round shapes run easily on filling lines and can be labeled at relatively high speeds. They can be accurately positioned in a spot-labeler via an indexing label lug on the bottle exterior. Round bottles have greater strength-to-weight ratios and better material utilization than irregular shapes.

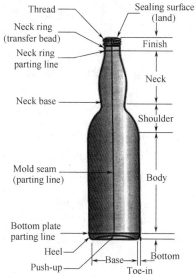

Figure 6.5 Glass bottle nomenclature

㉔ Square shapes, angular shapes, flat shapes, and sharp corners are more difficult to form properly and have many inherent problems. For example, flat **flasks** (Figure 6.4) are prone to having the two sides of the parison touch momentarily during transfer from the blank to the blow mold. *This results in spikes or in extreme cases a "bird swing" on the inside of the bottle.* Rectangular bottles still have a round finish, a factor requiring careful design to avoid stress points.

3.2 Finish and Closures

㉕ A bottle finish is defined as the part that will receive the closure. Bottle finishes are broadly classified according to diameter (expressed as the nominal inside diameter in millimetres), sealing method, and special features. Standards for finish sizes and tolerances have been set by the Glass Packaging Institute and are followed by both the bottle maker and the closure maker. Continuous-thread (CT), **lug**, crown, threaded-crown, and roll-on are common finish designs.

㉖ Closures are selected on the basis of cost, utility, and decoration. Particular closure requirements will dictate specific bottle finish designs. Stock closures should be used when cost is a key criterion. While glass-bottle-closure thread dimensions are similar to those for plastic bottles, the **thread profile** for glass has a curved or partially semicircular profile, whereas plastic bottle threads have flat lands. Care should be taken to match the bottle and closure-thread profiles.

3.3 Neck and Shoulder Areas

㉗ Neck designs have particular impact on filling, air displacement, and dispensing. Differences in fill level are more visible in long, narrow necks. **Headspace** is sometimes needed to provide for thermal expansion and to facilitate filling.

㉘ A **choke neck** is a manufacturing defect in which excess glass is distributed to the inside of the finish or opening. **Overpress** is a defect in which a small ridge of glass is formed on the sealing surface of the bottle finish.

㉙ The "upper shoulder" is the area directly below the neck. Blending of upper shoulder and neck is important to good design and efficient production. The "lower shoulder" is the integration point between the upper shoulder and the body. It is a vulnerable spot for abusive contact with other bottles and the origin of many handling and shipping fractures. Contact area with other bottles should be as large as possible.

3.4 Sides

㉚ The sides are the most generalized areas of the bottle. Labeling styles and means of preventing scuffing must be considered. Bottles are often designed with label panels that are recessed to prevent scuffing. The panel may have prominent base and shoulder ridges as part of the design.

㉛ In angular bottles, rounded corners rather than beveled ones are preferable for wrap-around or three-side labeling. **Spot labeling** is normally a one-or two-sided application, but four-sided labelers are available. Labeling of non-round shapes is typically slower than for round shapes.

3.5 Heel and Base

㉜ The heel is a high-abuse area. It should start as high from the base as possible, curving into the base to a suitable base diameter. The body-to-base curve should combine 3 radii. The largest radius blends body to heel, while the smallest blends heel into base.

㉝ The diameter of the base should be as large as possible within the constraints of good design. The center of the base is always domed inward (the "push-up") to ensure a flat, stable bottom that will not allow the bottle to rock. The circular bearing surface on which the bottle rests will usually have a stippled or knurled pattern so that the scratches that will inevitably occur during handling and usage do not weaken the bottle's body.

㉞ Ketchup bottles and other sauce bottles require that both heel and base be heavier and contoured to allow consumers to tap them safely and comfortably when expelling the contents.

㉟ Some wide-mouthed jar bases have been designed in stacking features. There are two types:
- Container base fits into recessed cap.
- Indented container base fits over cap.

㊱ A **heel tap** is a manufacturing defect in which excess glass is distributed to the heel area.

3.6 Stability and Machinability

㊲ Center of gravity and the base surface area will determine a bottle's stability. Stable bottle minimizes handling problems on both manufacturing and filling lines. Tall and narrow bottles present the most problems in manufacturing, packaging line handling and labeling due to the high center of gravity. Short bottles, usually with round or oval bodies, are an efficient type for machine handling and present minimal labeling problems. Examples of this type of bottle include baby food and cold cream jars.

㊳ As much as possible, bottles should be designed to be all-around trouble free to manufacture, fill, close, and ship. Some designs are inherently weaker or more prone to cause trouble in their filling and the distribution cycle than others.

3.7 Vials and Ampoules

㊴ **Vials** and **ampoules**, used mainly for pharmaceuticals and sera, are made from preformed tubing stock rather than by the blowing methods used for glass bottles. Ampoules are sealed glass containers with a constriction that has been treated to allow for easy fracture. This may be a controlled score, or it may be coated with a ceramic paint that causes a stress concentration in the constriction. Standard ampoule sizes are 1, 2, 5, 10, and 20 ml.

㊵ Serum vials are small bottles that are fitted with a rubber **septum** retained by an aluminum neck ring. The rubber septum is pierced by a needle **cannula** to withdraw serum. Unlike an ampoule, a vial can be accessed several times. Vials come in standard sizes of 1, 2, 3, 5, 10, and 20 ml.

㊶ **Tumblers** are wide-mouthed and (often) tapered containers for jams. **Carboys** are large bottles for bulk containment of acids or chemicals.

3.8 Carbonated Beverages

⑫ The pressure developed by a carbonated beverage depends on, amongst other factors, the amount of gas dissolved in the product. Beverage producers express this as the number of volumes of gas dissolved in a unit volume of the product. For example, if a 48 oz volume of carbon dioxide at standard conditions is dissolved in 12 oz of beverage, then the beverage is said to yield 4 gas volumes.

⑬ Carbonated beverage and beer bottles must withstand internal gas pressure and must be well capped. Internal pressure in a soft drink container may reach 0.34 millipascal (50 psi), while beer during pasteurization may reach 0.83 millipascal (120 psi). The stress on the glass causes a loss of bottle strength over time with the greatest losses occurring within the first week after filling. Bottle designs for pressurized products are always round in **cross section** and have gently curving radii in order to maximize bottle strength.

(3127 words)

New Words and Expressions

inorganic [ˌinɔrˈɡænik] *adj.* 无机的、无生物的
vitreous [ˈvitriəs] *adj.* 玻璃的、玻璃状的
silica [ˈsilikə] *n.* 二氧化硅、硅土
sodium [ˈsəudiəm] *n.* 钠
soda [ˈsəudə] *n.* 苏打、碳酸水
decolorizer [diːˈkʌləraizə] *n.* 脱色剂
colorant [ˈkʌlərənt] *n.* 着色剂
emerald [ˈemərəld] *adj.* 翠绿色的
Pyrex [ˈpaireks] *n.* 派热克斯玻璃（一种耐热玻璃）
retortable [riˈtɔːtəbl] *adj.* 耐蒸煮的
parison [ˈpærisən] *n.* 型坯
tong [tɔŋ] *n.* 钳子
lehr [liz] *n.* 玻璃韧化炉
pristine [ˈpristin] *adj.* 原始的
monolithic [ˌmɑnəˈliθik] *adj.* 整体的
monostearate [ˌmɑnəˈstiəˌreit] *n.* 单硬脂酸盐
cullet [ˈkʌlit] *n.* 碎玻璃
flask [flæsk] *n.* 烧瓶、长颈瓶
lug [lʌɡ] *n.* 支托、耳状物
headspace [ˈhedspeis] *n.* 顶部空间
overpress [ˌəuvəpres] *n.* 飞刺（玻璃制品的缺陷）
vial [ˈvaiəl] *n.* 小瓶、药水瓶
ampoul [ˈæmpuːl] *n.* 安瓿
septum [ˈseptəm] *n.* 隔膜
cannula [ˈkænjulə] *n.* 套管
tumbler [ˈtʌmblə] *n.* 平底玻璃杯

carboy ['kɑrbɔi] n. 用藤罩保护的大玻璃瓶
opacifying agent   遮光剂
chrome oxides   氧化铬、氧化铬绿
cobalt oxides   氧化钴
blow-and-blow   吹-吹法
press-and-blow   压-吹法
blank mold   初模
blow mold   吹模
titanium tetrachloride   四氯化钛
oleic acid   十八烯酸、油酸
plug gauge   测孔规
bird swing   瓶内粘丝
thread profile   螺纹牙形
choke neck   瓶颈阻塞
spot label   点标
heel tap   斜底、瓶底厚薄不均
cross section   横截面

Notes

1. *Silica sand fused with about 10% **sodium** compounds (usually carbonates) produces sodium silicate, or "water glass", a water-soluble glasslike form.* (Para. 3) 融合了10%钠组分（通常是碳酸盐）的二氧化硅砂子生产出的二氧化硅钠，即"水玻璃"——一种可溶于水的类似玻璃的形式。

2. *Along with Pyrex items, window glass, laboratory glassware, china and household glassware should not be included in glass collected for recycling.* (Para. 6) 除了派热克斯玻璃，窗户玻璃、实验室玻璃器皿、瓷陶器具和家用玻璃器具不应该和回收的玻璃包含在一起。"Along with"意指"除了"。

3. *This results in spikes or in extreme cases a "bird swing" on the inside of the bottle. Rectangular bottles still have a round finish, a factor requiring careful design to avoid stress points.* (Para. 24) 这导致了在瓶子内部的尖峰或者在极端情况下"瓶内粘丝"。矩形瓶仍是圆瓶口，这是一个需要精心设计以避免应力点的因素。"finish"是名词，指玻璃容器制造的"最后"部分即封口。

Overview Questions

1. What are the three principal constituents of glass?
2. What is annealing, and why is it necessary?
3. What problems might be encountered with a bottle that is wider at the top than at the base?

Lesson 7 Plastics In Packaging

1 Introduction To Plastics

① The term "plastics" can **be likened to** the term "metals" in the breadth of its application. "Plastic" describes the ability of something to be molded or formed. Historically this term referred to natural materials such as wax, clay, tar, rosin, and asphalt. With advances in chemistry, the term "plastic" began to describe **modified natural resins** and finally a large group of **synthetic** materials that could be formed into useful shapes. *The words "plastic" and "polymer" are used interchangeably, but plastic tends to describe finished parts; polymer tends to describe the raw material and is used by the scientific community.*

② Polymers are very large molecules. A water molecule has only three atoms, one oxygen and two hydrogen, whereas a typical polymer molecule contains hundreds, or more typically, thousands of atoms. A polymer is created when a large number of identical repeating monomer units are joined together to make a single large polymer molecule (from the Greek "polys", meaning many, and "meros", meaning parts). The "mer" is the smallest repetitive unit in a polymer and for this discussion is based on the carbon atom.

③ Today there are hundreds of identified "species" of synthetic polymer. Any of these is available in a range of molecular masses, most can be modified by the addition of other **monomers** and the properties of each can be dramatically influenced by processing conditions. In reality then, the choice in plastics is almost limitless.

④ Polymers can be grouped into two classes: "thermoplastic" and "**thermoset**" -terms that describe how polymers behave when heated. They also fall into two economic groups- "commodity polymers" (i.e., economical) and "engineering polymers" (i.e., costly). Only a small number of the available polymers are of practical significance for packaging, and practically all of those are in the commodity thermoplastics.

⑤ With very few exceptions, the essential raw materials used to make packaging plastics are derived from the **petrochemical industry**. The amount of petrochemicals diverted for plastics is quite small compared to that used for fuel and heating. Depending on the source, the estimated amount of petrochemicals used by all the plastics industries ranges from 1.5% to 3%. Packaging uses a fraction of this percentage.

⑥ Each plastic type has a unique structure and a proper chemical name. However, chemical names can be lengthy and so common industry usage is a mixture of trade names, common names, and abbreviations (Table 7.1). Spellings and abbreviations developed randomly, and today, the same polymer may be spelled and abbreviated in a number of different ways.

Table 7.1 Selected abbreviations and trade names for packaging polymers

Abbreviation	Generic Name	Common Trade Names and Alternatives
BOPP	Biaxially oriented polypropylene	
CTFE	Clorotrifluoroethylene	Aclar
CPET	Crystallized PET	
EEA	Ethylene-ethyl acrylate	Frequently grouped as an acid copolymer
EPS	Expanded polystyrene	
EVA	Ethylene-vinyl acetate	Also abbreviated EVAC
EVOH	Ethylene-vinyl alcohol	EVAL
HIPS	High impact polystyrene	
LDPE	Low-density polyethylene	
LLDPE	Linear low-density polyethylene	
OPP	Oriented polypropylene	
mPE	Metallocene polyethylene	
PA	Polyamide	Nylon, also abbreviated NY
PAN	Polyacrylonitrile	Barex. Also abbreviated AN
PEN	Poly(ethylene naphthalate)	
PC	Polycarbonate	
PE	Polyethylene	
PET	Poly(ethylene terephthalate)	Polyester, Mylar, Melinex
PETG	Poly(ethylene terephthalate) glycol	
PP	Polypropylene	
PS	Polystyrene	
PTFE	Polytetrafluoroethylene	Teflon
PVAC	Poly(vinyl acetate)	Also abbreviated PVA
PVC	Poly(vinyl chloride)	
PVDC	Poly(vinylidene chloride)	Saran (Dow trade name)
PVAL	Poly(vinyl alcohol)	Also abbreviated PVOH
None	Ionomer	Surlyn (DuPont trade name)

7 The casual use of spellings, abbreviations, and trade names can be confusing and sometimes technically incorrect. For example, **"Styrofoam"** is Dow Chemical's trade name for an expanded polystyrene material used primarily by the construction industry, but also used in other areas such as crafts. It is not correct to use the term "Styrofoam" to describe expanded polystyrene packaging materials. "Cellophane" was a trade name coined in the 1930s to market a regenerated cellulose product. It was the first clear plastic wrapping material used in quantity by the packaging industry. Up to the 1950s, if it was plastic and clear, it was quite likely cellophane. Today the packaging use of **regenerated cellulose** film is negligible. It is technically incorrect to refer to a plastic bag as "cellophane" because the possibility of it actually being made of this material is remote.

⑧ Most of the polymers listed in Table 7.1 are commodity thermoplastics. The majority are under two dollars per kilogram. Engineering plastics such as *polysulfone*, *acetal*, and silicone polymers are several orders of magnitude higher in cost.

⑨ A polymer's properties depend primarily on:

• **The elements that make up the polymer molecule.** By intuition alone, we would surmise that a polymer made up by hydrogen and carbon atoms alone—polyethylene, for example—would behave differently from one such as poly (vinyl chloride) that also included chlorine atoms.

• **The polymer molecule's polarity.** Depending on the participating atoms, molecules exhibit varying degrees of polarity. Degree of molecular polarity will influence such factors as melting point, coefficient of friction, solubility, barrier properties and adhesiveness.

• **The size or molecular weight of the molecule.** Most polymers are available in a range of molecular weights. Properties such as melting point, stiffness and solubility will change as molecular weight changes.

• **The molecule's shape.** *The shape of a molecule will determine how large numbers of them will fit together. Degree of crystallinity, clarity, barrier, melting point and other physical properties are affected by the molecule's shape.*

• **The polymer's thermal history.** Every thermoplastic needs to be melted in order to get it into a form that can be shaped, and so every plastic part has a thermal history. Changing the thermal history changes the plastic's performance characteristics.

• **Mechanical history.** A plastic's final properties also depend on its mechanical history—how it flowed in the molten state and how it was stressed when cold. Sometimes mechanical forces are used alone, but in other plastic forming methods, thermal and mechanical histories are combined to give still another variation on the plastic's basic properties.

2 Thermoplastic And Thermoset Polymers

⑩ A monomer can join to itself to form a complex polymer structure in a number of ways. The monomer units can simply join onto each other to form long chains as in a thermoplastic, or they can cross-link between the chains in a three-dimensional pattern, resulting in a thermoset plastic (Figure 7.1).

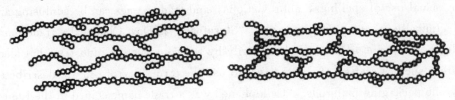

Figure 7.1 Thermoplastic polymer chains (left) are free to pass over one another at melt temperatures. Cross-linked thermoset polymer chains (right) can never come free of one another without destroying the material

⑪ Thermoplastic materials are fully reacted, or polymerized, products that, when subjected to sufficient heat, will soften. Pressure makes them flow and assume new forms, at which point they are cooled to form a useful shape. Scrap or defective shapes can be remelted and reprocessed. The easy formability and economic recovery of recyclable material make thermoplastics the material of choice for most packaging applications.

¶2 Thermoset plastics are polymers that have not fully completed their polymerization reaction, but do so when activated, usually by heat. The resulting product is cross-linked and will not be softened again by heat; therefore it cannot be reprocessed or reshaped. Thermosets such as **phenol**, **urea**, and **melamine** formaldehyde are occasionally used for specialty closures, but the use of thermosets for dimensional parts in the packaging industry is negligible. This discussion of plastic forming methods is limited to those methods used to shape thermoplastic materials.

¶3 The proper balance of product protection, containment, and appearance qualities, at an affordable cost, can be obtained by referring to the mechanical and chemical properties of the various polymers. Once a material is selected, the process or method for converting the selected polymer resin into a useful form must be determined.

3 Shaping Plastics

¶4 The method of converting polymer resin into the final package structure involves at least the following factors:
- Resin type.
- Geometry of the finished part.
- Number of units required.
- Dimensional tolerance requirements.
- Container wall thickness required to meet structural and economic requisites.
- Tooling (mold) cost and production life expectancy.

¶5 To relate these factors, the principal manufacturing methods must be understood. The following are the most common thermoplastic forming methods for packaging purposes:
- Extrusion (including profile extrusion, extrusion cast film and sheet, blown-film extrusion and co-extrusion).
- Injection molding.
- Extrusion blow molding.
- Injection blow molding (including injection stretch blow molding).
- Other less frequently used forming methods.

¶6 Thermoforming is a secondary forming process that further forms plastic sheet materials. In all thermoplastic shaping methods, the plastic must first be heated to a point where the material has a plasticity, or fluidity, appropriate to the intended molding method. A key attribute of any plastic material is its behavior at elevated temperatures. **Melt index** or melt flow rate is a common method of quantifying this behavior.

4 Plasticating Extruders

¶7 Polymer resins are received at a molding plant in the form of small granules or pellets, similar in appearance to rice. Regardless of the forming process, the first task is to heat and melt the polymer resin pellets into a flowable form. The plasticating extruder is a heavy barrel in which a screw rotates, driving pellets from the feed hopper at one end to the exit port at the other end (Figure 7.2). The work of driving polymer melt down the barrel provides most of the heat required to melt the polymer. Heater bands help maintain precise melt temperatures, and a cooling jacket keeps the **feed hopper** area cool.

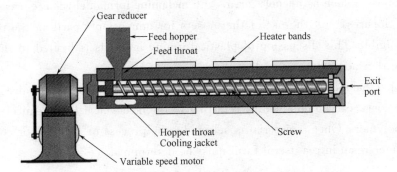

Figure 7.2 A pasticating extruder. Most plastics are colored by adding color concentrates along with polymer resin at the feed hopper

5 Extrusion

5.1 Profile Extrusion

⒅ A shape of constant cross-section (a profile) can be extruded by forcing polymer melt through a shaped orifice in a die placed over the end of the extruder (Figure 7.3). At its simplest, the die may be a metal plate with a round hole drilled into it, in which case the extrusion is a round rod of approximately the hole diameter. Tooling costs are low since the process takes place at relatively low pressures and neither the extruder nor the die needs to be substantial. Placing a suitable torpedo-shape in the die exit permits extrusion of a hollow pipe or tube. Extruded profiles are cut to length, and when hollow profiles are fitted with

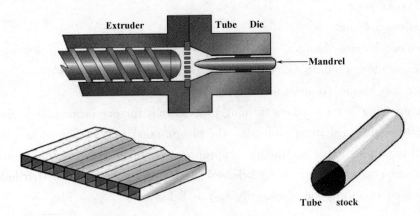

Figure 7.3 A profile extruder. Placing a torpedo-shaped mandrel in the die exit creates a tube. With other die and mandrel configurations, shapes resembling corrugated board are produced from the extruder

end closures, they can be made into versatile, inexpensive packages.

5.2 Sheet and Film Extrusion

⑲ Sheet extrusion is an application of profile extrusion but using a die with a slot orifice (The dies are also called "coat-hanger" or "T-shape" dies). The dies have a narrow opening between the die lips through which the plastic melt is extruded in a thin film. The film is immediately cooled and solidified on chill rolls (Figure 7.4). The dies can be several metres long. *Depending on thickness, the end product may be called "film" or "sheet". There is no clear division between the two. Film product formed by this process is referred to as "cast film"*.

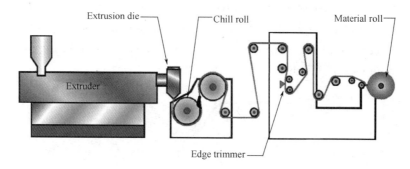

Figure 7.4   General layout of a cast-film extrusion line

⑳ Thicker sheet is used for thermoforming or is die-cut and folded into carton-type constructions similar to paperboard cartons. Thinner extruded films are used alone or laminated with other materials in a variety of flexible packaging applications.

5.3 Blown-Film Extrusion

㉑ Plastic film can also be manufactured by extruding the polymer through a circular die into a closed circular bubble and expanding the bubble with air. The material is extruded upward, and by pulling the inflated bubble upward and continuously extruding more plastic, a continuous seamless tube of thin film is created (Figure 7.5). Air flow along the outside of the bubble provides cooling and an air cushion. After the film has cooled, it is flattened in a **collapsing frame** and wound into rolls. The bubble can be wound up as a seamless tube, or it

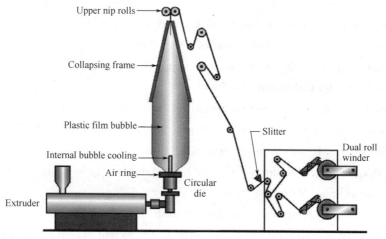

Figure 7.5   A blown-film line that produces a tubular film which is slit into two flat sheets

can be cut and sealed to make seamless plastic bags. Alternately, it can be slit and rolled into one or several flat film rolls.

㉒ In addition to devices that control size and shape, other process units, such as those for printing and embossing, gusseting, vacuum forming, slitting, and folding, can be introduced into the downstream system, providing more integrated product manufacturing lines. The blown-film process is used to make nearly all PE film as well as other films.

5.4 Orientation

㉓ The properties of cast and blown film and sheet can be improved by physically orienting the polymer molecules. Cast sheet is oriented in the machine direction by being pulled away faster than it is extruded, thus stretching it in the machine direction. This is usually accomplished by passing the cast film through a series of rolls, each roll rotating progressively faster than the previous roll (Figure 7.6, top).

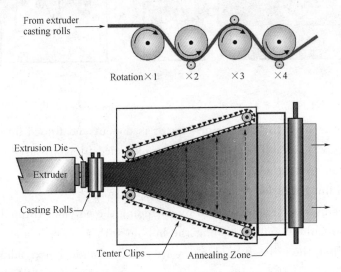

Figure 7.6 Orienting cast film. Machine-direction orientation (top) is done by pulling the film through progressively faster rotating rolls. Cross-direction orientation is done on a tenter frame (bottom), which stretches the film in the transverse direction

㉔ Cross-or transverse-direction orientation is done in a tenter frame (Figure 7.6, bottom). Clips traveling down diverging tracks grasp the film along each edge and stretch it in the cross direction up to about seven times its original cast width. Film oriented in two directions is said to have **"biaxial orientation"**.

㉕ Both machine-and cross-direction orienting are done at somewhat elevated temperatures. If the stretching and cooling processes are rapid, the film will retain some memory of its original dimensions. It will want to return to those dimensions if reheated; in effect, it will be a **"shrink plastic"**. Oriented films are heat-stabilized by keeping them at the elevated temperature (annealing) for a brief time.

㉖ Blown film is oriented by adjusting the inflation ratio and take-away speed relative to the tube-forming rate. Blown film cannot be oriented to the extent that is available with cast-film extrusion, however, the biaxial orientation of blown film is well balanced.

5.5 Co-Extrusion

㉗ Both cast-and blown-film extrusion dies can be designed to be fed from more than one extruder, thus producing a sheet composed of two or more different materials (Figure 7.7). Co-extrusion systems feeding as many as seven different layers through one die block have been made. Heavy polymer viscosity limits the mixing of the extruded layers, so they exist essentially as separate layers in the finished product.

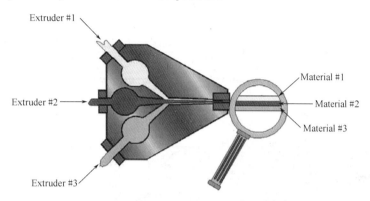

Figure 7.7 A co-extrusion, slot-orifice cast-film die. Similar multichannel dies can be made for blown-film extrusion

㉘ Most co-extrusion is done to combine the performance advantages of two dissimilar materials. For example, heat-sealable polyethylene is extruded onto materials that have poor heat-sealing characteristics, or a high-barrier polymer might be extruded between protective layers of a less costly low-barrier material. Co-extrusion is used to bury recycled plastics into hidden center layers or to produce sheets with decorative colored stripes or colored layers.

6 Injection Molding

㉙ Injection molding uses a powerful extruder with the capability to inject a precise amount of resin into a fully enclosed mold. Very high **hydraulic pressures** drive hot, relatively viscous molten material through the chilled passages of a part mold and fill the cavity before the plastic solidifies. The process requires substantial molds that will not flex or move under extreme temperature and pressure. An eight-cavity mold for margarine tubs may weigh upwards of a ton. This required mold mass and the complexity of the tooling make injection molding highest in tooling cost of the plastic forming methods.

㉚ Injection molding is the leading method of manufacturing closures, wide-mouthed thermoplastic tubs, jewel boxes, and other complex dimensional shapes. Because part dimensions are completely controlled by metal mold surfaces, injection molding gives the most dimensionally accurate part. Tooling sophistication, accompanied by newly developed high melt flow rate (low melt viscosity) thermoplastics, permits the manufacture of thin-walled plastic containers with wall-thickness in the order of $180 \mu m$ (0.007 in).

6.1 Injection Molding Machines

㉛ The extruder section of an injection molding machine must have provision for ejecting a precise amount, or "shot" of polymer melt into the mold as required. **Ram-screw-type machines** use a melt-conveyance screw designed to provide reciprocal as well as rotary motion, combining the function of screw and piston. Other machines have a separate piston for injecting the molten polymer (Figure 7. 8). The mold cavity is exactly in the form of the desired part. When the part has cooled, the mold opens and the part is ejected.

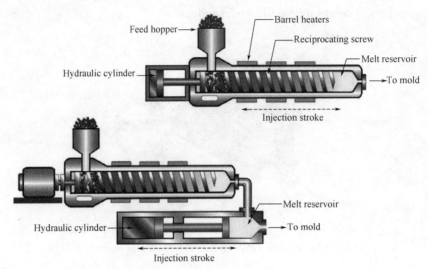

Figure 7. 8　"Shot" size can be metered either by a reciprocating screw (top) or by a separate screw or piston (bottom)

6.2　Co-injection Molding

52　Co-injection molding refers to methods of creating injection-molded parts that have multiple polymer layers. It is injection molding's counterpart to co-extrusion. Co-injection can be done by injecting the first material and then moving the mold core with the first material form into a second cavity. The second cavity has a clearance between it and the first material shape. The second material is then injected around the first.

53　Another system uses multiple **runners** that converge just before the **gate**. The materials flow into the mold cavity as concentric layers. Multilayer, injection-molded preforms for blowing into beverage bottles have been made using PET and EVOH in one version and PET and nylon in another.

7　Extrsion Blow Molding

54　Thermoplastic bottles are manufactured by one of two processes: extrusion blow molding (EBM) or injection blow molding (IBM). As the name implies, EBM combines extrusion with a blowing step. The majority of detergent, oil, and other household chemical bottles are made by extrusion blow molding.

55　Most plastics can be extrusion blow molded providing they have enough strength in the melt form (i. e., a low melt flow rate) to hold together when extruded into a parison. PE, PP, and PVC account for the majority of extrusion blow molded bottles.

56　EBM can accommodate many creative designs, including bottles with handles (handle-ware), two-part containers and containers with integral measuring chambers. Circumferential rings can be molded into a circular cross section to produce a flexible trunk or an accordion-type flexible section that can be used as a pump.

57 All EBM processes are based on a common underlying sequence (Figure 7.9). First, a hollow plastic tube, or parison, is extruded. While in a soft and formable state, the parison is captured between the mating halves of a bottle mold. Air is introduced into the hollow parison, stretching the deformable parison to conform to the mold walls. The newly formed bottle is held in the mold until it cools sufficiently to retain its shape.

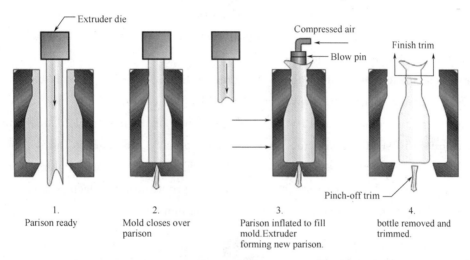

Figure 7.9 Typical extrusion blow-molding sequence

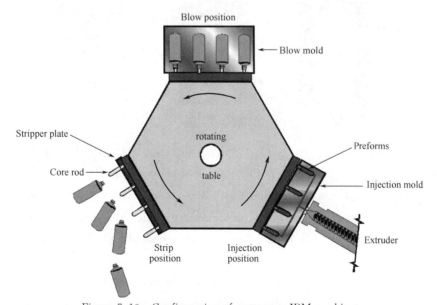

Figure 7.10 Configuration of a one-step IBM machine.
One core rod is shown empty for illustrations purposes

8 Injection Blow Molding

58 Injection blow molding (IBM) combines injection molding and blow molding. Instead of

extruding a parison, as in EBM, a parison or **"preform"** is injection molded. Injection molding of the preform allows more exact control over material distribution than available material when extruding a parison. After the preform injection cycle, the preform, still retained on the core pin, is transferred to the blow-molding station. The final blowing operation is similar to extrusion blow molding. When both injection and blowing are done on a single machine, the process is described as "one-step" (Figure 7.10).

39　Economics generally confine IBM to higher volume production, since two molds are required to make a container: the injection mold (s) to produce the preform and the companion blow mold (s) to blow the container. IBM allows great precision and complexity in the bottle's finish section. This is especially important with containers requiring close tolerance finishes and for **wide-mouthed** containers.

40　**Injection Stretch Blow Molding.** A variation of IBM, injection stretch blow molding (ISBM), uses a rod to stretch the preform during blowing (Figure 7.11). In a typical operation, at the point, that the core rod touches the bottom of the perform, a small amount of air is introduced to start the blowing process. When the core rod touches the bottom of the mold, the full volume of inflating air is introduced. This mechanical stretching orients the polymer molecules and improves stiffness and barrier properties. For deep bottles, the core rod ensures that the inflation is evenly centered in the mold.

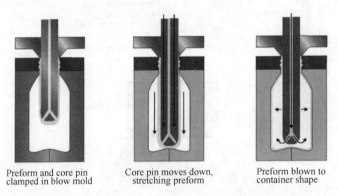

Figure 7.11　A major application for injection stretch blow molding is the manufacture of PET carbonated beverage bottles

41　ISBM is the favored process for making PET carbonated-beverage bottles. In this application, a two-step process is used: the preforms are made on a separate injection-molding machine and accumulated in bulk bins. They are then reheated and blown in a separate operation.

42　Both EBM and IBM can be used to produce narrow and wide-mouth bottle shapes. IBM, however, is able to produce bottle finishes that are more complex and with closer tolerances than could be done with EBM. Producing a close tolerance finish for a wide-mouth jar is particularly difficult for EBM. The absence of any trim and scrap with IBM is an advantage when manufacturing pharmaceutical and medical containers. Good manufacturing practice for these containers does not allow the inclusion of regrind material.

9 Thermoforming

9.1 Principle and Applications

㊸ Containers and other draw-formed packaging components can be readily manufactured from thermoplastic sheeting by a number of thermoforming variations. All variations heat the thermoplastic sheet material to the point where it becomes soft and pliable, but below the temperature at which melt flow might occur.

㊹ Most thermoplastic materials can be thermoformed, including single-polymer materials, co-extrusions, and laminated sheets. Multilayered laminated sheet can provide specialized physical and chemical properties that are not economically attainable by other means. Thermoforms made on automated, high-performance equipments may provide a possible alternative to injection-molded containers such as tubs.

㊺ Pliable plastic sheet can be formed by mechanical means, with vacuums, with pressure, or by any combination of these. In all cases, the relationship of the part surface area to the available sheet area determines the average material thickness. Material distribution is governed by the part's geometry and the particular method used to form the shape. Sheet gauge and mold accuracy are important.

㊻ Since molding temperatures and pressures are very low, thermoform molds are economical. It is not unusual for prototype molds to be made of wood or epoxy/aluminum. Production molds are either aluminum or **beryllium**/copper if exceptional thermal conductivity is required.

㊼ A sheet being formed should be heated to its maximum optimum temperature to reduce **residual stresses** set up when the material is stretched. Manual operations do not lend themselves to good, consistent qualities due to the variables of time-sequencing by "hand". Automatic, accurately timed, and thermally controlled equipment is preferred.

㊽ While thermoforming cannot make **narrow-mouthed** containers directly in the mold, two formed halves can be joined to create narrow-necked cylindrical containers or other partly or completely enclosed containers. The two halves can be joined by adhesive bonding, spin welding, or ultrasonic bonding.

㊾ The most common application of thermoforms is for various types of blister or clamshell display and product packaging. In most instances, these are formed from PVC, although PS and PET are also commonly used. Many **package inserts**, retaining devices, and countertop **point-of-purchase** displays are thermoformed from PS. Many medical and operating-room supplies are arranged in easily visible and readily accessible thermoformed trays.

㊿ Some food and product tubs are thermoformed rather than injection molded. There is an obvious **tooling cost** advantage; however, a thermoformed part may not have the precision required for consistent fit of, for example, a snap-on lid.

9.2 Thermoforming Methods

(51) Rotary or reciprocating matched die molding is the simplest thermoforming meth-

od. Matched dies are generally made from material with low thermal conductivity to prevent premature chilling of the heated sheet. The core part of the die simply pushes the softened plastic into the matching cavity half. Matched dies can be used only for shallow draws. Material distribution is poor.

52　Vacuum forming into a **cavity mold** or over a **plug mold** is the simplest form of vacuum molding (Figure 7.12). Vacuum holes are required in the cavity's lowest point. Vacuum forming into a cavity or over a plug has deficiencies similar to those of matched-die molding--limited draw and poor control over material distribution. Cavity molds allow for easier part removal, since the hot plastic shrinks away from the cavity when it cools, whereas the cooled plastic tends to tighten around a plug form.

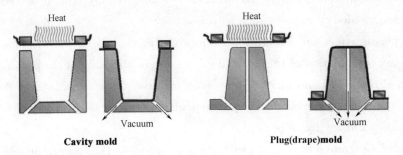

Figure 7.12　Simple vacuum forming over cavity and plug molds. The material is pulled to the mold shape when a vacuum is applied between the mold and sheet interfaces

53　Material distribution problems are reduced when several forming methods are combined. In plug-assist vacuum forming, the plug mold is above the sheet (Figure 7.13). The sheet is heated until it begins to sag, the plug moves the sheet into the mold, and a vacuum pulls it into conformity with the mold. This gives better material distribution to the corners than vacuum forming alone.

9.3 Billow forming

54　**Billow forming** uses air pressure to billow the sheet upward 50% to 75% of the anticipated mold draw (Figure 7.13). A plug pushes the billowed material into a cavity. A vacuum then pulls the intruded material to the cavity shape.

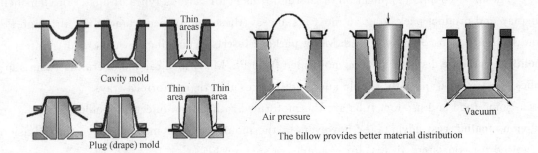

Figure 7.13　Plug assist vacuum forming. Billow forming inflates the plastic upward to produce a sheet of uniform thickness

10 Other Forming Methods

10.1 Pressblowing

55 Pressblowing is another way of combining the advantages of injection molding and blow molding (Figure 7.14). In pressblowing, polymer melt is injected from the extruder (step 1) into an injection mold head where the detailed head or finish section is formed. The injection mold head then lifts off the extrusion die nozzle, drawing with it a parison-like tube (step 2). The blow mold closes around the parison (step 3), and air expands the parison to the cavity dimensions. Lastly, the mold opens to eject the container and the container pinch-off is trimmed away.

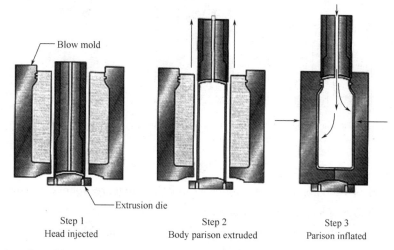

Figure 7.14 Pressblowing is another variation of injection molding and blow molding. A common application is the manufacture of collapsible plastic tubes

10.2 Rotational Molding

56 Rotational molding does not require an extruder. In this process, the polymer, usually a finely powdered polyethylene, is placed in a mold that is then heated while being rotated in two axes. The heat eventually melts the polyethylene, which flows and evenly coats the inside surfaces of the mold. While still rotating, the plastic is cooled. The mold is opened and the part is removed. An advantage of this process is the making of a hollow object with no openings.

57 Since rotational molding is not a pressure process, the molds need not be massive and are usually made from welded steel plate. Rotational molding is used to make very large bins and bulk containers. It is too slow and energy intensive for small plastic containers.

10.3 Compression Molding

58 Compression molding is primarily used to mold thermoset plastics. A measured charge of unpolymerized thermoset plastic is placed into the hot cavity of a mold. A mating core is brought down to squeeze the plastic into close conformity with the mold. The heat from the mold cures the plastic.

59 The compression molding technique has on occasion been used to mold thermoplastic

parts. The main difference is that the thermoplastic is melted in an extruder, and a measured shot is then ejected into a chilled mold cavity. As with thermosets, the core forces the molten plastic to conform to the mold profiles.

10.4 Blow-Fill-Seal Molding

60 Blow-fill-seal molding is similar to extrusion blow molding in that it starts with an extruded parison (Figure 7.15). However, as soon as the bottle is blown, product is introduced through a tube incorporated into the blowhead. Separate mold pieces then move in to form and seal off the finish.

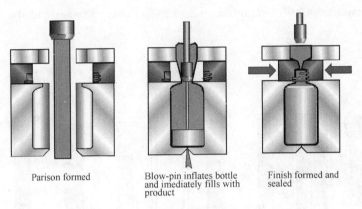

Parison formed　　Blow-pin inflates bottle and imediately fills with product　　Finish formed and sealed

Figure 7.15　Blow-fill-seal sequence

61 Blow-fill-seal is used to produce single-use bottles that are opened by breaking off the tip rather than removing a closure. It is particularly useful for sterile pharmaceutical preparations where the plastic forming temperature and the closure-less seal ensure sterility.

11　Recognizing Molding Methods

62 The process used to make a plastic part can be recognized by the part's nature and by an examination of the mold markings or lack of them. Extruded parts have no mold markings, since they are simply pushed through a shaped opening. Thermoformed parts also have no visible markings, since they are cut from a flat sheet. The process can only make containers that can be pulled from an open cavity or off a plug. There can be no undercuts.

63 An extrusion blow-molded bottle, by definition, must be a part that can be inflated. It will have a **pinch-off** across the bottom where the mold halves came together and cut off the excess parison. A faint **parting line** will be seen up the container sides.

64 Injection-molded parts will have a small bit of plastic at the gate point. Mold makers have become adept at hiding this gate mark, and sometimes it is difficult to find. Mold parting lines on simple tubs are usually put along an edge, which makes them hard to detect. Parting lines can be seen in parts that have undercuts or other features that require the mold to come apart in segments.

65 Injection blow-molded parts resemble extrusion blown parts except that instead of a mold pinch-off, there is a circular bull's-eye pattern on the container bottom. This is the residue of the gate mark from the initial injection molding of the preform. Faint parting lines up

the container sides can sometimes be seen.

(5093 words)

New Words and Expressions

synthetic [sin'θetik] *adj.* 合成的、人造的
monomer ['mɔnəmə] *n.* 单体
thermoset ['θə:məuset] *n.* 热固性
styrofoam ['stairə,fəm] *n.* 泡沫聚苯乙烯
polysulfone [,pɔli:'sʌlfəun] *n.* 聚砜
acetal ['æsə,tæl] *n.* 乙缩醛
surmise [sə'maiz] *v.* 猜测、推测
polarity [pə'lærəti] *n.* 极性
phenol ['finɔl] *n.* 苯酚
urea [ju'riə] *n.* 尿素
melamine ['meləmin] *n.* 三聚氰胺
mandrel ['mændrəl] *n.* 心轴
reservoir ['rezə,vɔr] *n.* 蓄水池
stroke [strəuk] *n.* 行程
runner ['rʌnə] *n.* 浇道
gate [geit] *n.* 浇口
preform [pri'fɔrm] *n.* 粗加工的成品
beryllium [bə'riljəm] *n.* 铍
be likened to　与……相比
modified natural resin　改性天然树脂
petrochemical industry　石油化工业
regenerated cellulose　再生纤维素
molecular weight　分子量
thermal history　受热历程
melt index　熔融指数
feed hopper　进料斗
gear reducer　减速器
variable speed motor　变速电动机
cooling jacket　冷却套管
profile extrusion　仿型（靠模）挤出
collapsing frame　倒人字夹板
nip roll　夹辊、压送辊
biaxial orientation　双向拉伸
shrink plastic　收缩塑料
hydraulic pressure　液压
ram-screw-type machine　柱塞螺杆型机器
hydraulic cylinder　液压缸

wide-mouthed 广口的
residual stress 残留应力
narrow-mouthed 窄口的
package insert 插页、药品说明书
point-of-purchase 购货点
tooling cost 加工成本
cavity mold 阴模
plug mold 阳模
billow forming 波浪成型
pinch-off 交错断裂
parting line 分型线、模缝线
BOPP 双向拉伸的聚丙烯
CTFE（Aclar） 三氟氯乙烯、三氟氯乙烯均聚物
CPET 结晶聚酯类
EEA 乙烯-丙烯酸乙酯、经常分组作为酸共聚物
EPS 发泡聚苯乙烯
EVA（EVAC） 乙烯-乙酸乙烯酯
EVOH（EVAL） 乙烯-乙烯醇
HIPS 高抗冲聚苯乙烯
LDPE 低密度聚乙烯
LLDPE 线性低密度聚乙烯
OPP 拉伸聚丙烯
mPE 茂金属聚乙烯
PA（NY） 聚酰胺、尼龙
PAN（AN） 聚丙烯腈、巴雷斯
PEN 聚萘二甲酸乙二醇酯
PC 聚碳酸酯
PE 聚乙烯
PET 聚对苯二甲酸乙二醇酯、聚酯纤维
PETG 聚（对苯二甲酸乙二醇酯）二醇
PP 聚丙烯
PS 聚苯乙烯
PTFE 聚四氟乙烯、特氟龙
PVAC（PVA） 聚醋酸乙烯酯、聚乙酸乙烯酯
PVC 聚氯乙烯
PVDC 聚偏二氯乙烯
PVAL（PVOH） 聚乙烯醇
None 无离聚物、沙林（杜邦商品名）

Notes

1. The words "plastic" and "polymer" are used interchangeably, but plastic tends to de-

scribe finished parts; polymer tends to describe the raw material and is used by the scientific community. (Para. 1) 单词"塑料"和"聚合物"可以交互使用，但在学术界里通常用塑料描述成品，聚合物描述原材料。

2. The shape of a molecule will determine how large numbers of them will fit together. Degree of crystallinity, clarity, barrier, melting point and other physical properties are affected by the molecule's shape. (Para. 9) 一个分子的形状将决定它们结合到一起的数量。结晶度、透明性、阻隔性、熔点和其他的物理性质会受到分子形状的影响。

3. Depending on thickness, the end product may be called "film" or "sheet". There is no clear division between the two. Film product formed by this process is referred to as "cast film". (Para. 19) 根据厚度，最终产品可以被称为"薄膜"或"片材"。两者之间没有明确的分界线。通过这个工艺过程成型的薄膜产品被称为"流延薄膜，平挤薄膜"。cast film 意指"流延薄膜，平挤薄膜"。

Overview Questions

1. Name four basic ways of forming plastics into semirigid shapes.
2. What is co-extrusion, and what advantages does co-extrusion offer?
3. To mold a plastic into a useful shape, it must first be heated and softened. What is the name of the machine that does this?
4. Why are combinations of vacuum, pressure, and mechanical assists used in thermoforming?

Lesson 8　Flexible Packaging Laminates

1　Laminates

① The purpose of a laminate is to combine the best of all properties-protection, aesthetics, machinability, and cost-into a single packaging structure. Laminates are made for the simple reason that there is no "super-substrate" possessing all desired properties for all applications. For example, polyethylene is economical and a good moisture barrier, but it is a poor oxygen barrier and may elongate if used to contain a heavy product. Polyester is a better barrier to oxygen, but it does not heat seal well and is more expensive.

② Flexible packaging based on laminates has been one of the major packaging growth areas in recent decades. Despite questions concerning the environmental status of multimaterial laminates, it is certain that these materials will continue to be a healthy part of the packaging mix simply because they are typically designed to replace a more material-or energy-intensive option.

③ A laminate is made by bonding together two or more selected material plies. Usually, but not always, these plies are in **roll-fed** form. The problem of creating an optimum flexible packaging material for an application can be resolved by selecting materials that have the individual desired properties and combining them into a single laminated structure featuring the most desirable attributes of the individual plies.

④ Individual laminate materials can best be examined by the quality or property they contribute to the final product:

- Structural properties such as physical strength, elongation, **abrasion resistance**, **puncture resistance**, and dead fold.
- Performance properties such as machinability, **sealability**, and environmental tolerance.
- Barrier properties against moisture, gas, odor, and ultraviolet light.
- Aesthetic properties such as clarity, opacity, feel, and metallic appearance.

⑤ These and other required performance characteristics must be delivered at minimum cost. Laminates are assembled from various combinations of paper, adhesives, plastic films, surface coatings, aluminum foils, and aluminum **metallized paper**. With the exception of aluminum, some of these materials have been discussed in previous chapters.

2　Aluminum Foil

⑥ Aluminum is made from **bauxite**, a claylike deposit containing aluminum oxides and silicates. Pure aluminum is a soft, silvery white, comparatively light metal (about one-third the weight of steel). The metal is ductile and **malleable** at normal temperatures and is a good conductor of heat and electricity.

⑦ Aluminum alloys containing small amounts of copper, zinc, magnesium, **manganese** and/or **chromium** have excellent strength properties. Alloys 1100, 1145, and 1235 are most commonly used for reroll stock. Alloy 3003, containing manganese, is used in applications such as pie plates, where good draw and greater stiffness are required. Aluminum and its alloys can be **reclaimed** readily at about 5% of the energy consumption required to refine the original ore.

⑧ By definition, **rolled aluminum** less than 152.4μm (0.006 in) is called foil. Foil is produced by being rolled from **ingots** or by continuous casting in-line with the furnace. All foil is supplied in 0 **temper**, the softest, most workable form.

2.1 Chemical Characteristics

⑨ When exposed to air, an aluminum surface acquires a natural, hard, transparent oxide layer that resists further oxidation. Aluminum foil's resistance to chemical attack depends on the specific compound or agent with which it contacts. Aluminum resists mildly acidic products better than mildly alkaline compounds such as soaps or detergents. Strong mineral acids will corrode **bare foil**. The mild organic acids generally found in food have little or no effect on aluminum.

⑩ Aluminum has high resistance to most fats, petroleum-based greases and organic solvents. *Generally, food products such as candies, milk, unsalted meats, butter and margarine are compatible with bare aluminum, as are many drug and cosmetic products.*

⑪ Intermittent contact with clean water has no visible effect on aluminum foil. However, in the presence of some salts and caustics, standing water can be corrosive. Hygroscopic products may cause corrosion if packaged in bare aluminum foil, particularly where a product contains salt or some mild organic acid. Applications that may subject aluminum to mild attack use coated, or laminated stock on the next-to-product surface. The decision as to whether to use a bare, coated or laminated surface in contact with a product must be based on reliable information and suitable testing.

2.2 Aluminum Foil in Flexible Packaging

⑫ Aluminum foil has many unique qualities that account for its widespread use in packaging.

Appearance	Aluminum foil has a bright, reflective metallic gloss that projects an exceptionally attractive and upscale appearance. In packaging, all reflective gloss surfaces are either solid aluminum foil or an aluminum metallized surface.
Barrier properties	Heavier foil gauges (>17μm/0.0007 in) are 100% barrier to all gases. As thickness is reduced, **pinholing** becomes more common. Typical **water-vapor transmission rate（WVTR）** for 0.00035 inch foil is 0.02 g or less/100 sq. in/24hrs. Many foil applications make use of foil's excellent gas-and light barrier properties. Food and nonfood products that are ultraviolet degradable can be protected by foil's opacity.
Dead fold	Foil has superior dead-fold properties. Dead fold is the ability of a material to hold the geometry of a fold. Wraps that must stay in place with-

out adhesive assistance and roll-up collapsible tubes require this property.

Friability　　**Unsupported foil** is easily punctured and torn, key properties when designing **unit-dose** and dispensing-tablet packages as well as various tear-away seals. An added benefit is the inherent tamper evidence.

Hygienic　　Aluminum foil can be easily sterilized. The smooth, metallic surface is nonabsorptive to contaminants. It is inert to or forms no harmful compounds with most food, cosmetic, or other chemical products.

Conductivity　　Microwave **susceptor** films, **electrostatic shielding**, and induction heat sealing are examples of applications that depend on aluminum's ability to conduct electricity and heat.

Formability　　Aluminum is a ductile metal. Heavier foils can be molded into trays and cups. Sheet stock is drawn into a variety of beverage and other can types. Solid aluminum slugs are impact extruded into collapsible tubes, cigar **humidor** tubes, and aerosol cans.

13　Foils are available in thicknesses of as little as 4μm (0.00017 in). The selection of foil alloys, gauges and tempers for bare conversion or for combining with other materials will be determined by the end use and conversion process requirements.

2.3 Foil Coatings

14　Packaging applications for plain foil are relatively limited and largely decorative. In most applications, aluminum foil is coated, and such coating may:

- Render the foil surface heat sealable.
- Increase foil's scratch or **scuff resistance**.
- Increase **tensile or burst strength**.
- Produce a specific surface (e.g., slip, nonslip, release, decorative).
- Improve adhesion of other coatings or printing inks.
- Enhance the water-vapor/gas-barrier properties of light-gauge foils.
- Increase foil's resistance to corrosive agents or products.
- Impart high gloss and three-dimensional depth to foil decoration or printing.
- Lubricate during converting or processing operations.

15　Coatings are employed to protect the package, the product or both. Coatings generally can be classified as decorative, protective, or heat sealing. In most instances, a coating is selected for one characteristic, but one coating may embody all three, as in a **tinted**, heat-sealable coating with high food-product compatibility.

16　Transparent lacquers and varnishes allow the brilliant reflective metallic **sheen** to show through. A transparent yellow lacquer would give the appearance of gold, while a transparent reddish orange would look like a copper alloy. Foil printed with an opaque ink has a particularly smooth, hard appearance.

17　The inherent heat sealability of many protective coatings is an added bonus. Effective heat-seal coatings provide strong, usually airtight seams and closures. Poly (vinyl chloride) heat-seal coatings are widely used with aluminum foil, but other formulations are also available. It is essential that the coating be compatible with all materials it contacts in order to protect and seal the contents.

⑱ Heat-seal coating does not appreciably add to a foil's bursting or tear strength in thicknesses under 25 μm (0.001 in).

⑲ Various polymeric coatings are applied to aluminum foils by extrusion or co-extrusion. *More typically, such coatings are used in combination with foil and other substrates such as papers, paperboards and plastic films to produce a multi-propertied laminated material.* Potential coating applications should be developed as individual cases. Selected materials and application weights should satisfy production criteria for coating and converting.

3 Vacuum Metallizing

3.1 The Metallizing Process

⑳ **"Vacuum metallizing"** refers to the depositing of a metal layer onto a substrate, carried out in a vacuum. Although many metals can be vacuum-deposited, only aluminum is used in packaging. The process, developed in the late 1940s, emerged as an important option for flexible packaging in the 1970s. Currently, packaging is the largest consumer of metallized papers and plastic films.

㉑ The initial role of metallized materials was for their decorative and aesthetic values. Recognition of the materials' functional properties has greatly extended packaging applications. Metallizing improves gas-and light-barrier properties and provides heat and light reflectance and electrical conductivity. The barrier properties achieved are a product of the thickness of the metal deposit and the properties of the substrate being metallized.

㉒ Most metallizing is done as a **batch process**. A typical system employs a horizontal tubular chamber, up to 2.1 m (84 in) in diameter and 2.8 m (110 in) long. Paper or film rolls are loaded onto unwind stations on one side of the chamber. The **web** is led down through tension rolls, under a large chilled roller, and through tension rollers to the **rewind roll** (Figure 8.1). A vacuum environment is needed to help aluminum vaporize at a lower temperature and to minimize oxidation of the metal vapor as it rises to the web.

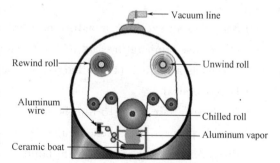

Figure 8.1 Cross section of a representative batch metallizer

㉓ Pure aluminum wire, approximately 3.2 mm (1/8 in) thick, is fed into machined metal "boats" located below the chilled drum. These are electrically heated to aluminum vaporizing temperatures. The boats rest in water-cooled **troughs** to extract radiant heat and minimize stray vaporization. Vaporized aluminum rises up and then condenses on the underside of the film or paper web as it passes over the chill roll. The thickness of the metal deposited is controlled by a combination of web speed, wire feed rate, and boat temperature.

㉔ Pattern metallizing is not possible with current technology; however, metallized pattern effects can be achieved by metallizing a preprinted film. Another process uses caustic solutions to remove the metallized layer selectively to create clear windows in a metallized film.

3.2 Vacuum-Metallizing Paper

㉕ Most metallized papers are quality virgin stocks that are clay-coated on one or two sides, depending on end-use criteria. The clay coating's weight varies, but generally, higher coating weights provide a smoother surface and higher reflective values in the metallized surface. All paper stock must be lacquer-coated before being metallized. The gravure-applied lacquer seals the paper surface, enhances surface smoothness, and promotes consistent metal adhesion.

㉖ Since the high vacuum within the chamber will **boil away** moisture, the paper's moisture content is normally reduced to below 5%. Remoisturizing is an essential production step after metallizing paper; this step re-establishes correct moisture content in the paper to improve its **resistance to curling**. Gravure coaters apply the primer coating for post-metallizing printing.

㉗ Vacuum-metallized paper gains the aesthetic appeal of a reflective aluminum surface, but does not gain barrier properties. A major market for metallized papers is for label stock.

3.3 Vacuum-Metallizing Films

㉘ Plastic film being metallized does not need to be sealed, smoothed, or dried, as a paper substrate does. In addition to adding decorative appeal, metallizing a plastic film significantly improves barrier properties to all gases. Although most plastic films can be metallized, OPP, PET, and nylon (PA) are the most commonly metallized packaging films.

㉙ OPP is the most widely metallized film. Snack-food packaging is the single largest application with potato chip bags being a major consumer. Metallization compliments OPP's properties:

- Serves as an excellent moisture barrier.
- Offers use temperatures up to 150°C (300°F).
- Serves as a limited oxygen barrier (which is significantly improved by metallizing).
- Is economical.
- Produces a fair metal-to-film bond.

Desirable properties also encourage PET's use. This material:

- Produces the best metal-to-film bond.
- Offers the best combination of oxygen-, moisture-and UV-light-barrier properties.
- Offers high use temperatures (up to 205°C/400°F).

㉚ Metallized **biaxially oriented nylon (BON) films** first gained market share in retail and institutional flexible packaging of ground coffee because of nylon's barrier qualities and resistance to abrasive coffee granules. Metallized PET is also used in this **niche** market because of current cost advantages. BON film:

- Is a good oxygen barrier.
- Has excellent tear, abrasion, and puncture resistance.

- Is hygroscopic (a modest moisture barrier).
- Is more costly than OPP or PET.

㉛ Metallized films are used alone in some applications but are usually a component in a laminated structure.

4 Other Inorganic Coating

㉜ An advantage of aluminum vacuum metallizing of plastic films is the significant increase in barrier properties. However, by its nature, the aluminum deposit makes the package opaque. This in itself is a benefit in some applications since it eliminates UV light degradation. However, in many instances, both a high gas barrier and high clarity are desired, as well.

㉝ Efforts continue in the development of other inorganic treatment to improve barrier properties and retain clarity at the same time. Significant advances have been made in depositing **silicone oxides** (**SiO_x**) and carbon (DLC or **diamond-like coating**). Both methods use plasma deposition technology rather than vacuum metallizing.

4.1 Silicone oxides

㉞ Like aluminum, SiO_x, (glass coatings) significantly improve gas barrier. Oxygen permeation values of 0.046 $cc/m^2/24$ hr (0.003 $cc/100\ in^2/24$ hr) have been reported. The significant advantage is that these coated films are clear. The cost of producing SiO_x coated films is still high, but a number of commercial applications, notably in the medical supply sector, are making use of the technology.

4.2 Carbon coatings

㉟ A number of companies have developed methods for depositing thin films of carbon over plastic substrates. The coatings are clear and claim to offer substantial barrier improvement over untreated plastic.

4.3 Nanocomposites

㊱ "**Nanocomposite**" technology describes a technique wherein extraordinarily fine particles of plate-like minerals such as clay are incorporated into the plastic. The mineral platelets themselves are not permeable to gases, so a permeating molecule has to go around each platelet. The overall effect is that a permeating molecule must work its way through a circuitous path that is several orders of magnitude greater than the actual thickness of the film. **Nanocomposite films** are said to be relatively clear.

5 Laminate Structural And Physical Properties

㊲ Structural properties are related to physical strength and performance. Physical strength is needed to hold the product. Relatively little strength is required to hold packaged dry soup, coffee, or confectionery products; more is needed when designing a 20 kg bag for industrial or bulk products.

㊳ For large bags, one might suppose that a material with high tensile strength, such as polyester, would be a good choice. However, material in a large bulk bag must be resistant to tear propagation. It should be reasonably flexible in the heavy gauges that would be used, and the large bag size **puts a premium on** material economy. Medium-density, low-density and linear low-density polyethylene better meet the requirements. The bags may be monolayer or could be coextruded with LLDPE, Surlyn or EVA as a heat-sealing layer. Additional layers might be incorporated if a high barrier is required.

㊴ A material with low elongation is desirable for heavy products; otherwise the weight would **distort** or **stretch** the package. Materials such as polyesters provide high tensile strength and low elongation. However, even though polyethylene has high elongation compared to polyester, the thick film gauges used for heavy products usually minimize this problem. Heavy-gauge polyethylene combinations provide the performance needed at a lower cost.

㊵ The property of "toughness"—the ability to resist puncture and abrasion—is needed for products that are abrasive or that have sharp edges. Nylon provides this, and frozen primal meat cuts are packed in nylon-based laminates for this reason. Some coffee brick pack laminates have a nylon layer to resist the abrasiveness of the coffee grounds.

㊶ Product weight or physical characteristics are not the only structural criteria. Retort pouches, for example, must undergo commercial sterilization that subjects the pouch to extreme temperature and pressure. End use and consumer preference may also dictate structural elements. Both a paperboard box and a polyethylene bag will hold fluid, but the paperboard box will stand by itself. Stiffer materials must be used where pouches must stand by themselves. **A gusseted pouch** bottom provides the required geometry to stiffen a flexible pouch.

5.1 Coefficient of Friction

㊷ Machinability is a composite of many properties, one of the most important being coefficient of friction (CoF). "CoF" refers to a material's ability to slide over itself, the product or a machine part during conveying, filling, collating, and casing operations. CoF can be reported as "static CoF" (i.e., the force needed to start an object moving from a standstill) and "kinetic" or "dynamic CoF" (i.e., the force required to maintain motion once it has been initiated). Static and kinetic CoF are almost identical for some materials, while in others static CoF can be significantly higher than kinetic CoF. Kinetic CoF is never higher than static CoF.

㊸ CoF can vary dramatically, depending on the surfaces being compared and the speed and temperatures at which the material is running. With numerous CoF test procedures available, always ensure that identical methods are compared and the appropriate test is used. CoF is critical for vertical form-fill-seal (VFFS) equipment and somewhat less critical on horizontal form-fill-seal (HFFS) equipment.

㊹ Generally, a low CoF allows for easy running of the web over stationary parts. However, low CoF can detrimentally affect material feeds where a **traction device** is used to pull film through the operation; low CoF can make for a slippery, unstable package. A low CoF, in the range of 0.12 to 0.2, would be considered slippery. A CoF range of 0.25 to 0.35 would be considered high.

5.2 Body and Dead-Fold Properties

㊺ Stiffness and "bulk" are usually obtained by incorporating economical paper layers into the laminate. Heavier aluminum foils, thicker—gauge plastics, and cellular plastics are also used to impart stiffness, depending on other properties that may be desired.

㊻ Laminated collapsible tubes must have some "body" or substance as well as the ability to "dead fold", so that they can be rolled up as the contents are ejected. Both paper and foil are used in the laminate layers of this construction.

㊼ Some laminate constructions are used as simple wraps, where the material's ability to conform to the wrapped object and stay permanently in position without a sealing medium is vital. Metal foils have the best dead-fold properties, followed by paper. Most plastic films have relatively poor dead fold. However, some PP, HDPE and PVC films have been designed to have reasonably good dead fold.

5.3 Tear Properties

㊽ Films such as polyester and polypropylene tend to propagate tears; that is, once a small cut or puncture has been made in the film, a tear readily propagates from that point. Other films, such as the polyethylenes, do not propagate a tear.

㊾ While a package should not tear open in transit, easy opening is a user benefit. Most plastic laminates have awkward tear properties. Paper or aluminum foil layers can enhance tear properties. In instances where a notch-sensitive material is being used, a small **nick** at the laminate's edge can help initiate the tearing action.

5.4 Thermoformability

㊿ Meat or cheese packs that have one part of the web drawn to a product-conforming tray require easy formability of the web stock. The web stock also needs good clarity for product viewing and good oxygen-barrier properties. This combination of properties is best found in nylon-based laminates (OPP does not thermoform and does not heat seal). The laminate would be coated, usually with PVDC to improve barrier qualities, and with an ionomer (Surlyn) heat-seal coating to seal through any fats on the seal line.

(51) Laminated materials are also used to thermoform rigid high-barrier cups and tubs for shelf-stable microwavable products. The material for these applications is in the order of 380μm (0.015 in) thick compared with the 10 to 50μm (0.0004 to 0.002 in) typical of flexible constructions.

5.5 Use Environments

(52) Laminate materials must be selected on the basis of the environment the package will experience. Low melting point adhesives and components cannot be used for boil-in-bag, ovenable or microwavable applications, as the package would come apart. Unoriented polypropylenes and some vinyls become brittle at low temperatures. Poly (ethylene terephthalate) (PET) and polycarbonate (PC) have the highest use temperature. For this reason, most dual-ovenable applications use PET as the base plies.

6 Flaxible Bags, Pouches And Sachets

(53) Every machine has its own characteristics and material requirements for optimum run-

ning. Since most webs are pulled through a machine, the material's **yield point**, i. e., the force at which permanent deformation takes place, is critical. Where **eye-mark** or other **register points** must be observed, recoverable elongation is also important. Other performance requirements will vary depending on the laminate purpose and application.

54 A significant proportion of flexible packaging is produced and filled on **vertical form-fill-seal (VFFS) machines** or **horizontal form-fill-seal (HFFS) machines** (Figure 8.2 and Figure 8.3). In a typical VFFS machine, the package material is unwound at the back of the machine and follows a vertical path over a **forming collar** where the flat material is shaped into a tube. A **longitudinal seal** is put into the tube, and product is introduced from the top. After product filling, horizontal sealing bars place a heat seal across the bag width, at the same time cutting across the center of the horizontal seal to separate one package, while leaving the tube sealed at the bottom ready to receive the next product dump.

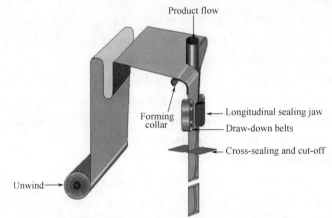

Figure 8.2 Material and product flow in a vertical form-fill-seal (VFFS) machine

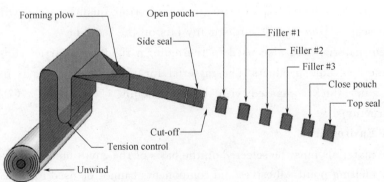

Figure 8.3 Material and product flow in a horizontal form-fill-seal (HFFS) machine

55 Most HFFS machines operate with the package material roll stock being fed horizontally over forming collars. Heat-seal bars seal off the appropriate-sized pouch, and after filling, seal the completed package. Depending on machine design, the formed pouch may be cut away from the parent web before or after filling. In some designs, laminate material may be formed directly around the product.

56 Both machine types can be used to fill liquid, powder, granular materials or small multiple products. VFFS equipment typically occupies a smaller **footprint**, but is more restricted in the number of filling or functional stations that can be grouped over the pouch. Horizontal machines can have a great number of operating stations grouped along the horizontal travel path of the pouch. This allows for multiple filling heads, steam purges, and other activities.

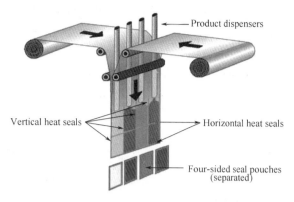

Figure 8.4 A multilane VFFS machine

57 Multiple-lane VFFS machines (Figure 8.4) mostly fill smaller pouches such as single-service condiments (mustard, ketchup, etc.) and sample sachets. Some high-volume larger pouches are also filled on multiple-lane machines. Since the machine is fed from two separate material rolls, there is the opportunity to use two different materials: an opaque back and a clear front, for example.

58 Pouches produced on different machines have different seal geometries (Figure 8.5).

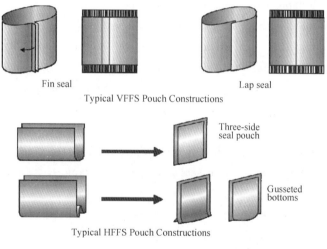

Figure 8.5 Vertical and horizontal form-fill-seal machines produce pouches with different seal geometries

59 Pouches made on a VFFS machine **are characterized by** a seal across the pouch top and bottom and a vertical seal across the center of the back. A **fin-style vertical seal** is easier to make since it brings together two inside heat-seal coated surfaces. However, it uses slightly more material, and the flattened seal is arguably not as aesthetically pleasing as a **lap seal**. A lap seal looks better and uses less material; however it requires that an inside surface be able to bond to the package's outside surface.

60 Pouches made on HFFS machines are characterized by a **three-sided sealed** edge. An advantage is that both front and back panels are free of seals. HFFS pouches can be easily accommodated to produce gusseted **stand-up pouches**.

61 Multiple-lane FFS pouches have **four sealed edges**.

7 Sealability

62 The majority of flexible packages are closed by heat sealing. The layer that provides this capacity is important, since **seal integrity** is absolutely essential, particularly for pharmaceuticals, foods, liquids, and fine powders. Heat-seal materials must be closely matched to the machine and the operating parameters being used to create the seal.

63 For heat seals, temperature, **dwell time** and sealing jaw pressure are the machine variables. However, a machine is set to run at design speed, and so dwell time is virtually a given. Jaw pressure is also fixed within narrow limits by the machine and is difficult to measure. In practice, most heat-seal control involves manipulating the temperature.

64 Waxes and wax blends are the most economical heat-seal materials, but they are not strong and have limited low-and high-temperature tolerance. They can be used only in noncritical applications.

65 Low-density polyethylenes are the most common heat-sealing mediums, although performance properties and cost varies widely depending on the specific type of polyethylene. These may be formulated with other additives or copolymers to improve selected adhesive qualities. Polyethylenes can be applied by extrusion, or a previously formed film can be bonded onto the laminate with an adhesive. In both instances, the film forms a layer over the entire laminate and contributes other properties as well as heat-sealability.

66 Where the seal requirements are more challenging, LDPE is replaced by more aggressive heat-seal mediums. The general order of performance and cost is as follows:

- LDPE.
- MDPE.
- PE/EVA (a soft film with a low heat-seal temperature).
- LLDPE (good **hot tack**, tough seal, wide sealing range).
- mPE.
- acid copolymers (for example, acrylics).
- ionomer (for example, Surlyn).

67 The ability to seal through contaminants such as grease or fat goes in the reverse order, with the best medium being ionomer.

68 Hot melts formulated to seal at temperatures as low as 60°C (140°F) can be used when LDPE's higher seal temperature (120°C/250°F) is not acceptable. Hot melts are easier to pattern-print and have easier seal initiation at lower temperatures and pressures. Ethylene-vinyl acetate is a common base resin used to make hot-melt heat-seal coatings.

69 Time/temperature/pressure data should be developed for all heat-sealing systems. Excessive combinations of time, temperature, and pressure tend to force melt from the seal area and create a weaker seal.

70 "Hot tack" is an important sealing-material characteristic, particularly on VFFS machines, where heavy product might be resting on a seal immediately after formation. A good

¶70 hot-tack seal will be able to resist peeling apart, even though it is still hot.

¶71 Cold seals are based on elastomeric materials that have a great tendency to adhere to themselves but not to other materials. Cold-seal adhesives are pattern-applied by the converter to the appropriate matching sealing surfaces on the web. In the packaging operation, machinery simply brings the two surfaces together and applies a slight pressure. Since time and temperature needs have been eliminated, sealing speed is very fast. Cold seals are useful in applications such as sealing chocolate wrappers, where the product is sensitive to elevated temperatures.

¶72 Laminate materials must be compatible with the product, and a laminate's internal plies must be adequately protected against aggressive product constituents. For example, some chemicals, particularly caustics, attack aluminum foils. Layers between the aluminum foil and the product must stop the penetration of any such ingredient to the foil interface. Aggressive flavor and perfume components can plasticize adhesive bonding layers, as well as detrimentally affecting other structural properties.

8 Barrier Properties

¶73 Almost every food or pharmaceutical application requires some form of barrier property". Barrier" is a nonspecific word that describes a material's ability to prevent gases from **permeating** through the volume of the material. The barrier qualities of a given material vary depending on the gas being permeated; it is rarely enough to say "we need a high-barrier material". The gas to which the barrier is required should be specified.

¶74 Oxygen and water vapor are the two most common barrier concerns, being the gases most responsible for product degradation (Table 8.1). Another class of permeants against which barrier properties are vital consist of the various volatile or essential oils associated with aroma and flavor. These are usually present in minute concentrations, and their presence is often critical to product quality.

Table 8.1 General oxygen-and moisture-barrier properties of various plastics

Material	Moisture Barrier	Oxygen Barrier
LDPE	fair	poor
HDPE	excellent	poor
EVAL	poor	excellent
PVDC	excellent	excellent
PA(nylon)	poor	good
PS	poor	poor
PET	fair	good
OPP	good	poor

¶75 Flavor and aroma constituents can be delicate or very aggressive, and retaining them is a packager's challenge. Fresh coffee aroma is a complex essential oil that can evaporate quickly or oxidize readily to leave a bland, **flat-tasting** beverage. Coffee packaging laminates must hold essential oils in and keep moisture and oxygen out. Additionally, vacuum-packing pulls the web tight against an abrasive granular material. This **places a premium on** toughness and abrasion resistance.

76　Barrier properties are usually reported as the rate at which a permeant passes through a given thickness and area over 24 hours and at atmospheric pressure. However, these data are expressed in many ways, and care must be taken to ensure that identical methods and units are being compared.

77　Oxygen permeability has traditionally been expressed in metric units:

$$\frac{mL(STP)mil}{m^2 \times d \times atm} \tag{8.1}$$

Where　STP—standard temperature and pressure; d—day; atm—atmospheric pressure.

78　Of the flexible packaging materials, only intact aluminum foil is potentially a 100% barrier to all gases. In the thinner gauges used for most packaging (as low as $7\mu m$/0.000285 in) foils suffer from pinholing (i.e., minute holes through the foil). Furthermore, foil is not durable to repeated flexing and can develop further pinholes during machining and shipping.

79　Water and oxygen permeation cannot be correlated. Other permeabilities do. In general, a good oxygen barrier will also be a good carbon dioxide barrier. In most cases, a good oxygen barrier will be a good barrier to many (but not all) organic vapors.

80　Metallizing plastic films significantly increases barrier to all gases. Poly (vinylidene chloride)(Saran) and poly (chlorotrifuoroethylene)(Aclar) are also good all-around barrier materials.

81　Ethylene-vinyl alcohol (EVOH) is used as a high oxygen barrier. However, EVOH absorbs water and must be sandwiched between materials that will keep moisture away from it. One common construction places the EVOH between two layers of polypropylene. Since EVOH and PP do not bond, PP films must be bonded to the EVOH with an adhesive or "tie" layer.

82　Barrier properties, as a general rule, **are inversely proportional to** film thickness on a fairly linear basis. Doubling the thickness of a given barrier layer will halve the permeation rate of the gas in question. Obviously, the required barrier property can be achieved with any material, provided it is thick enough. However, the oxygen barrier provided by $25\mu m$ (0.001 in) of PVDC would need about 25mm (1 in) of polyethylene to equal it—1,000 times thicker. Even though the cost of PVDC is two or three times that of polyethylene, it is a thousand times more effective as a barrier. For both practical reasons and economic reasons, the choice is a combination of the two.

83　In designing a barrier, the options must be evaluated and the best construction selected. Usually, the final analysis concerns the amount of barrier per dollar, tempered with machinability and other use properties.

84　Not all packages require a high barrier. In fact, some packages must have low barrier qualities. Red meats, for example, retain bright red coloration only when there is ready access to oxygen. Fruits and vegetables need to respire, ridding themselves of water and carbon dioxide while taking in oxygen. Products with high water activities would establish high humidities ideal for the propagation of microorganisms if they were packed in a high moisture-barrier package. Packaging films for these products must have permeabilities tailored to meet the individual respiratory requirements.

(85) A light barrier may be needed by products that can be harmed by light, usually the UV component. Fluorescent lighting in retail dairy cabinets degrade butter, and foil is frequently used, as it provides 100% dead fold and a light barrier.

9 Laminating Processes

9.1 Bonding Methods

(86) Individual layers, or plies, can be created or joined together by a number of techniques:

• ***Wet bonding*** refers to the use of solvent-or water-based adhesives. The solvents are dried or evaporated away after the plies are joined. Wet bonding can be used only if at least one of the plies is porous, to allow the solvents or water to escape. In practical terms, this means that at least one of the substrates must be paper.

• ***Dry bonding*** describes the process of applying solvent-carried adhesives to a surface and then drying away the solvents to leave a resinous adhesive surface. Sometimes this surface is tacky, reminiscent of household tapes. The second ply is nipped against the adhesive surface. In some formulations, the adhesive is softened to a fluid and tacky state, and the **curing process** is accelerated by passing the web between two heated nip rolls.

• ***Hot-melt bonding*** refers to the application of a hot-melt adhesive formulation. It is similar to dry bonding in the sense that no solvent needs to be eliminated, but has the advantage of not requiring a drying step. The substrates to be bonded need to be dimensionally stable at the application temperature, and which can limit the technique.

• *Extrusion* and *co-extrusion* use extruders to melt polymer resin, which is subsequently forced through a die and applied as a thin film onto a substrate. Extrusion coating applies the melt to a substrate. Extrusion laminating uses the melt to join two materials together. Co-extrusion feeds a die head from two or more extruders to produce multimaterial extruded film.

9.2 Laminating Machines

(87) Almost without exception, laminating is done as a **web-fed** rather than a **sheet-fed** process. Regardless of the machine, certain stations are common to all.

(88) A coating/laminating machine has an unwind stand for each of the raw web-fed materials that will be used in the process and a rewind stand to gather up the finished product. The unwind stands have tension-controlling devices so that materials are fed into the operating stations at a constant tension. A rewind-stand tension-controlling device ensures that the finished product is pulled out of the machine and laid up on the roll at a constant tension. Depending on the machine, additional tension-control zones may be placed within the machine.

(89) Tracking devices keep the material aligned with operational stations in the machine direction. Special rolls such as bowed rolls and herringbone rolls keep the material spread out flat while it is being processed.

(90) Coating/laminating machines have one or several coating stations at which additional material in a liquid state is applied to the base substrates. These may range from simple **glue applicators** to large extrusion dies attached to plasticating extruders. Materials are joined by pressure between two nip rolls. Depending on the coating process, the machine may have drying ovens either before or after the material-combining nip.

(91) The laminating process starts with a base web to which further layers are added until the desired construction is built up. Additional layers can be applied as fluid coatings, by bonding additional webs, by extruding and co-extruding, or by any combination of these. Base webs may be plain or may already have been printed, metallized, or otherwise treated.

9.3 Coating Stations

(92) Coating stations apply fluid materials such as varnishes, waxes, PVDC **emulsions**, and adhesives. There are literally hundreds of ways of putting a coating onto a web. The essential differences in the methods concern the way in which the coating is physically applied and the way the coating is metered. These factors, in turn, are related to the substrate type, its texture, and the coating's nature and purpose. Each method has its features and limitations. Gravure and extrusion coating predominate in flexible packaging.

(93) **Gravure coating** (Figure 8.6) applies and meters the coating from the cells of the gravure roll. Gravure rolls apply a precise coating thickness, regardless of variations in substrate thickness. Gravure does not work well with rough surfaces, where the small gravure cells may lose contact with the substrate surface. The coating must be of fairly low viscosity.

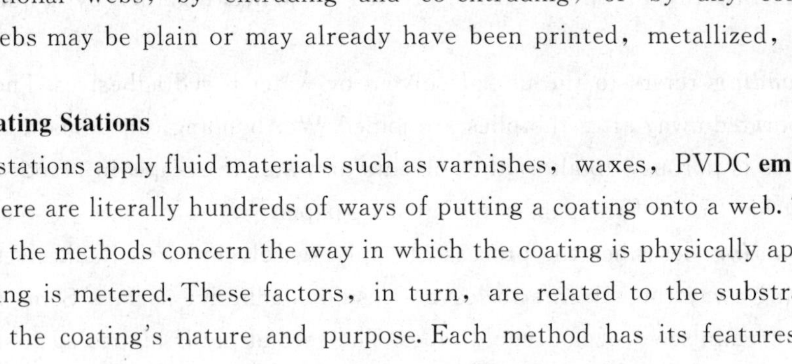

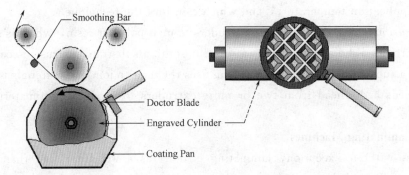

Figure 8.6 Gravure coating applies a predetermined amount of wet coating from engraved cells, shown enlarged at right

(94) Extrusion coating applies polymer melt to a substrate. In a typical application, polyethylene pellets are forced through an extruder barrel, where they are heated and softened to a syrup-like consistency. The viscous polyethylene is fed through a die with a long, narrow slot, forming a thin curtain that falls onto the substrate. At this point, it can be cooled to form a surface layer on the substrate, or a third material can be pressed against the still-mol-

ten polyethylene and bonded to the base substrate (Figure 8. 7).

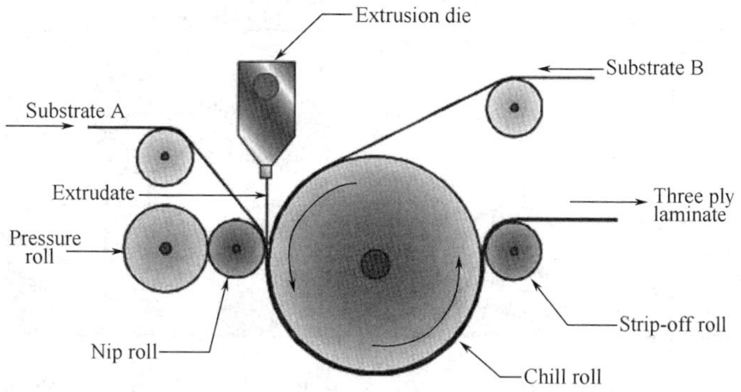

Figure 8. 7 Extrusion laminating uses an extruded melt to bond two materials

(95) Co-extrusion dies combine the outputs of two or more extruders so that the curtain of material exiting the extruder is actually several materials fused together.

(96) Wax lamination is an economical method of joining a paper or tissue support to aluminum foil. It is not a structurally and environmentally stable bond, so it has few other applications. The base material goes from the unwind roll to the heated-wax coating station. The second material is pinched against the hot wax at the combining nip and immediately chilled. After chilling, the finished laminate is rewound.

(97) Wet bonding (Figure 8. 8) requires that at least one substrate be porous enough to allow adhesive solvent or water to escape. Almost invariably this substrate is paper. Adhesive is applied to one substrate, the two substrates are nipped together, and the whole sent through an oven to set the adhesive.

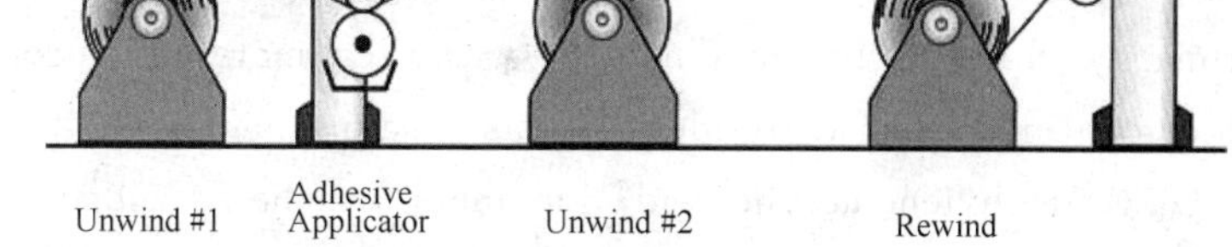

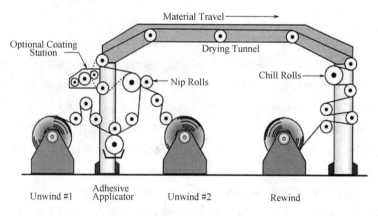

Figure 8. 8 A representative wet-bond laminator

(98) In dry bonding (Figure 8. 9) adhesive is applied to the substrate and then dried off solvents. The resinous adhesive is either tacky or can be activated and set by heat. The second substrate is joined by a heated nip against the first to effect the bond.

(99) With both wet and dry bonding, particular care must be taken to ensure that all traces of solvent are driven out. *Since laminators prefer to run their machines at top production speeds, there is always the danger that the "envelope will be pushed too far" and some vol-*

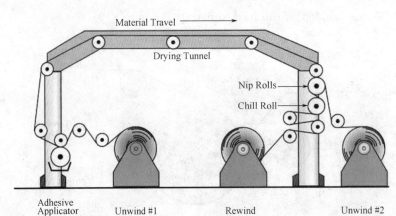

Figure 8.9 A representative dry-bond laminator. If both substrates are relatively impermeable, they cannot be nipped together until all adhesive solvents have been removed

atiles will remain in the laminate. These can give off-flavors to packaged food products. *Trapped solvents can cause later bubbling or blistering of the laminate. Quality laminators constantly check their production for volatiles using gas chromatography techniques.*

10 Specifying Laminates

100 Material plies are properly listed from the outside of the package to the inside.

101 In metric, a laminate's weight is given in g/m² (grammage) and in micrometres of thickness. Microns, used to describe micrometres, is not a recognized **SI unit**.

102 In inch/pound units, caliper may be quoted in thousandths of an inch, in mils, or in gauge: $1/1,000$ inch $= 1$ mil $= 100$ gauge (see Note 5).

$$0.0005 \text{ inch} = 50 \text{ gauge} = 1/2 \text{ mil} = 12.7 \mu m$$
$$0.001 \text{ inch} = 100 \text{ gauge} = 1 \text{ mil} = 25.4 \mu m$$
$$0.002 \text{ inch} = 200 \text{ gauge} = 2 \text{ mils} = 50.8 \mu m$$

103 Figure 8.10 shows examples of nine typical laminates containing aluminum foil as one of the components. The purpose of the foil component in all the examples is to provide a barrier to atmospheric gases and aroma constituents.

104 The identification LDPE in this figure includes LLDPE and various copolymers and blends based primarily on LDPE. In some instances, the seal material is proprietary or varies with application or converter. These are identified as "sealing medium". LDPE and other heat-seal layers (e.g., ethylene-acrylic acid; ionomer) can be extruded onto a substrate, or a previously manufactured film can be bonded on. Adhesive-bonded LDPE is somewhat stiffer than extrusion-applied LDPE. Bonded-on films are also used to build up thickness and provide better caliper control in thicker cross sections. A thick heat-sealing layer is desirable in such products as instant soup pouches where the thick polyethylene serves also to stop the dry hard noodles from cutting into the pouch stock. A vinyl heat-sealing film is used for lidding stock that will be applied to PVC single-service containers.

105 Retort-pouch laminates must withstand thermal-process stresses, and therefore the laminate's main structural component is polyester. It has low elongation, high tensile

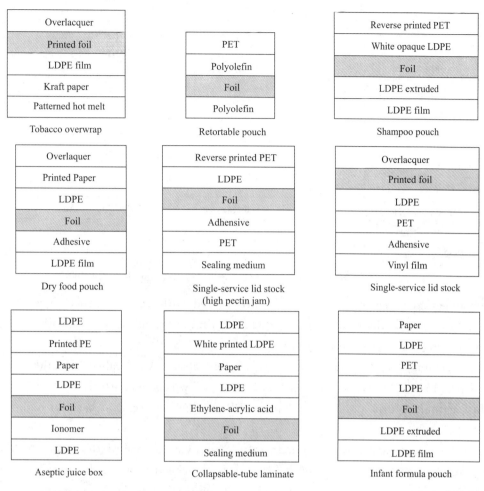

Figure 8.10 Examples of laminate constructions containing aluminum foil

strength and a high softening point. Aluminum foil provides the 100% barrier required for a shelf-stable package. The adhesive and a polypropylene-based heat-seal layer can withstand retort temperatures and pressures.

⑥ Laminates for aseptic juice packages usually consist of seven material layers. Paper provides body, while aluminum foil provides a barrier. The ionomer provides a superior bond to the aluminum metal.

⑦ The heat seal on a tobacco overwrap is a hot-melt adhesive applied in a pattern. This allows for heat sealing the overwrap and eliminates the need for a wet glue pot at the point of manufacture.

⑧ Pectin, a very aggressive component of some jams, will permeate LDPE and attack aluminum foil. The lidding stock for this product has a layer of PET to either side of the aluminum.

⑨ Figure 8.11 shows examples of six laminates that do not contain aluminum foil. The snack food laminate is typical of potato chip packaging. Barrier to moisture and oxygen is substantially increased by the metallized BOPP component. Laminates containing EVOH are

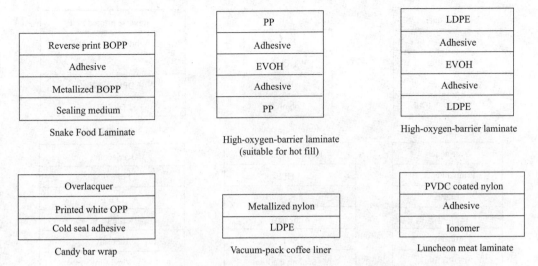

Figure 8.11 Examples of laminate constructions without aluminum foil

used where good oxygen barrier and clarity are desired. Although the figure shows outside LDPE and PP layers, these could also be nylon or PET, depending on what other laminate performance characteristics are needed.

　⑩ Heat sealing cannot be used near chocolate confections. A typical construction applies a pattern of cold-seal adhesive on the inside of the wrap. Two coated surfaces simply need to be brought together to form a seal. White opaque PP provides an excellent base for high-quality graphics.

　⑪ Luncheon meat and cheese laminates often use nylon as one of the layers. Nylon has good oxygen barrier properties (needed for high-fat luncheon meats and cheeses), and can be easily drawn by thermoforming into a product-conforming shape. The ionomer heat-seal layer is capable of sealing through any fatty contamination that might be over the seal area.

(7370 words)

New Words and Expressions

sealability [siːləˈbiliti] *n.* 密封性
bauxite [ˈbɔksait] *n.* 铝土矿
malleable [ˈmæliəbl] *adj.* 有延展性的、易适应的
manganese [mæŋgəˈniz] *n.* 锰
chromium [ˈkrɔmiəm] *n.* 铬
reclaim [riˈkleim] *v.* 回收再利用
ingot [ˈiŋgət] *n.* 铸块、锭
temper [ˈtempə] *n.* 韧度
pinholing [ˈpinˌhəuliŋ] *n.* 针眼、针孔
susceptor [səˈseptə] *n.* 感受器、基座
humidor [ˈhjuːmidɔː] *n.* 雪茄盒
tinted [ˈtintid] *adj.* 着色的
sheen [ʃin] *n.* 光辉、光泽

web [web] *n.* 卷筒材料
trough [trɔf] *n.* 水槽
niche [niʃ] *adj.* 有利可图市场的
Nanocomposite [ˌnænə(ʊ)ˈkɔmpəzit] *n.* 纳米复合材料
distort [diˈstɔːt] *v.* 扭曲、变形
stretch [stretʃ] *v.* 伸展
nick [nik] *n.* 裂口、刻痕
footprint [ˈfutprint] *n.* 脚印、占地面积
permeate [ˈpəmiet] *v.* 渗透、渗入
emulsion [iˈmʌlʃən] *n.* 乳剂、乳浊液
roll-fed 滚筒供料的
abrasion resistance 耐磨性
puncture resistance 抗戳穿性
metalized paper 敷金属纸
rolled aluminum 压延铝
bare foil 原铝箔
water-vapor transmission rate 水蒸气传输速率（WVTR）
unsupported foil 无载体铝箔
unit-dose 单元剂量
electrostatic shielding 静电屏蔽
scuff resistance 耐磨损性
tensile or burst strength 抗张或耐破强度
vacuum metallizing 真空镀敷金属
batch process 批量生产
rewind roll 收卷辊
unwind roll 放卷辊
caustic solution 苛性碱溶液
boil away 煮干
resistance to curling 耐折度
biaxially oriented nylon（BON）films 双向拉伸尼龙薄膜
silicone oxides 硅氧化物（SiO_x）
diamond-like coating 类金刚石涂层
nanocomposite film 纳米复合薄膜
put a premium on 重视、助长
gusseted pouch 有三角褶的袋子、折角袋
traction device 牵引装置
yield point 屈服点
eye-mark 眼标记、定位标
register point 定位点
vertical form-fill-seal（VFFS）machine 立式成型-充填-封口机

horizontal form-fill-seal（HFFS）machine　卧式成型-充填-封口机
forming collar　翻领式成型装置
longitudinal seal　纵封
be characterized by　具有……的特征
fin-style vertical seal　鳍形立式封口
lap seal　搭接封口、叠封
three-sided-seal pouch　三边封袋
stand-up pouch　直立袋
four sided seal pouch　四边封袋
seal integrity　密封完整性
dwell time　停留时间
hot tack　热黏性
flat-tasting　淡味
place a premium on　重视于、鼓励
be inversely proportional to　与……成反比例
wet bonding　湿法复合
dry bonding　干法复合
curing process　熟化工艺
hot-melt bonding　热熔复合
web-fed　卷筒材供料的
sheet-fed　片材供料的
glue applicator　上胶器
gravure coating　凹版涂布
SI unit　公制、国际单位制

Notes

1. *Generally, food products such as candies, milk, unsalted meats, butter and margarine are compatible with bare aluminum, as are many drug and cosmetic products.* （Para. 10）一般来说，食品如糖果、牛奶、新鲜肉类、黄油和人造黄油就像许多药物和化妆品一样，与纯铝兼容。"be compatible with"意指"与……兼容"。

2. *More typically, such coatings are used in combination with foil and other substrates such as papers, paperboards and plastic films to produce a multi-propertied laminated material.* （Para. 19）更典型地，这种涂层与铝箔和其他基材比如纸张、纸板和塑料薄膜结合起来使用可产生具有多种性能的复合材料。"in combination with"意指"与……结合起来"。

3. *Since laminators prefer to run their machines at top production speeds, there is always the danger that the "envelope will be pushed too far" and some volatiles will remain in the laminate.* （Para. 99）既然材料复合操作人员在最高生产速度下运行他们的机器，所以"增强包装技术系统的运转能力太快了的话"会存在危险，一些挥发物会留在复合材料里。"push the envelope"指"设备的性能范围，或功能极限"。

4. *Trapped solvents can cause later bubbling or blistering of the laminate. Quality lamina-*

tors constantly check their production for volatiles using gas chromatography techniques. (Para.99) 这些残留的溶剂会引起复合材料后来的起泡。优秀的材料复合操作人员经常使用气相色谱技术检查它们的生产过程是否产生易挥发物。"check … for" 意指"检查……是否含有……"。

5. *In the paperboard industry, 1/1,000 inch is a "point"; therefore, 1 point = 1 mil.* 在纸板企业里，千分之一英寸是一点，所以，1点＝1英寸的千分之一。

Overview Questions

1. What is the objective of combining various materials into a laminated packaging material?
2. Describe the difference between dry-bond laminating and wet-bond laminating.
3. What is the advantage of an HFFS machine over a VFFS machine?

Lesson 9 Closures

1 Selection Considerations

① **A closure** is a mechanical device that seals the contents within a container and can be removed to allow the contents to be dispensed. The term takes in **corks**, **stoppers**, **lids**, **tops**, and **caps**, made primarily from metal or plastic. Closures are applied to the "**finish**" of a glass, metal, or plastic container. The finish is that part of a container that receives the closure.

② The closure is a part of the container and shares the basic packaging functions of containing, preserving, and protecting the contents. *Closure specifications require the same upfront thinking as the container design itself: the designer must consider the contents and the end-use situations.*

③ There are many considerations when selecting a closure system:

• Closure materials must be chemically compatible with the **contents**. Particular attention must be paid to any **liner materials** or surfaces in direct product contact.

• Closure materials must be compatible with, and appropriate to, the container material.

• The closure must be easily handled on the filling line. This is especially important in high-speed production, where the closure must be located, correctly oriented, aligned with the container, placed on the container finish, and sealed in a fraction of a second.

• The closure is a possible path by which gases, contaminants, and organisms can enter the package or alternately, gases, moisture, or essential oils can leave the package. The closure must have adequate sealing qualities if these are factors.

• The method of creating a seal between closure and container surfaces must be carefully considered (e.g., inserted liner, flowed-in compound, molded-in seals, etc.).

• The need for **reclosability** affects closure choice. Some packages are opened once, the product is used, and the container is discarded. Other products may be used a little at a time over a long period. Reclosability should not be offered with a product that quickly becomes unfit for human consumption once the initial seal is breached.

• The closure must be "user friendly", with easy opening and, where applicable, easy and secure **resealability**. It is increasingly unacceptable to have a closure that requires a tool for opening.

• Economics are an ever-present aspect of any packaging design choice. Economics include not just closure cost, but also the cost of application and production losses.

• Closure convenience features such as applicators, measurement devices, and dispensers are sure to catch the eye of the convenience-minded consumer.

• As well as protecting the product, the closure must sometimes protect the consumer as well. Tamper-evident closures and child-resistant closures are obvious examples. Closures must not have sharp edges or corners and should not create **debris** that might be hazardous.

• Is the product aimed at a specific market segment? Products for seniors, for example, must allow for weaker, possibly **arthritic** fingers. Closures for hospital use are designed to maximum sterility confidence.

• Is the product dangerous or hazardous? Special closures have been designed that allow the safe measurement and dispensing of dangerous chemicals.

• Is there a particular persona that will provide a marketing advantage? Stock containers are used in many packaging applications. The only means of creating a unique appearance are the label graphics and the choice of a unique closure system.

2 Tamper Evident Closures

④ There is no officially recognized definition of what constitutes a tamper-evident (TE) closure. The U.S. **Food and Drug Administration** (FDA) published 21 CFR 211.132, **Tamper-Resistant** Packaging Requirements for Over-the-Counter Human Drug Products. The FDA definition reads: "Having an indicator or barrier to entry which, if breached or missing, can reasonably be expected to provide visible evidence to consumers that tampering has occurred."

⑤ Over-the-counter (OTC) drug products, products accessible to the public at point of sale, as well as **contact lens** solutions and some cosmetic products are the products mostly affected. There must be a clear statement prominently placed so that the consumer's attention is drawn to the particular feature.

⑥ The main approach to preventing tampering is to redesign the closure in some manner. This fixation on the closure as a defense against tampering is common to both consumers and legislators. It is not something that bothers the **tamperer**. Package types that are not normally considered vulnerable have been violated at points other than the closure. Professional packagers should never use the term **"tamperproof"**, since no practical package can **thwart** a determined tamperer.

⑦ **A case in point** was the Illinois person who laced 25 or so packages, including aseptic fruit drink packs, with **arsenic**, apparently using a **syringe**. In another incident a tamperer managed to get a dose of **cyanide** into a conventional single-serving yogurt tub. **In the final analysis** it is an often-ignored truth that we cannot protect ourselves entirely from the determined tamperer. It is also an ugly truth that the "urban terrorists" will always be there.

⑧ Understanding the nature of the problem, we realize that many legislated measures will not stop a determined tamperer. TE packaging may make it more difficult, or even very difficult, for the would-be urban terrorist to enter, but it rarely makes it impossible. This is not to say that such efforts should be abandoned. We should simply be aware that a tamper-evident closure does not guarantee our product against the determined tamperer.

⑨ Most tampering, fortunately, is simpler and is a **nuisance** rather than life-threatening. The casual extraction of a cookie while the parents are in another aisle, insertion of a dead insect, or exchanging salt for sugar are examples of this kind of activity. Properly sealed packages discourage this kind of impulse or nuisance tampering.

3 Tamper Evident Systems

⒑ Tamper-evident closures are important for those products prone to attracting would-be tamperers. Consumers need closures that demonstrate that the package contents have not been altered or interfered with. Some packages, such as aerosols, are inherently tamper evident or difficult to deliberately violate. Other packages must have features added that provide tamper evidence. These can be grouped into 12 recognized categories:

• Film wrappers. A transparent film with a distinctive design is wrapped securely around a product or product container. The film must be cut or torn to open the container and remove the product.

• Blister or **strip packs** (**Figure 9.1 and Figure 9.2**). **Dosage units** such as capsules or tablets are individually sealed in clear plastic or in foil. The individual compartment must be torn or broken to obtain the product.

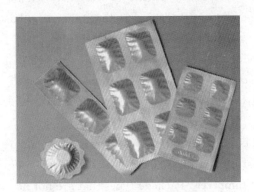
Figure 9.1 Blister or strip packs

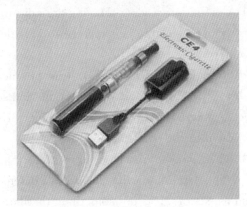
Figure 9.2 Blister card package

• Bubble packs (Figure 9.3). The product and container are sealed in plastic and mounted in or on a display card. The plastic or paper must be torn or broken to remove the product.

Figure 9.3 Bubble packs of pills

Figure 9.4 Tamper evident: Shrink sleeves, seals and bands

- Shrink seals and bands (Figure 9.4). Bands or wrappers with a distinctive design are shrunk by heat or drying to seal the cap-and-container union. The seal must be cut or torn to open the container and remove the product.
- Foil, paper or plastic pouches (Figure 9.5). The product is enclosed in an individual pouch that must be torn or broken to obtain the product.
- Bottle seals. Paper or foil with a distinctive design is sealed to the mouth of a container under the closure (Figure 9.6). The seal must be torn or broken to open the container and remove the product.

Figure 9.5　Pouch pack

Figure 9.6　Aluminum seals caps

- Tape seals (Figure 9.7). Paper or foil with a distinctive design is sealed over all carton flaps or a bottle cap. The seal must be torn or broken to open the container and remove the product.

Figure 9.7　Tape seal for a shipping box and a bottle closure

- Breakable caps. The container is sealed by a plastic or metal cap that either breaks away completely when removed from the container or leaves part of the cap attached to the container. The cap must be broken to open the container and remove the product.
- Sealed tubes. The mouth of a tube is sealed, and the seal must be punctured to obtain the product.
- Sealed cartons. All flaps of a carton are securely sealed, and the carton must be visibly damaged when opened to remove the product.
- Aerosol containers. Aerosol containers are inherently tamper resistant.
- Miscellaneous other systems.

4　Child-Resistant Closures

①　As with tamper-evident closures, reality demands that child-resistant packages never be referred to as **"childproof"**. Any closure that is devised for reasonable adult use will be openable by some children, either by accident and/or consistently. Hence, the correct term: "child resistant".

②　CR closures are under the jurisdiction of the **Consumer Product Safety Commission**. A list of substances covering drugs, household cleaning agents, pesticides, and other products defines which substances are regulated. CR closures serve the safety concerns of the marketplace. Although CR closures can certainly be a nuisance, their effectiveness is readily apparent. In 1995 participating poison control centers reported 1,024,094 exposures to harmful substances of children under five. Of these, only nine children died. Children will **ingest** virtually any thing. For example, in the mentioned study, 4,263 children drank illuminating/barbecue starter fluids. Since 1972, when CR closures were introduced, the incidence of accidental deaths has dropped 84% for all substances and 98% for aspirin alone.

③　*Stringent test protocols define which closures can be classified as CR. These protocols are described in 16CFR 1700.20 in the United States and are administered by the Canadian Standards Association in Canada.* The protocols are similar but not identical.

④　Briefly, the test is conducted using 200 children between the ages of 42 and 51 months. The children, working in pairs, are allowed 5 minutes to open the package. For those who are not able to open the package, a single nonverbal demonstration is given, and the children are allowed a further 5 minutes. The demonstration's purpose is to ensure that the package cannot be opened by the child once the "trick" has been observed. True child-resistant packages depend on a child's limited manual **dexterity**. Typically, this involves two dissimilar simultaneous motions or actions.

Figure 9.8　Push down and unscrew cap (child-resistant closure)

⑤　The CR effectiveness must be at least 85% without a demonstration and 80% after the demonstration. A separate adult test using 100 adults between the ages of 50 and 70 should have a 90% or better success rate. *Opening instructions are generally designed into the closure's **crown** or skirt, detailing a series of actions. Examples are:* "*line up the arrows and lift*", or "*lift and squeeze while twisting*". Figure 9.8

(1768 words)

New Words and Expressions

closure ['kləuʒə] *n.* 封闭物
cork [kɔːks] *n.* 软木塞
stopper ['stɔpə] *n.* 塞子

lid [lid] *n.* 盖子
top [tɔp] *n.* 顶盖
cap [kæp] *n.* 盖帽
finish ['finiʃ] *n.* 瓶口
content [kən'tent] *n.* 内装物
reclosability [rikləuzbiliti] *n.* 重新闭合
resealability [risiləbiliti] *n.* 重新密封
debris ['debriː] *n.* 碎片、碎屑
arthritic [ɑːr'θritik] *adj.* 关节炎的、关节炎患者
tamperer ['tæmpərə] *n.* 偷换者
tamperproof ['tæmpəpruːf] *adj.* 防偷换的
thwart [θɔːt] *v.* 阻挠
arsenic ['ɑːs(ə)nik] *n.* 砒霜、三氧化二砷
syringe [si'rin(d)ʒ] *n.* 注射器
cyanide ['saiənaid] *n.* 氰化物
nuisance ['njuːs(ə)ns] *n.* 讨厌的东西（人、行为）
blister ['blistə] *n.* 泡罩、气泡
ingest [in'dʒest] *v.* 摄取、吞咽
dexterity [dek'steriti] *n.* 灵活、灵巧
crown [kraun] *n.* 王冠
liner materials　衬料、里衬材料
Food and Drug Administration (FDA)　美国食品和药物管理局
tamper-resistant　阻抗偷换的
contact lens　隐形眼镜
a case in point　佐证、恰当的例子
in the final analysis　总之、归根结底
strip pack　条式包装
dosage unit　剂量单位
childproof　防止孩童的
Consumer Product Safety Commission　（美国）消费者产品安全协会

Notes

1. *Closure specifications require the same up-front thinking as the container design itself: the designer must consider the contents and the end-use situations.* (Para. 2) 封闭物在设计时要像容器的设计那样预先考虑其技术要求：设计者必须要考虑内装物及最终的使用情况。*up-front thinking* 表示"预先考虑"的意思，*end-use situations* 表示"包装物最终到达消费者手中时的使用情况"。

2. *Stringent test protocols define which closures can be classified as CR. These protocols are described in 16CFR 1700.20 in the United States and are administered by the Canadian Standards Association in Canada.* (Para. 13) 严格的测试协议规定了哪些封闭物可以归

类为儿童安全封闭物，这些协议在美国联邦法规 16 卷第 1700.20 条有描述，加拿大标准协会执行。16CFR 是美国联邦法案制定的强制性要求，1700.20 是对特殊包装的具体要求。

3. *Opening instructions are generally designed into the closure's crown or skirt, detailing a series of actions. Examples are:* "*line up the arrows and lift*", *or* "*lift and squeeze while twisting*". （Para.15）开启说明通常被设计在封闭物的冠盖或裙边上，详细描述了一系列的动作。例子有："对齐箭头后拉起"或"上拉后扭转挤压"。and 在这里表示"然后"，类似的还有"Push down and unscrew cap"，即"向下推然后拧开瓶盖"。

Overview Questions

1. List 10 considerations critical to the design or selection of an effective closure system.
2. Why should the common term tamperproof not be used?
3. List eight packaging systems that are considered to be tamper evident.
4. What is the design principle of a good child-resistant closure?
5. What is the relationship between tamper-evident closures and child-resistant closures?

UNIT 3

Packaging Dynamics and Distribution Packaging

Lesson 10 Shock, Vibration, and Compression

1 Shock

1 "Shock" is defined as an impact, characterized by a sudden and substantial change in velocity. For example, a dropped object gains velocity on its way down, which is rapidly lost when it meets the floor.

2 Shock is encountered in many places in the distribution environment:
- Accidental and deliberate drops during manual handling.
- Drops from **chutes**, conveyors, and other machinery.
- Falls from pallet loads.
- Sudden arrests on conveyors.
- Impacts occurring when vehicles hit **potholes**, **curbs**, or railroad tracks.
- Impacts occurring when a package is rolled or tipped over.
- Shock due to **rail shunting**.

1.1 Shock Resulting from Drops

3 Free-fall drop shocks, regardless of the cause, are identical in effect and can be treated as manual drops. Thousands of detailed observations have been made using instrumented packages or by direct and **discrete** visual monitoring to determine the typical manual handling patterns in a range of shipping situations. An example is given in Figure 10.1. This and other studies have shown the basic predictability of package handling. The details vary somewhat from study to study in the actual number of drops experienced, and the drop heights encountered. These reflect differences in handling modes and practices in the actual distribution environments.

4 Many such studies have been combined to provide generalized drop probability curves

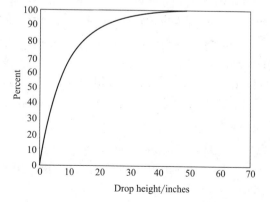

Figure 10.1 Cumulative percentage and drop height in next day air parcel delivery for United States Postal Service

such as those shown in Figure 10.2. This figure illustrates another predictable feature of **manual handling**: the lighter the package, the higher the probable drop height. That this should be so is simply human nature: lighter packages are more likely to be carried in groups and more likely to be tossed or thrown. Heavy packages cannot be thrown far and are usually handled more carefully to avoid personal injury.

5 Figure 10.2 or similar correlations are used to determine the probable drop height a given package should be designed to withstand. They are also the basis of preshipment test procedures and provide information for the development of protective packaging systems.

6 Fundamental lessons learned from such studies are as follows:

• The probability that a package will be dropped from a height greater than 1 metre (40 inches) is minimal.

• Packages receive many drops from low heights, while few receive more than one drop from greater heights.

• Skidded, wrapped, or otherwise unitized loads are subject to fewer drops than individual packages.

• There is little control over drop orientation with small packages. With larger packages, about half of the drops are on the base.

• A heavier package has a lower probable drop height.

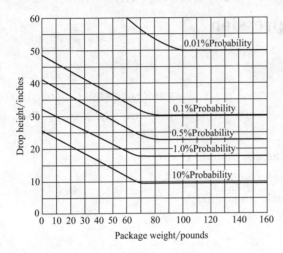

Figure 10.2 Generalized drop-height probability curves. The curves flatten out at the point where mechanical handling predominates

• The larger or bulkier the package, the lower the probable drop height.

• Handholds reduce the probable drop height by lowering the container relative to the floor.

• Cautionary labeling (fragile, this side up, handle with care) has only a minor effect. Cautionary labeling is no substitute for sound packaging practice.

• Address labels tend to orient the drop to a label-up position regardless of other instructions.

7 The usual result of drops and shocks fall into two categories:

• Damage to the package in such a manner that its protective or containment qualities are reduced.

• Damage to the product typified by bending, distortion, or, ultimately, breakage.

8 Typically, the greatest damage to the container is from edge or corner drops. The greatest damage to the contents is from a flat drop onto one of the faces. Shock can frequently damage the contents without adversely affecting the enclosing container. Usually this suggests the need for greater cushioning protection rather than an increase in the strength of the enclosing container.

1.2 Shock During Rail Transport

⑨ A special shock condition is experienced during railcar coupling. Boxcars are assembled into trains by moving individual cars at some speed against another car or cars. The average **shunting speed** is 8.4 kilometres per hour. This is the average speed; some of the impacts are at greater speeds (Figure 10.3).

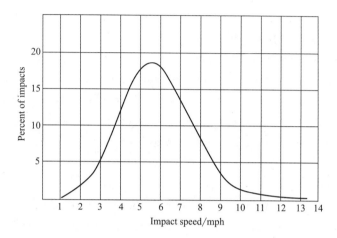

Figure 10.3 Distribution of rail coupling speeds

⑩ Shipper experience suggests that damage is greater for rail than for truck shipment. The high damage rates attributed to rail are probably due not so much to the actual shock forces as to load shifts and the effects of dynamic compression. Good loading and **bracing** and securing (**dunnage**) practices can substantially reduce rail damage. **Trailer-on-flat-car** (**TOFC**) shipping may be gentler than regular boxcar shipping because TOFC trains are not assembled by shunting; instead they are loaded after the cars have been assembled into a train.

1.3 Other Shock Conditions

⑪ Shock also occurs during the bumps and bangs, typical of mechanical handling and transport. Usually the degree of shock is not as great as that experienced during manual handling and free-fall drops. Generally, a package that will withstand manual handling shocks will survive mechanical handling.

⑫ "Repetitive shock" describes the low-frequency bouncing or rattling around that a product experiences if shipped as an unsecured load. The shock input of this "loose load" condition is low and not likely to cause typical shock damage. However, abrasion can occur, and if the product is in resonance with the input frequency, various forms of mechanical damage may develop. These, however, are better described as vibration-induced damage.

⑬ The Small-Parcel Shipping Environment (parcel post and parcel courier) represents a special environment. The packages are almost entirely under 20 kilograms, are not unitized, and undergo considerable manual handling, resulting in a high average number of impacts. This presents both user and operator with the most unfavorable situation possible. The use of highly automated conveying and sorting systems complicates matters further.

1.4 Quantifying Shock Fragility

⑭ Essential to the design of packages that will provide protection against shock damage is a knowledge of how "fragile" or "**sturdy**" the product is. A suitable cost-effective cushioning system is an engineered structure based on the product's quantified ability to withstand shock.

⑮ One way to quantify shock fragility is to measure the drop height at which damage starts to occur. However, quantifying shock fragility in terms of drop height is useful only if no additional protection is anticipated. This is helpful for products that may experience drops in their use environment, such as **cell telephones**, consumer electronics and **laptop computers**.

⑯ For products that are to be cushioned, packaging engineers use "**critical acceleration**", or "G", levels to describe an object's tendency to break when subjected to shock. An object will break if subjected to a force greater than its structure can bear.

⑰ Force is described by Newton's second law as
$$F = ma \tag{10.1}$$
where F—force; m—mass; a—acceleration.

⑱ Acceleration and deceleration are measures of the time rate of velocity change, and the forces are the same whether the object is accelerating or decelerating; only the direction changes. The most common measure of acceleration is g, or gravity, determined to be 9.81 m/s^2.

⑲ "G" is the ratio of observed acceleration to acceleration due to gravity:
$$G = \frac{\text{observed acceleration}}{\text{acceleration of gravity}} \tag{10.2}$$

⑳ A person encountering $2G$ would experience an acceleration twice that of normal gravity. The physical sensation would be a feeling of weighing twice as much as normal. At $3G$, a person would experience three times his or her normal weight. Since mass is constant for a given packaging problem, force is directly proportional to G.

㉑ If a 200gram coffee cup were dropped from 1 m, at the moment it reached the floor, its velocity would be 4.43 m/sec. If on hitting the floor it lost this velocity (decelerated) in 0.002 second, the deceleration could be calculated to be 2,200 m/sec^2. Expressing this as a ratio to normal gravity would give a G level of 224. At the moment of impact, the cup would, in effect, weigh 224 times normal-44.8 kilograms. Unless it was a very unusual cup, breakage could be guaranteed.

㉒ If the cup were dropped onto a sponge rubber pad, the impact velocity would remain the same. However, on impact the rubber pad would deflect, and the time over which the cup lost velocity would be extended. The deceleration would not be as severe, the stop not as abrupt. If the cup now stopped in 0.008 second, the G level would be 56. Another sponge layer might increase the deceleration time to 0.01 second, and the cup would experience 44 G. Adding still more layers would eventually reduce the G level to the point where the cup would not break. This would be one way of determining what cushioning protection the cup needed to protect it from a 1 m drop.

㉓ However, if the G level that would break the cup was known in advance, that is, if its fragility factor in G's was known, it would not be necessary to conduct the drop experiments; the cushioning needed could be determined by simple mathematics. It can be seen from the cup example that time is needed over which to dissipate the impact velocity and that this time is gained by the deflection of a resilient **cushioning material**. This is the basic principle of cushioning against shock.

㉔ A quick estimate of cushion material thickness can be made if the cushion material is treated as a linear, **undamped** spring. The deflection necessary to maintain a desired acceleration is calculated as follows:

$$D = \frac{2h}{(G-2)} \tag{10.3}$$

Where D = required deflection, h = anticipated drop height, G = fragility level (critical acceleration).

㉕ This formula provides the minimum distance over which the deceleration must take place in order not to exceed the critical acceleration. For example, for a product with a fragility factor of 40 G and an anticipated 1 m drop,

$$D = \frac{2 \times 1\text{m}}{40-2} = 0.053\text{m}(53\text{mm}) \tag{10.4}$$

㉖ The 53 mm deflection distance is the minimum stopping distance consistent with maintaining 40 G or less. Stopping in any lesser distance would raise acceleration to over 40 G and cause damage. The 53 mm deflection is the theoretical deflection distance, not the cushion thickness. To determine actual cushion thickness, it is necessary to know how far the proposed material will compress before reaching maximum strain, or **"bottoming out"**.

㉗ "**Static stress** working range" refers to the load per unit area that will cause a resilient material to deflect, but not to flatten out completely. Static load ranges can be found in supplier technical data sheets. With the correct static stress level, the cushion estimate can be completed. For this example, typical optimum strains for three commonly used cushion materials are on the order of the following:

Expanded polystyrene	40%
Polyethylene foam	50%
Polyurethane	70%

㉘ The theoretical deflection distance arrived at above can now be used to estimate the required thickness for the three different materials: 132 mm of polystyrene, 106 mm of polyethylene, or 76 mm of polyurethane.

㉙ More accurate estimates of cushioning thickness can be made using **dynamic cushioning curves** that are available for most cushioning materials. The information necessary to make these calculations using dynamic cushioning curves is:

- Product size and mass.
- Product fragility, expressed in G.
- Anticipated drop height.

50   To use a dynamic cushioning curve (Figure 10.4), locate the curve that crosses the desired critical acceleration line twice. The required foam thickness and the acceptable static load range can then be determined. The two points where the critical acceleration line is crossed represent the minimum and the maximum allowed static loads. Normally, a static load near the curve's minimum point would be selected, although designing with higher static loads would reduce cushion material area.

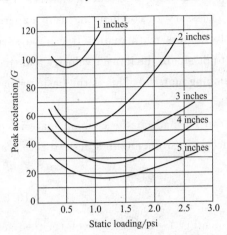

Figure 10.4   An example of a dynamic cushioning curve

51   Understanding shock and G will help you to understand many shipping damages. For example, a refrigerator shipped by rail has a compressor-motor assembly weighing 15 kilograms. The designer felt safe in securing this assembly to the frame with three fasteners capable of holding 120 kilograms, an ample safety factor. However, during shunting, the refrigerator experiences a 10 G shock and, during that brief moment, the motor behaves as if it had a mass of 150 kilograms. Since the three mounting fasteners can hold only a total of 120 kilograms, they may shear off.

52   The refrigerator sidewall, with a **bearing area** of 1.5 square metres, and the shipping box are able to distribute the load of a unit suddenly weighing 10 times more. With no external evidence of damage, the refrigerator is accepted at the receiver's dock and by the retailer. The problem is discovered only when a consumer plugs it in.

53   **Sophisticated** manufacturers know the G factor for all their products. In many instances, they will redesign products with low G levels, knowing that the savings in protective materials, and the goodwill generated by satisfied customers, will more than repay the cost of added engineering. Some typical G factor ranges are shown in Table 10.1.

Table 10.1   Typical fragility factor classes. A manufacturer would be advised to consider redesign of any product with a fragility level of less than about 30 G

G Factor	Classed as	Examples
15～25 G	Extremely fragile	Precision instruments, first-generation computer hard drives
25～40 G	Fragile	**Benchtop** and floor-standing instrumentation and electronics
40～60 G	Stable	Cash registers, office equipment, desktop computers
60～85 G	**Durable**	Television sets, appliances, printers
85～110 G	**Rugged**	Machinery, durable appliances, power supplies, monitors
110 G	Portable	Laptop computers, optical readers
150 G	Hand held	Calculators, telephones, microphones, radios

54   Fragility may be greatly dependent on how the force is transmitted to the product. *An egg on a flat surface has a fragility of 35 to 50 G, depending on the axis of impact.*

If the egg is supported in a conforming surface, its fragility can exceed 150 G (See Note 2).

1.5 Cushioning Against Shock

35 Any material that will deflect under an applied load can act as a cushioning material. By deflecting, the cushioning material attenuates the peak *G* level experienced by the product, compared with the shock pulse at the package surface (Figure 10.5). There are many materials capable of attenuating shock. Limitations imposed by the product, the process, or the environment will generally reduce the choices to a manageable few.

36 The cost of permanent molds, typically starting at $5,000, restricts the use of premolded shapes to **high-volume production**. Fabricated shapes, those cut and assembled from flat **planks**, are usually the choice for intermediate volumes. Loose fill, **foam-in-place**, and bubble pads are used for low-volume requirements or for those situations where nearly every shipment is unique.

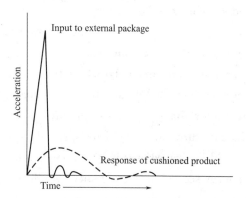

Figure 10.5 A cushioning material attenuates the initial shock pulse at the package's surface so that the product's response takes place over a longer period of time. The areas under the curves represent energy

37 Cushioning materials can be divided into those based on **cellulosic** material and on synthetic or polymeric materials. Cellulose-based cushioning materials are generally the most economical and include:

cellulose **wadding**	excelsior fill	corrugated **inserts**
molded pulp	**indented** kraft	newspaper

38 The shock absorption, **resiliency**, and cleanliness characteristics of these materials range from poor to fair. Some paper-based products are corrosive and should not be used with bare metal parts. Since all cellulose materials are hygroscopic, the risk of corrosion at high humidity is increased.

39 Many cellulose-based materials, particularly corrugated fiberboard, are quite **abrasive** and can **scuff** and polish finished surfaces. Corrugated fiberboard and rigid foams provide product protection by virtue of their own collapse. Their effectiveness is reduced after one major shock.

40 Polymeric-based cushioning materials include:

expanded polyethylene   expanded polypropylene
expanded polystyrene   polystyrene loose fill
air bubble sheet   expanded polyurethane foam
foam-in-place polyurethane

41 Polymeric materials have wide design latitude, and most polymeric cushioning materials can be produced in a range of densities and resiliences. As a class, they are clean materials

and have little or no corrosive properties. All plastic materials contribute to static problems unless they are specially treated. The resiliency of some plastic materials can change dramatically with temperature and altitude.

㊷ Cushioning polymers are not hygroscopic; however, some **open-celled foams** (typically polyurethane) will absorb liquid if they are wetted.

㊸ Loose fills are useful for random product packing but are difficult to get under large **overhangs** and are subject to **settling** during transport. In some instances loose fill can be recovered for reuse. In response to environmental issues, loose fills based on **popcorn** and expanded starches have been proposed. Popcorn fills will have the same shortcomings as other cellulose-based products, with the addition that they may attract **rodents** and other **vermin**.

㊹ Foam-in-place **urethane** cushioning material is made by mixing two reactive liquid chemicals (an **isocyanine** and a **glycol**) as they are ejected through an applicator. The two materials react almost immediately and begin to expand into a foam-like structure. During the foaming stage, the urethane is soft and pliable, but it quickly stiffens to a more semirigid state.

㊺ In use, urethane would be sprayed into the bottom of a box, and a protective plastic film would be **draped over** it. The object to be encased would be pressed into still soft, yielding foam. A second protective sheet would then be draped over the object and more foam applied to create the protective form's top half.

㊻ Foam-in-place urethane is versatile, and custom-made; form-fitting shapes can be easily fabricated. However, it is a labor-intensive process, generally used only in lower-volume or "one-off" applications.

2 Vibration

㊼ "Vibration" describes an oscillation or motion about a fixed reference point. The distance moved about the reference point is the **"amplitude"**, and the number of oscillations per second, expressed as hertz (Hz), is the frequency.

㊽ Vibration is associated with all transport modes, although each mode has its characteristic frequencies and amplitudes. Typically, the higher the frequency, the lower the amplitude. Frequencies above 100 Hz are of little concern to most packagers, because in most packaging situations, the product will become isolated (that is, its vibrational output will be less than the input received) at these higher frequencies. The most troublesome frequencies are below 30 Hz because they are the most prevalent in vehicles, and it is difficult to isolate products from them.

㊾ Vehicle vibrations come from many sources. Truck vibrations (Figure 10.6) occur predominantly at the natural frequencies of the load on the suspension system, of the unsprung mass of the tires against the suspension system, and of the trailer and body structure. They are excited by the condition and irregularities of the roadbed, the engine and **drive train**, tire and wheel imbalance, and the dynamics of the lading, or freight.

㊿ During vibration, an object is constantly being accelerated and decelerated as it moves through the vibration cycle. Since an acceleration can be described by its G level, one dimen-

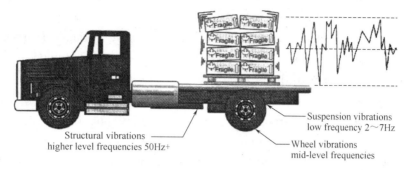

Figure 10.6 Typical source of truck-bed vibrations

sion of vibration is its G level. Vibrations at identical frequencies will have greater G levels at greater displacements or amplitudes.

2.1 Vibration Damage Due to Relative Motion

⑸ Vibrational damage can take several forms. Scuffing and abrasion can occur wherever one part is free to move against another. Designs affording point contact and those that allow movement against corrugated fiberboard surfaces are particularly prone to abrasion. Abrasion is sometimes aided by external substances, often dust from the product itself, for example, detergent powders. Scuffing is particularly **objectionable** on labels and graphics.

⑿ *Reducing or eliminating relative motion **lessens** this type of vibrational damage. Tight shipping case dimensions, particularly in the vertical axis, are preferred wherever this is compatible with top load compression of product and package.* Good bottle design incorporates **recessed** label areas. In other instances, hard surface **varnishes** will protect graphics. Some printing inks are soft and particularly prone to scuffing. In designs where scuffing is liable to be a problem, these inks should be avoided or protected.

⒀ Soft, nonabrasive plastic or cellulose wraps are used to protect finished and painted parts. Coatings are available that will reduce the abrasive character of corrugated board surfaces. Wax coating is the simplest and most economical of these.

⒁ Many particulate products sift or settle when vibrated. This leaves an open void at the tops of boxes and bottles, which the consumer invariably interprets as an underfill. Many products are shipped inverted so that the settling and compaction take place against the container top. When the consumer opens the container, the product is usually seen to fill the container almost level with the top.

2.2 Vibration Resonance

⒂ The spring/mass relationship between an input vibration and the response of a mass can have three outcomes:

      Output＝input   direct coupling
      Output＞input   resonance
      Output＜input   isolation

⒃ "Resonance", the condition where a vibration input is amplified, is the key packaging concern. In some cases, the end result is an output that is greater than, and out of all proportion to, the input. For example, at resonance, a 10 mm input amplitude might be amplified to produce a movement of 25mm or 40 mm.

57   A property of all spring/mass systems is that they have a unique frequency at which they will go into resonance. Resonance occurs whenever the **forcing (input) frequency** is the same as the **natural frequency** of the product and/or the package system. Resonance exists not only for the total assembly but also for parts or subsections within the total structure.

58   For protective packaging purposes, all resonance points should be located and quantified. This is done by subjecting the product to a range of frequencies and observing the frequencies at which a resonance condition occurs. *For packaging purposes, a typical resonance search might sweep the frequencies between 3 Hz and 100 Hz at 0.5 to 1.0* **octave** *per minute (refer to ASTM D 999).* The search should be done in all axes.

59   Identifying resonance damage is usually straightforward once the principles are understood. As with shock, resonance damage can occur without any visible external signs of abuse.

60   The energy developed on the output side during a resonance condition can do many things:
- Fatigue and finally fracture metal cans and **pails**.
- Flex and crack delicate circuits on circuit boards.
- Disintegrate or otherwise alter the texture of food products.
- Separate and settle granular components in a food product or settle loose protective fill.
- Aggravate scuffing and abrasion problems by several orders of magnitude.
- Cause individual containers or components to bang into one another.
- Disturb pallet patterns or dunnage (load-securing) systems.
- Initiate stack resonance.
- Unscrew bottle caps and threaded fasteners.

61   The greatest vibrational input in a typical truck is directly over the rear wheels and **tailgate**. If damage is restricted to or most severe in this section of the vehicle, vibrational inputs are almost certainly the cause. Vibrational inputs are also usually the source if damage seems to occur only in the product layer next to the pallet or on the top layers.

62   Damage caused by resonance vibration can be difficult to resolve. The problem is complicated in that all cushioning materials are resilient and, while they are acting to attenuate shock, they are also acting as a spring in response to vibrational input. For many applications, it has been found that redesigning the product to eliminate critical resonance points is the most cost-effective method of decreasing damage. The last resort is the design of a **vibration-isolation cushioning system**.

2.3   Stack Resonance

63   Occasionally, entire loads go into a stack resonance condition, where each succeeding container goes into resonance with the previous container until the entire stack is bouncing, creating conditions of extraordinary destructiveness (Figure 10.7).

64   For example, if a truck bed is moving at an amplitude of 5 mm and the bottom container goes into resonance, it might acquire an amplitude of 10 mm. The input into the second container in the stack is now 10 mm and, when it goes into resonance, its movement might be

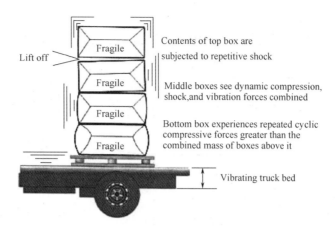

Figure 10.7 In stack resonance, the entire stack is bouncing, creating a destructive condition

20mm. The amplitude is multiplied as it goes up the stack until the top container actually bounces off the top of the load.

65 The dynamic load on the bottom container in such a system can be several orders of magnitude greater than the actual weight resting on it. The top container is subjected to extremes of repetitive shock and vibrations of considerable amplitude. Since the top layer in a unitized load is essentially weightless for short periods of time, small side loads (such as a bump from the side) will cause it to "float", or move. In a **stretch-wrapped** load, this tendency for the top layer to move can lean or skew the entire pallet load to one side. Individual boxes will sometimes interlock into the adjacent pallet load, making it difficult to pick up a pallet without disrupting or damaging the neighboring loads (Figure 10.8).

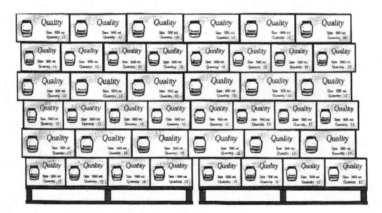

Figure 10.8 Skewing of a load may cause boxes in adjacent loads to interlock

2.4 Isolating Vibration

66 Materials used to isolate vibration are for the most part the same as those used to isolate shock. An ideal vibration-isolation material provides isolation in the 3 to 100 Hz range, since these are the predominant frequency ranges that cause damage during transport. However, cushioning materials, like all other springs, also have characteristic resonance points. Vibrational response curves are available for many resilient materials.

67 A material with the characteristics shown in Figure 10.9 could be used effectively to iso-

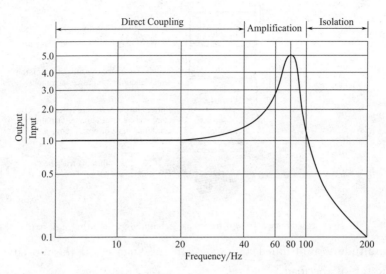

Figure 10.9 Vibration response curve for 175pound, C-flute pad at 0.5 psi static load.
The three conditions-direct coupling, amplification, and isolation-can be clearly identified

late vibrational inputs over 100 Hz. The amplification between 40 and 100 Hz is not necessarily a problem, provided the product has no response in this zone. If by bad choice this is where the product resonates, damage is almost certain. Below 40 Hz there is direct coupling, and the product will not see any worse vibration than the input level.

68 A properly selected isolation material resonates at an input frequency that is less than half of the product's resonance frequency. For example, if a product has a major resonance at 48 Hz, the isolation material should resonate at less than 24 Hz.

3 Compression

3.1 Static and Dynamic Compression

69 Most products are stacked during warehousing and shipping. It is important for a container to be able to bear static compression loads safely without damage to product or container. Static compression is determined by mechanically applying a load at a slow rate or by conducting **dead-load** stacking tests.

70 "Dynamic compression" describes a condition where the compression load is applied at a rapid rate. Dynamic compression is experienced during **clamp-truck** operation, rail shunting, and stack resonance as well as during normal transit conditions. According to railroad authorities, longitudinal dynamic compression forces, developed by the movement of **back-load** against the **bulkhead** of a railcar, can exceed 1,400 lb/sq. ft in a standard boxcar at typical shunting speeds.

71 Most standard laboratory compression tests are relatively dynamic in nature. Fiberboard containers are able to bear dynamic loads almost as high as the determined compression strength, but only for a very short duration.

72 **Apparent compression strength** is affected by the load application rate. Rapidly loading a viscoelastic material will give a higher apparent strength than when the material is loaded slowly. To ensure that data are comparable among laboratories, load rates have been stand-

ardized to 12.7mm/min, ±2.5 mm/min.

3.2 Compression Strength and Warehouse Stack Duration

73 The warehouse condition is one of static loading over time. However, quick answers are usually needed for predicting safe warehouse stack duration or for evaluating new container designs. The laboratory compression test is completed within minutes; it is a dynamic test.

74 **Compression strength** (a dynamic value) is not the load that can safely be applied in the warehouse. Stacking strength (a static value) for a given situation can be estimated from

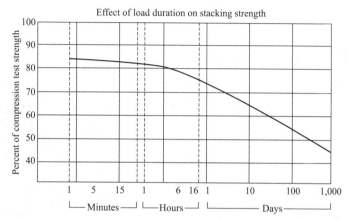

Figure 10.10 The compression strength of corrugated board falls off with time. This graph relates dynamic compression (laboratory) to static compression (warehouse)

Figure 10.10, which relates dynamic and static loading conditions. The initial part of the curve in the figure shows what the container will bear under dynamic, or short-term, load application. If 85 kg of product were stacked on a box with a compression strength of 100 kg load, the box would fail after about 10 minutes. The same container would fail in about 10 days if loaded to 65 kg. If the container is to last 100 days, the stack load should not exceed 55% of the dynamic compression-test value.

3.3 Compression Strength and Humidity

75 Paper is a hygroscopic material, and the moisture content of paper can range from about 3% to over 20%, depending on the humidity of the surrounding atmosphere. A change in relative humidity from 40% to 90% can result in a loss of about 50% of a corrugated container's stack strength. Corrugated containers destined for very humid conditions need excess stack strength to allow for this loss.

76 Since the physical properties of paper change dramatically with moisture content, all paper tests are conducted at standard temperature and humidity conditions. Container compression strength is normally given as load to failure at 50% R.H. and 23°C.

77 Figure 10.11 contains a chart for estimating the compression strength of corrugated board at different moisture levels. To use the chart, make a line from the compression strength to the board moisture content. Mark the point where the line crosses the pivot line. Now project a line from the new moisture content through the pivot line mark, and read the new compression strength.

3.4 Other Factors Influencing Box Stack Strength

78 In cases where the contents do not contribute to the compression strength of a container, com-

pression strength is mostly a function of the wall **perimeter**, with the greatest contribution made by the four corners (Figure 10.12). A box will fail at loads far below the measured compression strength if the loads are applied unevenly at points away from the corners or in a concentrated area. Thus, in addition to the total compressive load, one must also consider the load per unit perimeter length and the load distribution.

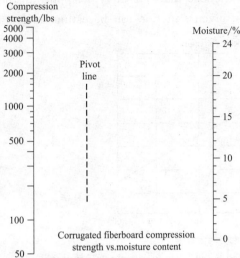

Figure 10.11 Chart for estimating compression strengths at different board moisture contents

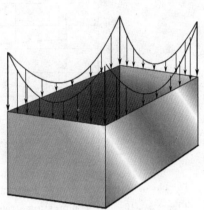

Figure 10.12 Distribution of load-bearing ability around a box perimeter. The length of the arrow is proportional to the load-bearing strength at that point

⑲ Laboratory compression tests are conducted on new, undamaged containers. Consider, however, the effect on stack strength of clamp-truck handling creases container side panels, or of a cord used to unitize or secure the pallet load that cuts into the edges of all corner boxes. Such damage, precisely at the container's most important point, dramatically reduces available load-bearing ability.

⑳ Higher initial compression strength is needed where warehousing follows a long journey or **rough handling**, since the containers will have experienced attrition factors that will have an accumulated effect on load-bearing ability. Lower initial compression strengths can be used only in those instances where the product has a short distribution cycle.

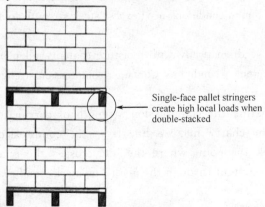

Figure10.13 Crushing loads from a single-face pallet

㉑ Compression strengths are normally measured as the full bearing area on both the container top and bottom. However, most pallets are decked with boards, and therefore the bottom container does not have full support over its base. Without this full support, a proportion of the stack strength is lost. **Single-face pallet stringers** produce a much greater unit area load on the topmost container of the lower pallet (Figure 10.13). Such loads may be 3 or 4 times greater than that assumed for the bottom container in a stack, resulting in unanticipated damage.

82　Most shipping containers are designed to provide maximum vertical stack strength, since this is the common warehouse condition. Dynamic compression by clamp trucks and rail shunting is in the longitudinal direction, normally the container's weaker axis.

83　The best possible use of container load-bearing ability is when boxes are stacked directly on top of each other in a vertical column. Unfortunately, column stacking is the least stable technique, and other stacking patterns are used to provide better load stability or cube utilization. Each palleting pattern has a different total stacking strength.

84　Allowing boxes to overhang the edge of a pallet leaves the load-bearing walls of the boxes suspended in midair. Pallet overhang is often deliberate but is rarely a good idea. Inadvertent overhang can occur internally because of pallet board geometries relative to the container size. Typical loss of available compression strength to overhang is shown in Figure 10.14.

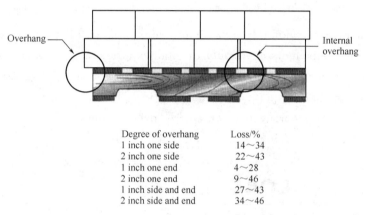

Degree of overhang	Loss/%
1 inch one side	14～34
2 inch one side	22～43
1 inch one end	4～28
2 inch one end	9～46
1 inch side and end	27～43
2 inch side and end	34～46

Figure 10.14　Effect of overhang on compression stack strength

3.5　Contents' Effect on Compression Strength

85　Contents sometimes increase apparent compression strength even though they do not apparently support a load. The usual reason is that the contents prevent the container sidewalls from buckling inward, thus delaying the failure point.

86　Often times, the contents are counted on to provide some portion of the load-bearing ability of a shipping container. In the instance of **shrink-wrapped** trays, the contents must bear all of the stacking load. The asymmetrical nature of the oil bottle shown in Figure 10.15 can result in overhang of the major load-bearing bottle wall segment. The available bottle compression strength is a fraction of the measured value.

87　It is generally assumed that stack forces are acting on a vertical wall of a corrugated box. However, flexible primary packages and bag-in-box systems containing liquids or semi-solids exert varying degrees of hydrostatic pressure perpendicular to the vertical container wall. Bag-in-box systems and many semi-solids such as cheese and premixed grouting compounds develop hydrostatic forces that reduce compression strength.

88　Hydrostatic pressure appears as an outward **bulge** slightly below the midpoint and is greatest at the bottom of the enclosing box. Supporting walls become **bowed** rather than per-

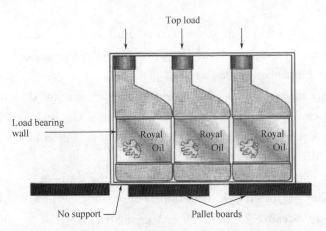

Figure 10.15 Overhang of asymmetrical supporting contents can have a major impact on a container's calculated ability to hold a load

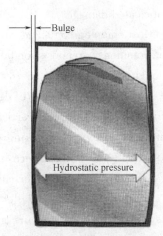

Figure 10.16 Bag-in-box systems reduce compression strength by bowing out the side walls

pendicular, with a corresponding loss of compression strength (Figure 10.16).

⑧⑨ Where various components contribute to total compression strength, good design calls for the individual components to act collectively. Maximum strength is gained when all components have the same failure point. For example, if a plastic bottle and a partition are expected to contribute to a corrugated container's overall compression strength, the three should be sized so that they fail as a single unit (Figure 10.17).

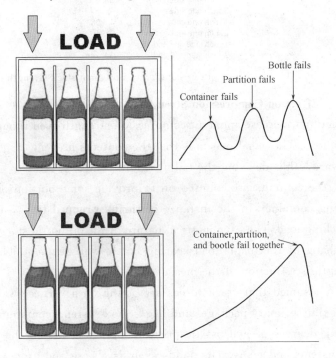

Figure 10.17 The highest compression strength in a multicomponent system is achieved when all components act together. Top example shows separate failures, while the bottom example shows simultaneous failure at a higher compression

3.6 Plastic Bottle Stacking Factors

⑩ Plastic bottles are expected to contribute compression strength in some shipper designs. This practice is acceptable provided that plastic's viscoelastic, or "creep", property is taken into account. With plastic, as with corrugated board, the dynamic compression strength must be related to static warehouse conditions.

⑪ Stack duration for polyethylene bottles can be estimated using the bottle load ratio shown in Figure 10.18; this ratio consists of the expected load over the compression strength. A bottle with a compression strength of 10 kilograms and loaded with 3.1 kilograms would have a load ratio of 0.31. It could be expected to last about 180 days under this loading.

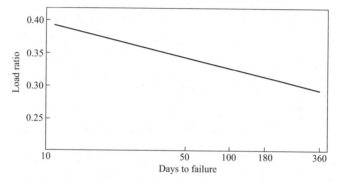

Figure 10.18 Warehouse stacking and load ratios for polyethylene bottles

⑫ The design of any proposed plastic container should be carefully reviewed. Features such as sharp corners, edges, or small-radius curves are to be avoided since they act as **stress concentrators** and promote flex cracking. Containers that are expected to contribute to load bearing should be designed with this in mind. This usually means circular cross sections, large **finish surfaces** to spread the load, and shallow-angled transitions to distribute the load from the finish to the container walls.

3.7 Stacking and Compression

⑬ Stacking strength is a key requirement of most transport packages. Stacking strength is defined as the maximum compressive load (pounds or kilograms) that a container can bear over a given length of time, under given environmental/distribution conditions without failing.

⑭ The ability to carry a top load is affected by the structure of the container and the environment it encounters, and the ability of the inner (primary) packages and the **dividers**, corner posts, etc. to sustain the load. Compression strength is related to stacking strength but is actually quite different.

⑮ The simplest and most common corrugated transport packages are regular slotted containers (RSCs, Box Style 0201) in which the corrugation direction is typically vertical-parallel to top-bottom stacking forces. Since the early 1960s, we have been able to estimate the compression strength of regular slotted containers with reasonable accuracy and precision.

96 Compression strength of regular slotted containers is a function of:
- Perimeter of the box (two times length plus two times width).
- Edge crush test of the combined board.
- Bending resistance of the combined board.
- **Aspect ratio** ($L : W$) and other factors.

97 When we know the above variables, we can estimate the compression strength through an equation known as the McKee formula.

$$BCT = 2.028 \times (ECT)^{0.746} \times [(D_x \times D_y)^{0.254}]^{1/2} \times BP^{0.492} \qquad (10.5)$$

Where BCT—RSC top-to-bottom box compression strength, kN/m² (lbf/in² or psi); ECT—edge crush test, kN/m (lbf/in); D_x, D_y— **flexural stiffnesses** of combined board in the machine direction and cross direction, kN/m (lbf/in); BP—inside box perimeter, m (in).

98 The McKee formula can only be applied to RSCs, and only those with a perimeter-to-depth ratio no greater than 7 : 1.

99 McKee also created a simpler formula based on caliper of the combined board instead of **bending stiffness**:

$$BCT = 5.87 \times ECT \times (T \times BP)^{1/2} \qquad (10.6)$$

Where T—caliper of combined board, m (in).

100 It provides accuracy close to the original equation and is easier to use, both in testing and mathematically. McKee's work was based on averages. Individual boxes will vary above and below the predicted value.

101 The ability to predict the compression strength of a container is a considerable tool, but it is even more powerful to take a compression requirement, back out an ECT requirement and use it to determine appropriate board combinations.

102 Solving for ECT, the simplified McKee formula is:

$$ECT = BCT/[5.87 \times (T \times BP)^{1/2}] \qquad (10.7)$$

3.8 Distribution Environment and Container Performance

103 The ability of a container to perform in distribution is significantly impacted by the conditions it encounters throughout the cycle. Some of these conditions are difficult for the packaging engineer to influence, including stacking time and relative humidity. Others are determined by handling and unitizing packages; for example, pallet patterns, pallet overhang, pallet deck board gaps and excessive handling.

104 We can now estimate the impact of these conditions on container strength. If the original box compression strength is known (determined in the lab using a dynamic compression tester), we can factor it by generally accepted multipliers to arrive at an estimated maximum safe stacking strength (Table 10.2).

Table 10.2 Stacking strength multipliers that compensate for storage time, humidity and palletizing conditions

		Compression Loss	Multifliers	
Storage time under load	10 days-37 percent loss		0.63	
	30 days-40 percent loss		0.6	
	90 days-45 percent loss		0.55	
	180 days-50 percent loss		0.5	
Relative humidity, under load (cyclical RH variation further increase compressive loss)	50 %-0 percent loss		1	
	60 %-10 percent loss		0.9	
	70 %-20 percent loss		0.8	
	80 %-32 percent loss		0.68	
	90 %-52 percent loss		0.48	
	100 %-85 percent loss		0.15	
Pallet Patterns			Best Case	Worst Case
Columnar, aligned	Negligible loss			
Columnar, misaligned	10~15 percent loss		0.9	0.85
Interlocked	40~60 percent loss		0.6	0.4
Overhang	20~40 percent loss		0.8	0.6
Pallet deckboard gap	10~25 percent loss		0.9	0.75
Excessive handling	10~40 percent loss		0.9	0.6

3.9 Estimating Required Compression Strength
① **Compression Requirement**

[05] If the compression strength and distribution environment is known, the effective stacking strength of any given RSC can be reasonably estimated. Similarly, if the distribution environment, container dimensions and flute profile are known, a compression requirement can be estimated. This can be of great value, because once a compression requirement is determined, the ECT requirement can be determined (and, therefore, board combination options as well).

[06] *The minimum dynamic (in-lab) compression strength required to provide safe stacking performance throughout that container's expected life cycle (given time, environmental/distribution conditions).*

[07] In order to integrate the calculated compression requirement into manufacturing specifications, customers and box manufacturers must agree on the nature of its use: long term average, average of a five (or more) box sample or absolute individual box minimum value.

[08] Typical compression requirement determination only considers the static (warehouse) portion of the distribution environment. In some instances the compression loading on the bottom box in a stack or unit may be greatest in the dynamic (transportation) portion of the environment. Containers in motor freight transport routinely see dynamic loading forces ranging from less than 0.5 to greater than 1.5 G's. It is very important to consider top loads and

shock and vibration inputs in transportation.

② **Example of Calculations**

A box of 0.5m × 0.25m × 0.30m (outside dimensions) will have 12 kg, stacked 2.7 m high in the warehouse. Boxes will be arranged in an interlock pattern and will be required to hold the load for 180 days at 80% R.H.. The pallets are in good condition; there will be no overhang. What should the required compression strength of the box be?

1. Determine maximum number of boxes above bottom box: 2.7/0.30 − 1 = 8
2. Determine load on bottom box: 8 × 12 kg = 96 kg
3. Determine Environmental Factor by multiplying together all factors that apply:

180 days	0.50
80% R.H.	0.68
Interlocked stack	0.50
Multiplier product (environmental stacking factors) =	0.17

4. Determine required box compression strength:

$$BCT = \text{anticipated load/stacking factor} = 96 \text{ kg}/0.17 = 564 \text{ kg} \quad (10.8)$$

Now that the actual compression strength is know, this value can be plugged into the McKee formula (10.7), and the required edge crush test value of the corrugated board can be calculated.

③ **Compression Solutions**

Following are a variety of approaches to increasing compression and stacking strength. The most efficient and cost-effective approach will depend on the product, package size and distribution environment.

• Stronger liners and medium (s): Edge crush and box compression are dependent on the stiffness of both liners and medium measured as **ring crush**.

• Load sharing: This is a technique where the product and/or primary package carry some portion of the static and dynamic stacking loads. To maximize the benefit, all participants must "load up" simultaneously. To optimize this requires load versus deflection compression testing of the completed packages, as well as the product and all package components.

• Increase the number of corners: Corner or angled bends reinforce the walls of the corrugated structure and increase compression strength. For example, using the same amount of combined board, a **hexagonal** or **octagonal** cross-section will provide more compression strength than a rectangular one.

• Change corrugation direction: Designing the corrugation direction to be parallel with the load is commonly accepted practice and is typically the approach that yields greatest top-to-bottom container strength. A greater percentage of strength is derived from the corners than the walls. One possible exception is when using small-flute combined board, such as E-flute and smaller. Horizontal corrugation tends to make corners more rigid in small-flute combined board.

• Dimensions

- Depth: Compression strength typically decreases as depth increases, until depth reaches 15 to 16 inches, after which loss rate diminishes.

- Length to width: Aspect ratio affects compression strength. Compression strength is typically greatest with ratios of 1∶1 to 1.5∶1.

- Perimeter: More board typically means more strength, although actual contribution-per perimeter inch-falls as the container becomes larger.

- Panel size: Gains from an increased perimeter are at least partially offset by performance losses from overly large panel sizes. These spans are less impacted by the natural reinforcement from corners and are more subject to bowing, or the bulging or distortion forces created by the product inside.

- Multiwall corrugated fiberboard: Double wall and triple wall fiberboard can provide greater compression strength than single wall fiberboard of similar combined basis weight. This is primarily due to enhanced bending stiffness and greater caliper.

- Partitions, inserts and interior packaging: Whether separate or integral, these forms can provide significant compression strength. This is especially true when optimized for top-to-bottom fit. Others yield improvement by reinforcing other load-bearing panels, keeping them vertical under the load.

- Lamination: Whether laminating combined board to combined board, or adhering multiple mediums or liners together, lamination can provide tremendous gains in package performance, improving both edge crush and bending stiffness.

- Treatments, **impregnations** and coatings: These are sometimes added to strengthen the components; other times they are used to preclude moisture and its detrimental effect on compression strength.

(7256 words)

New Words and Expressions

chute [ʃu:t] *n.* 斜槽、滑道
pothole ['pɔthəul] *n.* 壶穴
curb [kə:b] *n.* 路边
discrete [dis'kri:t] *adj.* 不连续的、离散的
bracing ['breisiŋ] *n.* 支撑、支柱
dunnage ['dʌnidʒ] *n.* （在回收再用包装容器内不作为缓冲垫的）填充物、衬板（如填补空隙、定位、隔开等）
resonance ['rezənəns] *n.* 共振、共鸣
sturdy ['stə:di] *adj.* 强健的、坚定的
undamped ['ʌn'dæmpt] *adj.* 无阻尼的
polyurethane [,pɔli'juərəθein] *n.* 聚氨酯
sophisticated [sə,fistikeitid] *adj.* 先进的
benchtop ['bentʃtɔp] *n.* 台式
durable ['djuərəbl] *adj.* 持久的、耐用的
rugged ['rʌgid] *adj.* 结实的、粗糙的

attenuate [ə'tenjueit] v. 削弱、衰减
plank [plæŋk] n. 厚木板、支架
cellulosic [ˌselju'ləusik] adj. 纤维质的
insert [in'sə:t] n. 插入物、添加物
wadding ['wɔdiŋ] n. 软填料、纤维填料
indented [in'dentid] adj. 锯齿状的、犬牙交错的
resiliency [ri'ziliənsi] n. 弹性、回弹
abrasive [ə'breisiv] adj. 研磨的
scuff [skʌf] v. 刮伤、划伤
overhang [ˌəuvə'hæŋ] n. 伸出、悬空
settling ['setliŋ] n. 沉淀物
popcorn ['pɔpkɔ:n] n. 爆米花
rodent ['rəudənt] n. 啮齿动物
vermin ['və:min] n. 害虫、寄生虫
urethane ['jurəθein] n. 氨基甲酸乙酯、尿烷、聚氨酯
isocyanine [aisəu'saiənin] n. 异花青
glycol ['glaikəul] n. 乙二醇
amplitude ['æmplitju:d] n. 振幅
objectionable [əb'dʒekʃənəbl] adj. 引起反对的、讨厌的
recess [ri'ses] v. 使凹进
varnish ['va:niʃ] n. 清漆、凡立水
octave ['ɔkteiv] n. 倍频
pail [peil] n. 桶、提桶
tailgate ['teilgeit] n. （卡车等的）后挡板
bulkhead ['bʌlkhed] n. 隔壁、防水壁
perimeter [pə'rimitə] n. 周长、边长
stringer ['striŋə] n. 纵梁、长梁
bulge [bʌldʒ] n. 突出量、凸出部分
bowed [bəud] adj. 弯曲成弓形的、弯如弓的
divider [di'vaidə] n. 分割物、间隔物
hexagonal [heks'ægnəl] adj. 六角形的、六边形的
octagonal [ɔk'tægnə] adj. 八角形的、八边形的
rectangular [rek'tæŋgjələ] adj. 矩形的
impregnation [ˌimpreg'neiʃən] n. 注入、浸渗
rail shunting　铁路调车
manual handling　人工搬运
shunting speed　调车速度
trailer-on-flat-car（TOFC）　平板拖车
cell telephone　手机
laptop computer　笔记本电脑

critical acceleration 临界加速度
cushioning material 缓冲材料
bottoming out 触底、从低点回升
static stress 静应力
dynamic cushioning curve 动态缓冲曲线
bearing area 承载面积
high-volume production 大量生产
foam-in-place 现场发泡
molded pulp 纸浆模塑
expanded polystyrene 发泡聚苯乙烯
air bubble sheet 气泡垫
open-celled foam 开孔泡沫
drape over 披上
drive train 传动系统
forcing (input) frequency 激励频率
natural frequency 固有频率
vibration-isolation cushioning system 隔振缓冲系统
stretch-wrapped 拉伸裹包的
dead-load 静载
clamp-truck 钳式卡车、夹抱车
apparent compression strength 表观（视）抗压强度
compression strength 抗压强度
load-bearing ability 承载能力
rough handling 野蛮装卸（同 abusive handling）
single-face pallet 单面托盘
shrink-wrapped 收缩裹包的
stress concentrator 应力集中点
finish surface 瓶口表面
aspect ratio 长宽比
flexural stiffness 弯曲刚度
bending stiffness 弯曲刚度
ring crush 环压强度

Notes

1. *An egg on a flat surface has a fragility of 35 to 50 G, depending on the axis of impact. If the egg is supported in a conforming surface, its fragility can exceed 150 G.* (Para. 34) 位于平面上的鸡蛋具有 35～50G 的脆值，这取决于鸡蛋的冲击轴线。若鸡蛋被支撑在一个合适的表面上，其脆值可达 150G。

2. *The explanations for shock provided in this text are simplified. Proper consideration of shock and shock protection takes into account not only peak G but also velocity change. These two factors are usually represented by a "damage boundary curve". The*

proper method of quantifying shock fragility is through the use of a shock test machine. This device is capable of providing a shock pulse of an accurately defined amplitude, duration, and shape.（Para. 34）本课关于冲击的解释简单化了。冲击和冲击防护的合理考虑不仅有峰加速度，还有速度变化量。这两个因数通常由"破损边界曲线"来表示。冲击脆值定量化的合理方法要借助冲击试验机。该实验设备能够产生准确定义的幅值、持续时间和形状的冲击脉冲。

3. *Reducing or eliminating relative motion lessens this type of vibrational damage. Tight shipping case dimensions, particularly in the vertical axis, are preferred wherever this is compatible with top load compression of product and package.*（Para. 51）减少或消除相对运动可减轻这种振动引起的破损。使运输包装箱尺寸更紧固，特别是将纵轴尺寸作为首选。这是因为在纵向，产品与包装承受最大的压力荷载。

4. *For packaging purposes, a typical resonance search might sweep the frequencies between 3 Hz and 100 Hz at 0.5 to 1.0 octave per minute (refer to ASTM D 999).*（Para. 57）对于包装来说，常用的共振扫频可在 3～100 Hz 范围内，按照每分钟 0.5～1.0 Hz 的倍频扫频（参考 ASTM D 999 标准）。倍频指的是工作频率与基准频率的比值关系。ASTM D 999 指 Standard Test Methods for Vibration Testing of Shipping Containers（运输容器振动测试的试验方法）。

5. *The minimum dynamic (in-lab) compression strength required to provide safe stacking performance throughout that container's expected life cycle (given time, environmental/distribution conditions).*（Para. 105）在整个容器的预期生命周期中，需要最小的动态抗压强度（实验室测得）来提供安全堆码性能（给定时间、环境/流通条件）。

Overview Questions

1. What fundamental lessons can be learned from such studies about shock?
2. What shock conditions are there during the distribution environment?
3. Why does the coffee cup have lower acceleration dropped onto a sponge rubber pad than directly onto the floor?
4. How does the humidity influence the compression strength of the corrugated paperboard container?
5. How to calculate the compression strength of regular slotted containers based on the McKee formula?

Lesson 11 Mechanical Shock

1 Introduction

① Throughout the distribution system packages are **manhandled** and **mishandled** in various ways. They are dropped, thrown, kicked and otherwise roughly abused. Packages may fall from conveyors or **forklifts** and crash to the floor. They are also subjected to a variety of vehicle impacts: trucks starting, stopping, hitting chuckholes and railroad crossings, railcar humping, jolting and other moderately violent actions. In each instance the package suffers an impact with another object: floor, truck bed, pallet, bulkhead or another package. This impact results in a mechanical shock to both objects.

② A mechanical shock occurs when an object's position, velocity or acceleration suddenly changes. Such a shock may be characterized by a rapid increase in acceleration followed by a rapid decrease over a very short period of time. The acceleration versus time plot for most shocks is very complex, as shown in Figure 11.1(a). Shocks are more easily understood if we replace the complex history of Figure 11.1(a) with the smooth approximation shown in Figure 11.1(b). A package shock may typically be 20 milliseconds (0.020 seconds) long and have a magnitude or "height" of 150 G's.

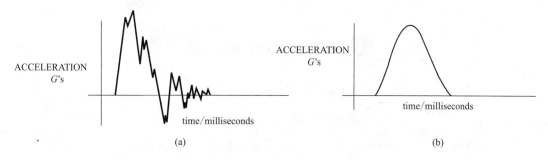

Figure 11.1 Representation of mechanical shock

③ A mechanical shock may be thought of as a multiple of weight applied suddenly and lasting only a fraction of a second. If someone were to receive a 200 G shock, for instance, he would feel as if he weighed 45,000 pounds for an instant. *This would feel like nothing more than a minor jolt due to the natural shock absorbing muscle and tendon system of the body (if, however, the shock were to last longer, the consequences would be considerably worse).* Mechanical shocks to packages, however brief, may cause damage.

To understand and estimate the potential damage a shock may cause, we need to know both the magnitude of the acceleration and the duration of the shock.

2 The Freely Falling Package

④ We have learnt that the motion of a freely falling body. The following equations respectively describe the length of time, t, it takes a package to fall from a drop height, h, and the downward velocity at which it will be traveling a moment before impact, v_I, the **impact velocity**:

$$t = \sqrt{\frac{2h}{g}} \tag{11.1}$$

$$v_I = \sqrt{2gh} \tag{11.2}$$

⑤ Such a fall is shown in Figure 11.2. From experience, we know that a small object like a package impacting a large object like a floor will bounce. The amount of bounce depends on the nature of the objects involved. A bean bag dropped on a concrete floor will not rebound, whereas a "superball" will rebound nearly to the point from where it was dropped. Likewise, a package will rebound a little or a lot depending on the nature of the package and the surface it hits.

⑥ The **coefficient of restitution**, e, describes the **rebound velocity** as a function of the impact velocity (See note 2):

$$v_R = e v_I \tag{11.3}$$

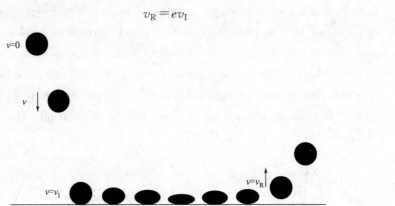

Figure 11.2 A falling package

The coefficient e takes on values between zero and one, with typical values falling in the 0.3 to 0.5 range. Note that if e were to be greater than one, the rebound velocity would exceed the impact velocity and the package would bounce higher than the drop point.

⑦ The total **velocity change**, Δv, for an impact is defined as the sum of the **absolute values** of the impact and rebound velocities:

$$\Delta v = |v_I| + |v_R| \tag{11.4}$$

or, since we know e and v_I:

$$\Delta v = (1+e) v_I = (1+e)\sqrt{2gh} \tag{11.5}$$

Because $0 \leqslant e \leqslant 1$

$$\sqrt{2gh} \leqslant \Delta v \leqslant 2\sqrt{2gh} \tag{11.6}$$

From Equation (11.6) we may estimate the velocity change resulting from an impact to be somewhere between the value of the impact velocity and twice that value.

⑧ The velocity change is also numerically equal to the area beneath the **shock pulse** as shown in Figure 11.3.

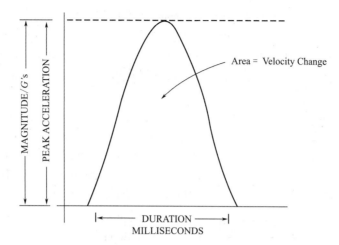

Figure 11.3   The relationship among shock parameters

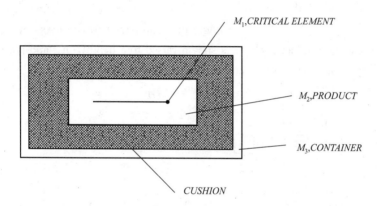

Figure 11.4   A simple product-package system

Package damage is related to the three factors involved in mechanical shock:
Peak Acceleration.
Duration.
Velocity Change.

Knowing any two of the factors allows us to estimate the third.

3   Mechanial Shock Theory

Mindlin, in considering distribution breakage of radar tubes during World War II, reduced the product-package system to its simplest form. Shown in Figure 11.4, the product-package system consists of four basic components: the outer container, the cushion, the product, and a critical element. The critical element is the most fragile component of the product (the filament of the radar tube, for instance). It is the part that is most easily damaged by mechanical shock.

This model may be further reduced to something we may investigate mathematically, shown in Figure 11.5. This model and the following analysis will help in understanding the potentials for product shock damage. For our product-package model:

M_2 represents the mass of the product.

M_1 represents the mass of the critical element or CE.

M_3 represents the mass of the outer container.

k_1 is the linear spring constant of the sprint-mass system representing the critical element.

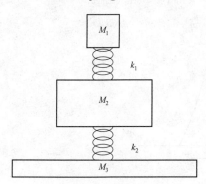

⑩ k_2 is the linear spring constant of the cushion system (See note 3).

⑪ For simplicity we will ignore the mass of the outer container and assume that it provides no spring action. We will also assume that the cushion has no mass or **damping** and suffers no permanent deflection from a shock. Also, we will assume that the product-package system impacts a perfectly rigid floor and that:

$$M_1 \ll M_2$$

Figure 11.5 A spring-mass model for the product-package system shown in Figure 11.4

That is, the mass of the critical element is negligible compared to the mass of the product.

In Figure 11.6, the product-package is shown raised to a height, h, and poised to drop. The critical element may be ignored for the moment as we investigate the shock to the product.

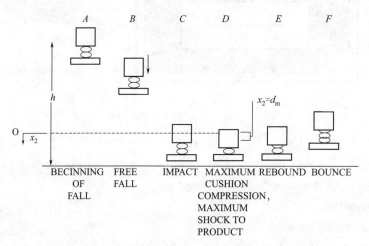

Figure 11.6 The impact of a product-package system

At point A, the product is ready to fall. The **potential energy** at this moment is:

$$PE = M_2 g h = W_2 h \tag{11.7}$$

where the weight of the product is given by W_2 ($W_2 = M_2 g$).

⑫ The product-package system is allowed to fall. At point C, the outer container makes contact with the impact surface. At this point, before the product begins to compress the cushion, we set up a reference line so that we may measure the product's downward displacement, x_2, on the cushion. Also at point C, the **kinetic energy** of the system is given by:

$$KE = \frac{1}{2} M_2 v_1^2 \tag{11.8}$$

where the impact velocity is given by Equation (11.2).

Substituting:
$$KE = \frac{1}{2}M_2(2gh) = M_2 gh = W_2 h \tag{11.9}$$

we see that the kinetic energy at this moment is equal to the initial potential energy of the system.

In the moments following the initial contact of the container with the impact surface the cushion must absorb this kinetic energy. In this situation the amount of energy absorbed by the linear cushion is:

$$E = \frac{1}{2}k_2 x_2^2 \tag{11.10}$$

where x_2 is the downward displacement of the product on the cushion.

At point D the cushion has absorbed all of the system's kinetic energy and the downward velocity of the product has slowed to zero. The maximum deflection, d_m of the cushion occurs at this point:

maximum dynamic compression, d_m = maximum value of x_2

Equation (11.10) then becomes:

$$E_{max} = \frac{1}{2}k_2 d_m^2 \tag{11.11}$$

Since the maximum amount of energy absorbed by the cushion must equal the system's kinetic energy at impact, we may equate expressions (11.9) and (11.11):

$$W_2 h = \frac{1}{2}k_2 d_m^2 \tag{11.12}$$

Equation (11.12) may be rearranged to produce an expression for the maximum dynamic compression:

$$d_m = \sqrt{\frac{2W_2 h}{k_2}} \tag{11.13}$$

The static deflection of the product on the linear cushion is:

$$\delta_{st} = \frac{W_2}{k_2} \tag{11.14}$$

so Equation (11.13) may also be written as:

$$d_m = \sqrt{2h\delta_{st}} \tag{11.15}$$

The maximum force P_{max} exerted upward by the cushion against the product occurs when $x_2 = d_m$:

$$P_{max} = k_2 x_2 = k_2 d_m = k_2 \sqrt{\frac{2W_2 h}{k_2}} = \sqrt{2k_2 W_2 h} \tag{11.16}$$

The maximum acceleration (or deceleration) experienced by the product, G_m, may be found from the relationship (See note 4):

$$G_m = \frac{P_{max}}{W_2} = \frac{\sqrt{2k_2 W_2 h}}{W_2} = \sqrt{\frac{2k_2 h}{W_2}} \tag{11.17}$$

where G_m is unitless, but is understood to be in unit of 1 g.

Again, recalling the expression for δ_{st}, an alternate form of Equation (11.17) may be written as:

$$G_m = \sqrt{\frac{2h}{\delta_{st}}} \qquad (11.18)$$

Notice that in Equations (11.17) and (11.18), G_m is proportional to the square root of the drop height:

$$G_m \propto \sqrt{h} \qquad (11.19)$$

⑯ This means that if we double the drop height, the magnitude of the shock is not doubled, but rather increased by a factor of $\sqrt{2}$ (about 1.4). Equation (11.17) may be rearranged to produce an expression for k_2:

$$k_2 = \frac{W_2 G_m^2}{2h} \qquad (11.20)$$

Equation (11.13) may also be rearranged to produce an equivalent expression:

$$k_2 = \frac{2W_2 h}{d_m^2} \qquad (11.21)$$

Equations (11.13) and (11.17) may also be manipulated to give us a simple expression for d_m. Multiplying both sides of Equation (11.13) by 1:

$$d_m = 2h\sqrt{\frac{W_2}{2k_2 h}} = \frac{2h}{\sqrt{\frac{2k_2 h}{W_2}}} = \frac{2h}{G_m} \qquad (11.22)$$

giving us an expression for the maximum dynamic compression as a function of the maximum acceleration and the drop height. Several examples will illustrate how these equations may be used.

⑰ **[Example 1]** Find d_m and k_2 for a cushion which will limit the maximum acceleration of an 82 lb product to 62 G's in an eleven foot drop.

Solution:

Here we are given that:

$W_2 = 82$ lb, $h = 11$ ft = 132 in, $G_m = 62$

From Equation (11.22):

$$d_m = \frac{2h}{G_m} = \frac{2 \times 132 \text{in}}{62} = 4.26 \text{in}$$

From Equation (11.20):

$$k_2 = \frac{W_2 G_m^2}{2h} = \frac{82 \text{lb} \times (62)^2}{2 \times 132 \text{in}} = 1194 \text{lb/in}$$

We may conclude that an 82lb. product cushioned so that k_2 is equal to 1194 lb/in and falling eleven feet will experience a shock of 62 G's and depress the cushion 4.26 inches. Obviously, enough cushion must be present to absorb a deflection equal to or greater than d_m.

⑱ **[Example 2]** A one hundred pound product measures 10in × 10in × 10in and can sustain up to a 50 G shock without damage. It is equally sensitive to shock on all six faces. If a maximum drop height of five feet is expected in distribution, what is the **modulus of elasticity** of the required cushion material? Assume that the working length of the cushion material is fifty percent of cushion's total thickness.

Solution:

We are given that:
$$W_2 = 100\text{lb}, h = 5\text{ft} = 60\text{in}, G_m = 50$$

First, we need to calculate d_m from Equation (11.22):
$$d_m = \frac{2h}{G_m} = \frac{2 \times 60\text{in}}{50} = 2.4\text{in}$$

We are to assume that for the cushion:
$$\text{Working length} = 50\% \text{ total thickness}$$
$$WL = 0.5\ TT$$

In order to prevent "bottoming out":
$$WL \geq d_m$$

Setting the maximum dynamic compression equal to the working length:
$$d_m = WL = 0.5\ TT$$
$$TT = \frac{d_m}{0.5} = \frac{2.4}{0.5} = 4.8\text{in}$$

The cushion must be 4.8 inches thick.

⑲ More importantly, the modulus of elasticity, E_2, is related to the spring constant (See note 5):
$$k_2 = E_2 \frac{\text{Area}}{\text{Thickness}}$$

And k_2 for our problem from Equation (11.20) is:
$$k_2 = \frac{W_2 G_m^2}{2h} = \frac{100\text{lb} \times (50)^2}{2 \times 60\text{in}} = 2083\text{lb/in}$$

so
$$2083\text{lb/in} = E_2 \frac{\text{Area}}{\text{Thickness}}$$

The thickness of each cushion is 4.8 inches and the bearing area is equal to the area of each face of the product, 100 in².
$$2083\text{lb/in} = E_2 \times \frac{100\text{in}^2}{4.8\text{in}}$$
$$E_2 = 100\text{lb/in}^2$$

⑳ We also may want to know the size of the container that this ten inch cubic product and its cushions need for distribution. If we place a 4.8 inch thick cushion on each of the product's six faces, each dimension will be increased by the thickness of two cushions:
$$10\text{ in} + 4.8\text{ in} + 4.8\text{ in} = 19.6 \text{ inches}$$

The inner dimensions of the container are then:
$$19.6\text{in} \times 19.6\text{in} \times 19.6\text{in}$$

Before leaving this section, there is one more interesting relationship to cover. We expect a maximum acceleration of G_m to produce a displacement of d_m. Under normal conditions the one g acceleration of gravity produces a displacement of δ_{st}. Therefore, we would expect the relationship below to hold:
$$\delta_{st} = \frac{d_m}{G_m}$$

Therefore by substituting:

$$\frac{d_m}{G_m} = \frac{\sqrt{\frac{2W_2 h}{k_2}}}{\sqrt{\frac{2k_2 h}{W_2}}} = \frac{W_2}{k_2} = \delta_{st}$$

we see that it does.

㉑ In this lesson we have explored the relationships of G_m, d_m and k_2. Even though our model is based on a theoretical linear spring we now know:

• G_m is related to the drop height, the springiness of the cushion and the weight of the product.

• The stiffer the spring or cushion, the larger the value for G_m.

• The higher the drop height, the larger the value for G_m ($G_m \propto \sqrt{h}$).

• The heavier the product, the smaller the value of G_m as long as the spring or cushion is working.

4 Shock Duration

㉒ Packages and products typically receive mechanical shocks lasting somewhere between one and fifty milliseconds in the distribution system. The time length or duration of the shock is referred to as τ.

We may approximate the displacement of the product on the cushion, x_2, at any time during the impact with the function:

$$x_2(t) \approx d_m \sin(\omega_2 t) \tag{11.23}$$

shown in Figure 11.7, where $\omega_2 = 2\pi f_2$, and

$$f_2 = \frac{1}{2\pi}\sqrt{\frac{k_2 g}{W_2}} \tag{11.24}$$

the natural frequency of the product (M_2) on the cushion (k_2). The period of the free vibration of the product on the cushion, T_2, may be seen from Figure 11.8 to be twice the shock duration:

$$\frac{1}{f_2} = T_2 = 2\tau \tag{11.25}$$

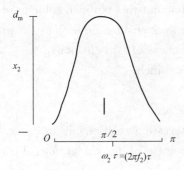
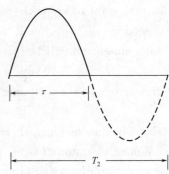

Figure 11.7 Shock duration and maximum displacement

Figure 11.8 Shock duration and the natural period of vibration

㉓ For any shock of known period τ, we may calculate the **equivalent shock frequency** (See note 6):

$$f_2 = \frac{1}{2\tau} \tag{11.26}$$

Equations (11.24) and (11.26) may be rearranged to give us an expression for the shock duration:

$$\tau = \frac{1}{2f_2} = \frac{\pi}{\sqrt{k_2 g / W_2}} = \pi \sqrt{\frac{W_2}{k_2 g}} \tag{11.27}$$

5 Shock Amplification And The Critical Element

㉔ To adequately understand the shock phenomenon and ultimately protect a product from shocks encountered in distribution, we must compare the response of the critical element within the product to the shock occurring to the product during impact. In the last two sections, expressions for G_m and τ were developed. Now we look at what happens to the critical element as a result. We will define G_e as the maximum acceleration experienced by the critical element, resulting from a shock to the product of magnitude G_m and duration, τ, as shown in Figure 11.9.

㉕ Analogous to our approach to **vibration magnification** we may define an amplification factor, A_m relating the input and output shock levels:

$$A_m = \frac{G_e}{G_m} \tag{11.28}$$

The maximum acceleration at the critical element is simply the product of this amplification factor and G_m:

$$G_e = A_m G_m \tag{11.29}$$

During the impact the amplification factor is given by the expression:

$$A_m(0 \leqslant t \leqslant \tau) = \frac{\frac{f_1}{f_2}}{\left(\frac{f_1}{f_2} - 1\right)} \sin \frac{2N\pi}{\left(\frac{f_1}{f_2} + 1\right)} \tag{11.30}$$

where N is an integer, f_2 the equivalent shock frequency and f_1 the natural frequency of the critical element (See note 7). Just after impact the amplification factor is:

$$A_m(t > \tau) = \frac{2\left(\frac{f_1}{f_2}\right) \cos\left(\frac{f_1}{2f_2}\right)}{1 - \left(\frac{f_1}{f_2}\right)^2} \tag{11.31}$$

㉖ The largest value of A_m calculated from Equations (11.30) and (11.31) must be chosen for Equation (11.29). This depends on the relationship of f_1 and f_2:

① If $\frac{f_1}{f_2} < 1$, $f_1 < f_2$ or $\frac{1}{2}T_1 > \frac{1}{2}T_2 = \tau$, i. e. if the duration of the shock is less than one half the CE's natural period of vibration, the maximum amplification occurs just after the impact is over, and,

$$A_m = A_m(t > \tau) \text{[Equation(11.31)]}$$

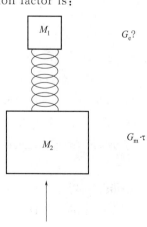

Figure 11.9 Shock transmission

② If $\frac{f_1}{f_2}>1$, $f_1>f_2$ or $\frac{1}{2}T_1<\frac{1}{2}T_2=\tau$, i.e. if the duration of the shock is longer than one half the CE's natural period of vibration, the maximum amplification occurs during the shock, and,

$$A_m = A_m(0 \leqslant t \leqslant \tau)[\text{Equation}(11.30)]$$

The maximum amplification values are given in Table 11.1 for frequency ratios between 0.01 and 11.0. Notice that as the f_1/f_2 ratio becomes large the value for A_m approaches 1, reflecting the "direct" transfer of the shock to a stiff element. For small values of the frequency ratio:

$$A_m = 2\left(\frac{f_1}{f_2}\right), \text{for } \frac{f_1}{f_2} \leqslant 0.20 \tag{11.32}$$

[**Example 3**] Suppose we have a product which contains a critical element, as shown in Figure 11.10. The natural frequency of vibration for this critical element is 38 Hz. A shock pulse is applied to the product. The shock pulse may be approximated by a **half sine**, the peak acceleration is 100 G's and the duration is 25 milliseconds. What is the shock to the critical element?

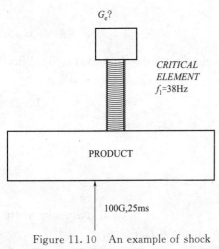

Figure 11.10 An example of shock amplification

Solution:

① $\tau = \dfrac{25\text{ms}}{1000\text{ms/sec}} = 0.0250\text{sec}$

② $f_2 = \dfrac{1}{2\tau} = \dfrac{1}{2 \times 0.025\text{sec}} = 20\text{Hz}$

③ $\dfrac{f_1}{f_2} = \dfrac{38\text{Hz}}{20\text{Hz}} = 1.90$

④ $\dfrac{f_1}{f_2} = 1.90$

corresponds to an A_m of 1.747 (from Table 11.1)

$$A_m = 1.747$$

⑤ $G_e = A_m \times G_m = 1.747 \times 100G\text{'s} = 174.7\ G\text{'s}$

We see that the critical element experienced a shock of about 175 G's as a result of the 100 G shock to the product.

In order to predict damage or evaluate the cushion effectiveness we need to know the highest acceleration level a critical element can withstand without damage. This is defined as G_s, the safe level; the maximum value of G_e to occur without damage to the element. Shock accelerations at the CE exceeding G_s will result in damage. An example will illustrate this approach.

[**Example 4**]   Suppose using a shock machine programmed to produce half sine shock pulses of 20 milliseconds duration, we find that a product can withstand 100G's. At higher accelerations there is damage to a critical element. The critical element has a natural frequency of 80 Hz and the entire product weighs four pounds. In distribution we will be using a cushion for which $k_2=750$ lb/in and we expect a maximum drop height of four feet. Will it be safe?

Solution:

First, we must determine G_s using test data.

① Calculate the equivalent shock frequency of the shock machine:
$$\tau = 20\text{ms} = 0.020\text{sec}$$
$$f_2 = \frac{1}{2 \times 0.020\text{sec}} = 25\text{Hz}$$

② Find the ratio of the element's f_1 to the shock machine's f_2:
$$\frac{f_1}{f_2} = \frac{80}{25} = 3.2$$

Then look up the amplification value from Table 11.1:
$$A_m = 1.451$$

③ The 100 G's (G_m) produced by the shock machine produces a response at the CE of:
$$G_e = A_m G_m = 1.451 \times 100 \text{ G's} = 145.1 \text{ G's}$$

Therefore, $G_s = 145.1$ G's.

Now we look at what will happen in distribution.

④ The shock in a 4 foot drop is:
$$G_m = \sqrt{\frac{2k_2 h}{W_2}} = \sqrt{\frac{2 \times 750 \text{lb/in} \times 48\text{in}}{4\text{lb}}} = 134 G's$$

⑤ The shock duration (See note 8) on the cushion is:
$$\tau = \frac{\pi}{\sqrt{\frac{k_2 g}{W_2}}} = \frac{\pi}{\sqrt{\frac{750 \times 386.4}{4}}} = 0.0117 \text{sec}$$

Then
$$f_2 = \frac{1}{2\tau} = 42.83 \text{Hz} \approx 43 \text{Hz}$$

⑥ $\frac{f_1}{f_2} = \frac{80}{43} = 1.87$, and from Table 11.1, $A_m = 1.752$.

⑦ So the shock at the CE during distribution is:
$$G_e = 1.752 \times 134 \text{ G's} = 234.8 \text{ G's} > 145.1 \text{ G's} = G_s$$

Because the G_e in distribution will be larger than the safe level, damage will occur and we may conclude that this cushion is inadequate to protect the product in a 4 foot drop.

6 Horizontal Impacts

㉙ Up to this point we have dealt strictly with vertical impacts. But what happens when we are confronted with a horizontal impact, as in the Example 5?

[**Example 5**] A 40 pound product is encased in linear cushioning material, for which $k=200$ lb/in. The product within its cushioning is securely fastened to the floor of a railcar. The railcar is coupled to the rest of the train at 10 miles per hour. If the working length of the cushion materials is 40% of the total length, what is the minimum thickness of the cushion needed to prevent "bottoming out" against the outside container? What are G_m and τ at the product?

Solution:

We may solve this problem by computing a vertical drop height which produces the same impact velocity as the horizontal impact.

$$v_I = 10\,\text{miles/hr} = \frac{10\,\text{miles} \times 5280\,\text{ft/mile} \times 12\,\text{in/ft}}{3600\,\text{sec/hr}} = 176\,\text{in/sec}$$

From Equation (11.2), we have:

$$h = \frac{v_I^2}{2g} = \frac{(176\,\text{in/sec})^2}{2 \times 386.4\,\text{in/sec}^2} = 40\,\text{in}$$

Now we may compute d_m, G_m and τ as if we had a vertical drop from 40 inches:

$$d_m = \sqrt{\frac{2W_2 h}{k_2}} = \sqrt{\frac{2 \times 40\,\text{lb} \times 40\,\text{in}}{200\,\text{lb/in}}} = 4\,\text{in}$$

$$G_m = \sqrt{\frac{2k_2 h}{W_2}} = \sqrt{\frac{2 \times 200\,\text{lb/in} \times 40\,\text{in}}{40\,\text{lb}}} = 20\,\text{G's}$$

$$\tau = \pi\sqrt{\frac{W_2}{k_2 g}} = \pi\sqrt{\frac{40\,\text{lb}}{200\,\text{lb/in} \times 386.4\,\text{in/sec}^2}} = 0.071\,\text{sec} \qquad \text{(See note 9)}$$

Since $WL = 0.40\,TT$, we may compute the required cushion thickness:

$$d_m = WL = 0.40\,TT$$

$$TT = \frac{d_m}{0.4} = \frac{4\,\text{in}}{0.4} = 10\,\text{in}$$

Some useful expressions may also be generated for G_m and d_m following the reasoning above.

$$G_m = \sqrt{\frac{2k_2 h}{W_2}}, \text{ and } h = \frac{v_I^2}{2g} \text{ from Equation (11.2).}$$

Substituting for h,

$$G_m = \sqrt{\frac{2k_2}{W_2}\frac{v_I^2}{2g}} = v_I\sqrt{\frac{k_2}{W_2 g}} \tag{11.33}$$

From $\delta_{st} = \dfrac{W_2}{k_2}$, Equation (11.33) becomes:

$$G_m = \frac{v_I}{\sqrt{\delta_{st} g}} \tag{11.34}$$

The natural frequency is given by:

$$f_2 = 3.13\sqrt{1/\delta_{st}},\ \delta_{st}\ \text{in inches (See note 10)}$$

so

$$G_m = \frac{v_I f_2}{3.13\sqrt{g}} = \frac{v_I f_2}{61.53} \text{ (if } v_I \text{ and } f_2 \text{ are in in/sec and Hz)} \tag{11.35}$$

Likewise, an expression for d_m may also be generated:

$$d_m = \sqrt{\frac{2W_2 h}{k_2}} = \sqrt{\frac{2W_2 v_I^2}{k_2 2g}} = v_I\sqrt{\frac{W_2}{k_2 g}} = v_I\sqrt{\frac{\delta_{st}}{g}} = \frac{v_I}{\sqrt{g}}\frac{3.13}{f_2} = 0.16\frac{v_I}{f_2} \tag{11.36}$$

Table 11.1 Amplification factors (half sine pulse)

f_1/f_2	A_m	f_1/f_2	A_m	f_1/f_2	A_m	f_1/f_2	A_m	f_1/f_2	A_m	f_1/f_2	A_m
.01	.020	.82	1.397	1.66	1.768	2.50	1.625	3.85	1.300	6.90	1.169
.02	.040	.84	1.419	1.68	1.767	2.52	1.620	3.90	1.289	7.00	1.167
.03	.060	.86	1.441	1.70	1.767	2.54	1.615	3.95	1.279	7.10	1.164
.04	.080	.88	1.462	1.72	1.765	2.56	1.610	4.00	1.268	7.20	1.160
.06	.120	.90	1.482	1.74	1.764	2.58	1.605	4.05	1.258	7.30	1.157
.08	.160	.92	1.501	1.76	1.762	2.60	1.600	4.10	1.247	7.40	1.153
.10	.200	.94	1.520	1.78	1.761	2.62	1.595	4.15	1.237	7.50	1.149
.12	.239	.96	1.538	1.80	1.759	2.64	1.590	4.20	1.227	7.60	1.145
.14	.279	.98	1.555	1.82	1.757	2.66	1.585	4.25	1.217	7.70	1.140
.16	.318	1.00	1.571	1.84	1.755	2.68	1.580	4.30	1.207	7.80	1.135
.18	.357	1.02	1.586	1.86	1.753	2.70	1.575	4.35	1.198	7.90	1.131
.20	.396	1.04	1.601	1.88	1.750	2.72	1.570	4.40	1.188	8.00	1.126
.22	.435	1.06	1.614	1.90	1.747	2.74	1.565	4.45	1.179	8.10	1.120
.24	.474	1.08	1.627	1.92	1.745	2.76	1.560	4.50	1.170	8.20	1.115
.26	.512	1.10	1.640	1.94	1.742	2.78	1.555	4.55	1.160	8.30	1.110
.28	.550	1.12	1.651	1.96	1.739	2.80	1.550	4.60	1.151	8.40	1.104
.30	.588	1.14	1.662	1.98	1.735	2.82	1.545	4.65	1.142	8.50	1.099
.32	.625	1.16	1.672	2.00	1.732	2.84	1.540	4.70	1.133	8.60	1.093
.34	.662	1.18	1.682	2.02	1.729	2.86	1.535	4.75	1.125	8.70	1.087
.36	.698	1.20	1.690	2.04	1.725	2.88	1.530	4.80	1.116	8.80	1.082
.38	.735	1.22	1.699	2.06	1.722	2.90	1.525	4.85	1.108	8.90	1.076
.40	.771	1.24	1.706	2.08	1.718	2.92	1.520	4.90	1.099	9.0	1.070
.42	.806	1.26	1.714	2.10	1.714	2.94	1.515	4.95	1.091	9.1	1.075
.44	.841	1.28	1.720	2.12	1.710	2.96	1.510	5.00	1.083	9.2	1.080
.46	.875	1.30	1.726	2.14	1.706	2.98	1.505	5.10	1.098	9.3	1.083
.48	.909	1.32	1.732	2.16	1.702	3.00	1.500	5.20	1.112	9.4	1.087
.50	.943	1.34	1.737	2.18	1.698	3.05	1.488	5.30	1.124	9.5	1.090
.52	.976	1.36	1.742	2.20	1.694	3.10	1.475	5.40	1.134	9.6	1.092
.54	1.008	1.38	1.746	2.22	1.690	3.15	1.463	5.50	1.143	9.7	1.094
.56	1.040	1.40	1.750	2.24	1.685	3.20	1.451	5.60	1.150	9.8	1.097
.58	1.071	1.42	1.753	2.26	1.681	3.25	1.438	5.70	1.157	9.9	1.098
.60	1.102	1.44	1.757	2.28	1.676	3.30	1.426	5.80	1.162	10.0	1.100
.62	1.132	1.46	1.759	2.30	1.672	3.35	1.414	5.90	1.167	10.1	1.101
.64	1.162	1.48	1.761	2.32	1.667	3.40	1.402	6.00	1.170	10.2	1.102
.66	1.191	1.50	1.763	2.34	1.663	3.45	1.390	6.10	1.172	10.3	1.102
.68	1.219	1.52	1.765	2.36	1.658	3.50	1.379	6.20	1.174	10.4	1.103
.70	1.246	1.54	1.766	2.38	1.654	3.55	1.367	6.30	1.175	10.5	1.103
.72	1.273	1.56	1.767	2.40	1.649	3.60	1.356	6.40	1.176	10.6	1.103
.74	1.299	1.58	1.768	2.42	1.644	3.65	1.344	6.50	1.175	10.7	1.102
.76	1.325	1.60	1.768	2.44	1.639	3.70	1.333	6.60	1.175	10.8	1.102
.78	1.349	1.62	1.769	2.46	1.635	3.75	1.322	6.70	1.173	10.9	1.101
.80	1.373	1.64	1.768	2.48	1.630	3.80	1.311	6.80	1.172	11.0	1.100

(3710 words)

New Words and Expressions

manhandle [ˈmænˌhændl] *v.* 粗暴地对待、野蛮装卸（同 mishandle）
forklift [ˈfɔːklift] *n.* 叉车
duration [djuəˈreiʃən] *n.* 持续时间
damping [ˈdæmpiŋ] *n.* 阻尼
impact velocity　冲击速度
coefficient of restitution　恢复系数
rebound velocity　回弹速度
velocity change　速度改变（量）
absolute value　绝对值
shock pulse　冲击脉冲
peak acceleration　峰值加速度
potential energy　势能
kinetic energy　动能
modulus of elasticity　弹性模量
equivalent shock frequency　等效冲击频率
vibration magnification　振动放大（因子）
half sine　半正弦

Notes

1. *This would feel like nothing more than a minor jolt due to the natural shock absorbing muscle and tendon system of the body （if, however, the shock were to last longer, the consequences would be considerably worse*）(Para. 3) 由于人体天生具有对冲击起吸收作用的肌肉组织及肌腱系统，这种感觉只不过是一个轻微的震动（然而，如果这个冲击持续的时间较长，后果会相当严重）。"*nothing more than*" 意指"仅仅、只不过"。

2. for a lightly damped system, $e \cong 1 - \pi c/(2\sqrt{km})$, where c is the damping coefficient (Para. 6).

3. We will use a simple linear spring to represent the cushion to simplify our analysis. Other nonlinear models may be more appropriate, but also entail complex mathematics (Para. 10).

4. At the point of maximum compression there is no motion, the upward and downward forces are in equilibrium (Para. 15):
$M_2 \ddot{x}_2 + P_{max} = 0$, $M_2 \ddot{x}_2 = -P_{max}$, $\frac{W_2}{g}\ddot{x}_2 = -P_{max}$, $W_2 \frac{\ddot{x}_2}{g} = -P_{max}$, $W_2 \left|\frac{\ddot{x}_2}{g}\right| = P_{max}$, $W_2 G_m = P_{max}$ where $G_m = \left|\frac{\ddot{x}_2}{g}\right|$

5. $\sigma = E_2 \varepsilon$, where σ—normal stress; ε—normal strain; A—area; $\sigma = \frac{W_2}{A} = \frac{k_2 d_m}{A}$, $\varepsilon = \frac{d_m}{TT}$ (Para. 19).

6. The equivalent shock frequency and the natural frequency of the product on its cushion are numerically equal (Para. 23).

7. $f_1 = \frac{1}{2\pi}\sqrt{\frac{k_1 g}{W_1}}$. This frequency is usually determined experimentally by vibrating the product and observing the resonant frequency of the critical element (Para. 25).

8. $g = 9.8$ m/sec^2 = 32.3 ft/sec^2 = 386.4 in/sec^2 (Para. 28).

9. We have assumed here that only one of the two side cushions is acting on the product. If both side cushions act, staying in contact with the product during impact, the expression for τ is: $\tau = \pi\sqrt{\frac{W_2}{2k_2 g}}$ (Para. 30).

10. $f_2 = \frac{1}{2\pi}\sqrt{\frac{k_2 g}{W_2}} = \frac{\sqrt{g}}{2\pi}\sqrt{\frac{k_2}{W_2}} = \frac{\sqrt{386.4}}{2\pi}\sqrt{\frac{k_2}{W_2}} = 3.13\sqrt{1/\delta_{st}}$ (Para. 31).

Overview Problems

1. An article weighing 12 lb can sustain a maximum shock of 80 G's ($G_m = 80$). Find k_2 for a linear cushion which will limit the acceleration of the product to 80 G's in a 36 inch drop. What is the maximum deflection of the cushion? Find the duration (τ) of the shock pulse sustained by the article.

2. The damage occurred to a critical element; the natural frequency of this critical element is 60 Hz. The product is packaged with linear springs for which $k_2 = 1010$ lb/in. A maximum drop height of 60 inches is expected in the distribution system.

    (1) What is the frequency of the shock produced by the shock machine?
    (2) What is the amplification factor for the test situation?
    (3) What is G_s, the safe acceleration level for the critical element?
    (4) What is the magnitude of the acceleration experienced by the product in the distribution system?
    (5) What is the duration of the shock experienced by the product in the distribution system?
    (6) What is the amplification factor in the distribution system?
    (7) Does the cushion material provide adequate protection for the product (and critical element)?

Lesson 12 *Lansmont* Six Step Method

① The function of a **cushioned package** is to provide a buffer between the product and the world of distribution and handling. To design this interface we must be able to determine the types and severity of the hazards that the package will encounter. These may encompass many things, but the most powerful include drops which occur during handling, vibration of the transportation vehicle, and compressive loads encountered during warehousing.

② Once we have identified what inputs to expect from the shipping environment, and to what extent the unpackaged product can withstand these, we can go about making up the difference between these two levels with a cushioned package system.

③ Ideally, the package system will provide enough protection to exactly match the requirements of the product and **distribution environment.** There are, however, two pitfalls which may occur if a systems approach to package design is not adopted. In the first situation, the package falls short of the protection requirements and a significant amount of damage occurs during shipment. This **"under-packaging"** is fairly obvious to detect, but is avoidable and easily corrected with changes to the method of shipment, package design, product design, or combinations of the above. In the second situation little or no damage occurs, but the product is "over packaged". In effect, the package is providing more protection than is required. Just as "under-packaging" wastes money through damaged product and loss of customer good will, "over-packaging" siphons money directly from a company's bottom line.

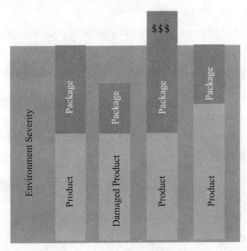

Figure 12.1 The general concept of product/package design

④ Try to visualize the general concept of product/package design as the bar chart depicted in Figure 12.1. The shaded background can be thought of as the level of environmental **intensity or severity** for a given distribution channel. The product has some inherent ability to withstand this abuse, however it usually is not rugged enough to make it through shipment on its own. The role of the package, therefore, is to make up the difference between what the environment has to offer and what the product can withstand.

⑤ The ideal case, as depicted by the first product/package system bar, is where the package exactly makes up the difference between the product **ruggedness** and the environmental inputs. If the package falls short, as depicted by the second product/package system bar, "under packaging" has occurred and damage in shipment will most likely result. If the package provides too much protection, as depicted by the third product/package system bar, "over packaging" has occurred and money is being wasted on protection which is not required. In certain instances it will actually be cheaper to ruggedize the product rather than put an expensive package around each unit. This product improvement is depicted in the fourth product/package system bar.

1 Step 1 Define The Environment

⑥ An essential step to designing a cushioned package system is to determine the severity of the environment in which it will be shipped. The general idea is to evaluate the method of distribution to determine the hazards which are present and the levels at which they are present. These may include such things as accidental drops during handling, vehicle vibration, shock inputs, temperature extremes, humidity levels, and **compression loads.** This text will focus on the areas of shock and vibration, but it is important that other areas also receive proper consideration during the package design process.

⑦ It would be nice to follow every package through the distribution environment and observe what actually happens to it. Usually, however, we must accept another approach. The next best thing to being there is using some sort of recording device to monitor the package and/or the vehicle during shipment. Provided we do this enough times, we begin to gain some sort of statistically valid information which can be used to describe that particular channel of distribution. The events will obviously change from trip to trip, but in general we have an idea of what to expect. This is the best approach to gaining information about a specific distribution channel. Probably the most widely used approach, however, is to study available published data. The difficulty here is that the data is usually outdated, and was not originally recorded from the environment through which you will actually be shipping your package. In general, however, it may provide the guidelines and rules of thumb necessary for the package design process.

⑧ The importance of this environmental information cannot be over stressed. This information will eventually become the part of the package design requirement and if not described correctly the package may appear to fail in distribution even though the design goals were met. In addition, over packaging may result if the actual inputs are lower than those chosen for the design goal.

1.1 Shock Environment

⑨ Shocks may result from many types of events, but it is generally agreed that the most severe shocks a package will receive occur during handling operations. These include the times when a package is dropped while being loaded or unloaded from a vehicle, sorted or staged for further distribution, or when bulk is being made or broken. It is important, therefore, to identify the **drop height** from which the package will be expected to fall.

⑩ Of course not all packages are handled exactly the same way, even when shipped by the same carrier over the same route. Some packages may never be dropped, while others will fall from a height many times higher than anticipated. Some may fall on the bottom, and others on a side, the top, a corner, or an edge. What this means is that there is a certain inherent variability with the manner in which packages are handled.

① Figure 12.2 describes drop height for a particular package in terms of its probability of occurrence over a given distribution route. This chart indicates that low level drops occur frequently, while very high level drops are rare. Although this is just an example plot, the general thrust of the data is valid. Many small drops can occur during normal handling when the package is picked up, set down and just plain jostled around. Large drops, however, usually only result from accidents such as a package falling off the top of a stack or loading platform.

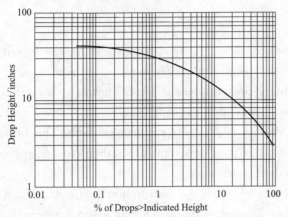

Figure 12.2 Drop height probability chart (example only)

② Drop height information tied to probability of occurrence is the most accurate way to theoretically design a package and tailor it to meet a certain damage rate. For example, if we wanted to design a package that would arrive with less than 1% of the products damaged, based upon Figure 12.2 we would select a design drop height of about 32 inches. If, however, we were willing to accept a 4% damage rate, then we could reduce the design drop height to 20 inches. On the other hand, if we insisted on having a damage rate less than 0.1% then our design drop height would jump up to 42 inches. *This type of evaluation allows trade-offs between damage costs and packaging costs. In most cases a certain amount of damage is acceptable because of the expense associated with trying to protect each and every unit from the highest level event.*

③ Although designing a package with this type of information is obviously the most informed approach, rarely is this type of detailed data available for your particular package and channel of distribution. The next best approach is to fall back upon some **the general rules of thumb** which have been developed in the packaging industry over the years. These include data like the table 12.1 which is presented in **ASTM D-3332**. This table describes drop height as a function of package weight and indicates that light packages will fall farther because they can easily be picked up and tossed about. Heavy packages, on the other hand, usually require mechanical handling and therefore will not be dropped as far.

Table 12.1 Drop heights as a function of package weights

Package weights	Type of handling	Suggested drop test heights
0~20 lbs	one man throwing	42 inches
20~50 lbs	one man throwing	36 inches
50~250 lbs	two men carrying	30 inches
250~500 lbs	light equipment	25 inches
500~1000 lbs	light equipment	18 inches
over 1000 lbs	heavy equipment	12 inches

1.2 Vibration Environment

¶4 It is virtually impossible to travel in a vehicle without experiencing some form of vibration. The rotation of engine and wheels induce vibration to the frame. Inconsistencies in the travel medium cause the suspension system to respond and the frame to flex. These inconsistencies may be semi periodic in nature such as expansion joints in a road or rail joints in train tracks, or they may be purely random occurrences such as potholes or railroad crossings. In any event, all these types of vibration become mixed together to form a composite input to the package.

¶5 The vibration encountered in the distribution environment is very complex in nature, consisting of intermixed frequency excitations emanating from a variety of sources. This type of vibration is often considered random in terms of the time domain because it is almost impossible to predict what will happen at any one instant. Yet in the frequency domain, a vehicle may display a very distinct signature which allows for the determination of the frequencies and levels which are present.

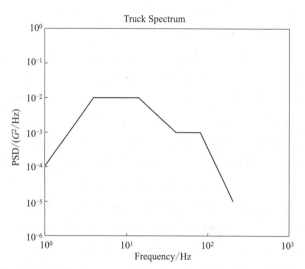

Figure 12.3 Power spectral density characteristics for truck

¶6 Figures 12.3 thru 12.5 are derived from **ASTM D-4728** and display the **power spectral density** (PSD) characteristics for several types of vehicles. These plots define the vibration in terms of the average power associated with each frequency. It should be noted that these particular plots do not represent any one trip, but rather encompass the general characteristics of the vehicle type.

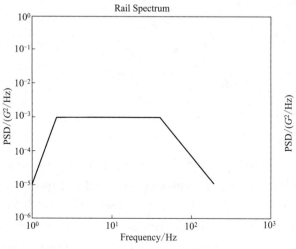

Figure 12.4 Power spectral density characteristics for rail

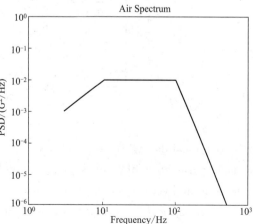

Figure 12.5 Power spectral density characteristics for airplane

⑰ It is generally thought that **steady state vibration** occurs at relatively low levels during shipment. **ASTM D-4169** suggests that a 0.5 G **sinusoidal** input over a frequency range of 3 to 100 Hz can used as a rule of thumb to predict how a package will perform during truck shipment.

2 Step 2 Product Fragility Analysis

⑱ Just as the weight of the product can be measured using a scale, the product ruggedness can be measured with dynamic inputs. A **shock machine** is used to generate a **damage boundary curve**, and a vibration system is used to map out the natural frequencies of a product.

2.1 Shock: Damage Boundary

⑲ The damage boundary theory is a testing protocol which determines, in an engineering sense, which shock inputs will cause damage to a product and which will not. There are two parts of a shock which can cause damage, the acceleration level and the velocity change. The velocity change, or the area under the acceleration time history of the shock, can be thought of as the energy contained in a shock. The higher the velocity changes the higher the energy content. There is a minimum velocity change which must be achieved before damage to the product can occur. This level is called the critical velocity change. Below the critical velocity change, no damage occurs regardless of the input acceleration level. In essence, there is not enough energy in this region of the damage boundary to cause harm to the product. Exceeding the critical velocity change, however, does not necessarily imply that damage results. If the change in velocity occurs in a manner which administers acceptable doses of acceleration to the product, the velocity change can be very large without causing damage. However, if the critical velocity and the critical acceleration are both exceeded, damage occurs.

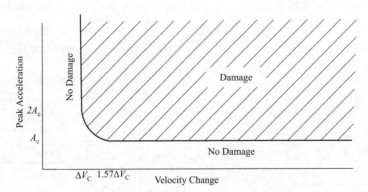

Figure 12.6 Typical damage boundary curve

⑳ A typical damage boundary curve can be found in Figure 12.6. It is a plot of the damage causing parameters of a shock pulse and defines the region where certain combinations of acceleration and velocity change will cause damage. If the combination of acceleration and velocity change fall in the clear band outside the damage region, no damage occurs. For example, if the velocity change of the input is below that of the product's critical velocity change, then the acceleration level of the input can be in the $100G$, $1000G$, $10,000G$, or even infinite without causing damage. In practical terms, to achieve these very high accelerations means that the duration of the shock must be very short. If we think of it in a graphical sense, we have a certain amount of area or velocity change which we can re-arrange to produce a shock pulse. If we decide to make the pulse very tall, i. e. very high in acceleration, then because of the limited amount of area that the pulse can contain it must also be very short in duration. In fact, the duration is so short that the product cannot respond the acceleration level of the event, only the energy input. Because the input velocity change did not exceed the critical velocity change of the product, no damage has occurred.

㉑ When the velocity change of the input exceeds the critical velocity change of the product, however, the only way to avoid damage is to limit the input acceleration to a level below that of the product's critical acceleration. This is usually one of the functions that a cushioned package performs. It translates the high acceleration events experienced on the outside of the container to lower acceleration events experienced inside at the unit.

㉒ To determine a damage boundary requires running two sets of tests. A step velocity test is used to determine the product's critical velocity change and a step acceleration test is used to determine the critical acceleration.

① **Step velocity test**

㉓ To run the step velocity test, the unit is fixed to the table of the shock machine and subjected to a short duration pulse with a relatively low velocity change content. It is important that the duration of the event be so short that the unit cannot respond to the acceleration level of the shock, only the velocity change. Most commercial machines intended for use in this type of testing produce a half sine pulse 2～3 ms in duration. Following the input, the test unit is examined for functional, physical and aesthetic damage. If none has occurred, it is given an input with a slightly larger velocity change component, but with roughly the same duration. This process is repeated until damage to the unit has occurred. The last non-failure input defines the critical velocity change for the unit in the orientation in which it was tested.

㉔ The critical velocity change can be equated to an equivalent free fall drop range (EFFDR). These numbers describe how far the unpackaged unit can fall onto a rigid surface before damage will occur. The process for calculating the EFFDR is demonstrated in Figure 12.7. In essence, it describes the ranges of heights that the bare product may fall before damage will occur, based upon the type of surface it impacts. Undoubtedly, if you drop the unit onto something soft, you can drop it farther without damage than you could if you dropped onto something

EFFDR CALCULATION

$$h = \left(\frac{V_C}{1+e}\right)^2 \times \frac{1}{2g}$$

Coefficient of Restitution

$0 \leqslant e \leqslant 1$

Realistic Limits of e

$0.25 \leqslant e \leqslant 0.75$

Acceleration Due to Gravity

$g = 386.1 \text{in/s}^2$

High End of Range

$$h = \left(\frac{V_C}{1+0.25}\right)^2 \times \frac{1}{2 \times 386.1}$$

Low End of Range

$$h = \left(\frac{V_C}{1+0.75}\right)^2 \times \frac{1}{2 \times 386.1}$$

Figure 12.7 The process for calculating the EFFDR

hard. This is why a range of heights is defined, rather than one specific height above which damage will occur and below which damage will not occur.

② Step acceleration test

㉕ For the step acceleration test, a new unit is fixed to the table of the shock machine and given a low acceleration **square wave pulse** with a relatively large velocity change content. The velocity change of the input must be at least 1.57 times the critical velocity change defined in the step velocity test. This will ensure that the test is conducted past the knee of the damage region. This knee occurs where the product shifts from being velocity-sensitive to being acceleration-sensitive. Following the input, the test unit is examined for damage and if none has occurred the unit is subjected to a slightly higher acceleration level with roughly the same velocity change. This process is continued until damage to the unit has occurred. The last non-failure input defines the critical acceleration for the unit in the orientation in which it was tested.

㉖ The designer now has all the information necessary to set the shock protection requirements of his product. The critical velocity change shows the designer the maximum drop height the bare product can be subjected to before damage will occur. If that drop height is less than the design drop height specified in Step 1, a package or interface material is necessary. If a package system is necessary, it must transmit less than the critical acceleration to the unit, when dropped from the design drop height.

㉗ In a rigorous testing program, damage boundary curves are generated for each orientation of the unit. To do this requires damaging 2 units per orientation, one for the step velocity test and one for the step acceleration test. Rarely are this many units available for destructive testing in the **prototype** stage of a product's life, the most beneficial time to do this type of work. Compromises are often made to limit the number of units which must be damaged. In certain situations it is possible to perform the testing only along the three **orthogonal axes**. In addition, the unit can often be repaired between tests, so that one unit and a few spare parts may be used to perform all of the testing.

㉘ It should be noted that the square wave used to determine the critical acceleration provides conservative results. In general, a square wave of a given acceleration and duration is the most severe **waveform** possible. It contains not only the fundamental frequency associated with the pulse duration,

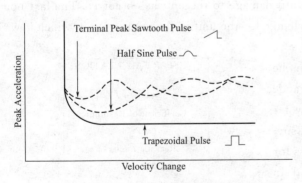

Figure 12.8 Damage boundaries of different wave forms with same peak acceleration and same velocity change

but also all the higher harmonics associated with the quick rise and decay of the waveform. What this means is that square wave can cause damage, at a given acceleration level and velocity change, while other waveforms do not. Figure 12.8 displays examples of the damage regions which various waveforms may produce. As can be observed form this plot, the square wave encompasses all damage regions produced by other waveforms. This is useful to the package designer, because in the early stages of development it is not known what shape waveform the package will transmit to the product. By using a square wave for this test, we gain confidence in our final package design. We know that if the unit can pass a square wave of a given velocity change and acceleration level, we can pass any waveform that our package may produce of the same velocity change and acceleration level.

㉙ In addition to the engineering reasons, there are also the economic and practical reasons for using the square wave to determine the critical acceleration. As can be noted from Figure 12.8, the square wave produces a flat horizontal line to bound its damage region while other waveforms tend to make scalloped upward sloped shapes. *Because of the square wave's flat line, we can define the critical acceleration with just one set of tests. This is something that can be done in an afternoon. To define the actual shape of the damage region for other waveforms, requires performing tests along the entire length of the velocity change axis. This approach can damage hundreds of units and require weeks or even months to complete. Although it may provide more precise results for a given waveform, it says nothing about how other waveforms may cause damage.* Usually the expense and effort to define the critical acceleration in this manner is not warranted because no practical benefits are gained.

2.2 Vibration: Resonance Search & Dwell

㉚ It is generally agreed that damage due to vibration is unlikely except at those frequencies where the product is most sensitive. The identification of those frequencies, therefore, becomes critical in designing a package system. *The purpose of the bare product vibration testing is to identify the natural or* **resonant frequencies** *of the critical components within the product.*

㉛ To run the vibration test, the unit is secured to the table of a **vibration test machine** and subjected to a low level sinusoidal input over a broad frequency range. The product can be observed for resonances either visually, audibly, or fitted with response **accelerometers** attached to its critical components. If the unit has been instrumented, the table input and component responses are monitored throughout the test.

㉜ Typically, the ratio of the component response to the table input acceleration is plotted as a function of frequency and is called a transmissibility plot (Figure 12.9). The **transmissibility ratio** (response divided by input) reaches a peak at the natural or resonant frequency of the component. The plotting of this ratio comprises the resonance search portion of the testing.

㉝ Once the product resonant frequencies

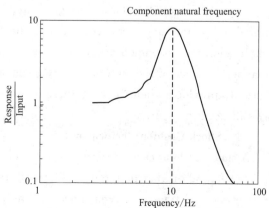

Figure 12.9 Transmissibility plot

have been determined, the vibration system is tuned to those frequencies and the product is forced to dwell there for a predetermined length. This will identify those frequencies which are prone to induce damage or fatigue.

㉞ If the unit is likely to be shipped in more than one axis, the vibration sensitivities of the product in those axes should be investigated as well.

3 Step 3 Product Improvement Feedback

㉟ Based upon the results of the fragility tests, it may be desirable to strengthen or ruggedize the product rather than ship each one inside an expensive package. Trade-offs between product cost, product reliability, and packaging costs should be identified and ranked for effectiveness. Often times it is possible to raise the fragility level of a product with minor modifications or design changes. This may add a slight cost to each product, but if the packaging requirements drop significantly the total system price goes down.

㊱ The ability to get product modifications implemented can vary widely depending upon the atmosphere within each company, and the position of the individual trying to get it accomplished. For some companies, this type of feedback to product designers is a formal step used in developing all new products. This allows the product to become more reliable, of better quality, and also keeps packaging costs to a minimum. In other situations, particularly when packaging is being developed by an outside supplier, it can be almost impossible to convince a company that making changes to the product is in their own best interest. It is, however, still important to present these ideas and take the role of educator where needed. If possible, try to identify the trade-offs between minor product changes, reliability and repair costs, and packaging expenses.

4 Step 4 Cushion Material Performance Evaluation

㊲ Material performance data should generally be available from the manufacturer of the material. In certain instances, however, it may be necessary to generate this type of data. This involves examining both the shock absorbing and vibration transmission characteristics of the materials.

㊳ It should be noted that the data generated by these methods is applicable to the cushion material only, and may not necessarily be the same as the response obtained in a complete pack. In addition, specimen area, thickness, loading rate and other factors will affect the actual performance of the material in any given situation. What this means is that the data can be used to provide a scientific best guess for the initial package design, but some fine tuning may still need to occur.

4.1 Shock Cushion Performance

㊴ **A shock cushion curve** describes the material in terms of the deceleration transmitted to an object falling on that material at different static loadings. One cushion curve is generated for each material type, material thickness and drop height combination.

㊵ The test procedure is basically one of dropping a platen of specified weight from a known drop height onto a cushion of predetermined bearing area and thickness. The deceleration experienced by the platen at impact is monitored and recorded by an accelerometer. Five drops from a particular drop height are performed on a sample at a given static stress loading. The average of the deceleration readings from the last four of these drops is the value used in plotting each cushion curve point. By adding weights to the platen, the static stress on the cushion material can be changed. Through a series of tests at various static loadings, data is generated and presented in the form of cushion curves (Figure 12.10). A minimum of five static loadings are tested to plot each curve, with a new sample being used at each loading.

④ It should be noted that these curves are "best fit" curves. This is because they are generated from averaged drop data at each static stress point on specific cushion samples. There is a certain inherent variability in the manufacture of the material as well as the

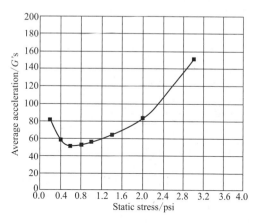

Figure 12.10 A shock cushion curve (10-inch drop height, 2inch thickness)

judgment involved in drawing smooth curves from "non-classical" data. The effects of sample variability, averaged data and curve fitting cannot be ignored, and therefore, the curves must be properly interpreted. Most cushion curves tend to have what is generally referred to as a "smile" shape (Figure 12.10). At low static loadings, the materials transmit relatively high accelerations. In this area, the impacting object does not have sufficient force to deflect the material, thus the material does not act much like a cushion. As the static loading increases, the transmitted accelerations tend to drop. In this region, the object now is able to deflect the material and cause it to work like **shock absorber**. At higher static loadings, the object deflects the material so far that it bottoms out. This is what causes the transmitted acceleration to rise along the right end of the curve.

4.2 Vibration Cushion Performance

⑫ The amplification/attenuation curve defines the frequencies at which a cushion material will amplify vibrational input and the frequencies at which it will filter out or attenuate the vibration. One amplification/attenuation curve is generated for each material type and material thickness combination.

⑬ To run the test, a block is monitored with a response accelerometer and loaded with weight until it reaches the desired static stress level when resting on a cushion sample. One cushion sample is placed below the test block and another is placed above it. This whole configuration is then placed in a corrugated container and secured to the table of the vibration test machine. A resonance search test is performed and a transmissibility plot of the cushion response is generated. The weight in the test block is then changed to obtain the next desired static stress loading and the test is repeated with fresh cushion samples. This process is repeated until the desired range of static loading has been explored. A minimum of five static loading test points are used to generate an amplification/attenuation curve.

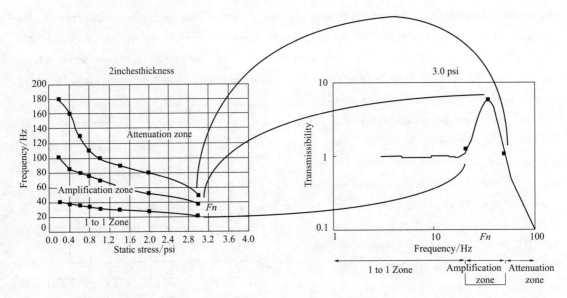

Figure 12.11 Transmissibility plot (3.0 psi)

㊹ Once all the transmissibility plots have been generated, the data is plotted on the amplification/attenuation curve as shown in Figure 12.11. The amplification/attenuation curve describes the vibration performance of the material as a function of static loading and can be thought of as a "top view" of a series of related transmissibility plots.

㊺ *In general, the shape of the amplification/attenuation curve slopes downward as static loading increases. This results from the basic characteristics of spring-mass systems. As static loading increases, the amount of weight supported by a given area of cushion increases. Since the cushion/spring characteristics have not changed, the natural frequency of the system tends to decrease.*

5 Step 5 Package Design

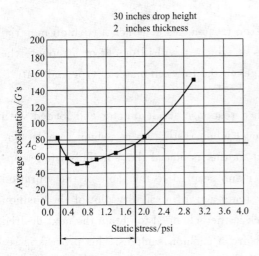

Figure 12.12 Determine the proper static stress range

5.1 Shock: Package Design

㊻ First, gather together cushion curves for the selected cushion materials. It is important to check that the drop height at which the cushion curves were generated is the same as the design drop height selected in Step 1. Next, locate the critical acceleration level determined in Step 2 on each of the cushion curves. Draw a horizontal line across the plots through this point. Any portion of the curve which falls below critical acceleration line indicates the static loading range where the material should transmit less than the critical acceleration (Figure 12.12).

5.2 Vibration: Package Design

㊆ For vibration consideration, we need to collect amplification/attenuation curves for the selected materials. Locate the lowest product natural frequency on each of the curves and draw a horizontal line across the plot. Any portion of the line which extends into the attenuation zone indicates the static loading range where the material should attenuate vibration at the frequencies where the product is most sensitive (Figure 12.13).

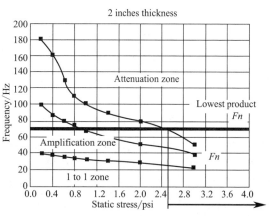

Figure 12.13 Determine vibration range

㊇ Once the static loadings which appear to provide adequate shock and vibration protection have been identified, material and thickness selections can be made. The actual static loading which is chosen for the package is dependent upon several factors, however designing at the highest possible static loading means using less material. When other considerations such as compressive creep are important, designing at the lowest possible static loading may be warranted. Figure 12.14 displays the method for calculating the amount of material which must be used around a product to reach a desired static loading.

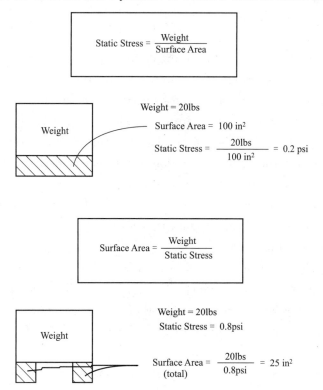

Figure 12.14 Calculate the amount of material

5.3 Package Design Considerations

㊾ The package design must be able to achieve both the shock protection requirements and the vibration protection requirements of the unit. This can sometimes present a challenge since the best design from a shock standpoint is rarely the best design from a vibration standpoint, and vice versa. Often times, due to material limitations, compromises need to be made. When this is the case, intelligent decisions can be based upon the facts and techniques used for Steps 1 and 2. We know, for instance, that vibration is a certainty. We will encounter vibration no matter which method of shipment is used because it is inherent to vehicles as they travel. Drops, on the other hand, have a certain probability associated with them. Not all packages will be dropped and certainly not all will be dropped from the design drop height. In addition, the critical acceleration of the unit was determined in a conservative manner. The package system will most likely not transmit a rectangular waveform to the product. Thus the product should be able to withstand somewhat higher accelerations than were predicted by the step acceleration test because the actual transmitted waveform is less damaging. What this means is that it is usually best to lean toward vibration protection when compromises need to be made. Of course this will depend upon the individual situation, however, in general this is the optimum approach.

6 Step 6 Test The Product/package System

㊿ Once the package design is completed, the prototype package system is tested to ensure that all design goals were met.

6.1 Shock: Package Testing

㈤ The package must be able to fall from the design drop height, set in Step 1, and transmit less than the critical acceleration to the unit. An accelerometer mounted to a rigid portion of the product near the center of gravity can be used to monitor the acceleration level transmitted through the cushioning material into the product. A rigid location is selected so that the input to the product as a whole can be observed during the impact, and thus directly compared to the results of the Step Acceleration test. It may be desirable to monitor additional locations during the tests, such as certain critical components, however the only effective way to evaluate package performance is to monitor a rigid location. Packaging itself does not directly change the response of the product to a given input, but it can modify the input which eventually reaches the product.

㈤ **Flat drops** are usually thought to be the most severe drops possible in terms of the acceleration level transmitted to the product. Flat drops focus all of the input along one axis of the unit and little energy is lost to the crushing of corners or edges of the package, or to package rotation. Flat drops, therefore, are used to measure the performance of the package system.

㈤ Corner and edge drops, however, often cause damage to the structure of the package which similar flat drops do not. These types of drops are often used as part of a test sequence to verify the package's ability to hold together during shipment.

6.2 Vibration: Package Testing

[54] Ideally, the package system will attenuate or filter out vibration at those frequencies where the unit is most sensitive. To accomplish this most effectively, the package system should have a natural frequency less than one half that of the product's lowest natural frequency.

[55] To run the package response vibration tests, the product is again fitted with a response accelerometer attached at a rigid location. The product is then placed in the prototype package system and secured to the table of a vibration test machine. The package is subjected to a low level sinusoidal input over the same frequency range which was used to test the bare product. In this test, however, it is the response of the package system rather than the product components which is monitored. A transmissibility plot of the package response is generated and used to verify that the package is working properly.

[56] Once the package resonant frequencies have been determined, dwell tests are performed at those frequencies. In addition, dwell tests may be performed at each of the product resonant frequencies. Random vibration tests may also be conducted in which the vibration system is programmed to mimic the real world motions of transportation vehicles.

(5506 words)

New Words and Expressions

cushion ['kuʃən] *n. v.* 衬垫、缓冲
ruggedness ['rʌgidnis] *n.* 强度、坚固性
sinusoidal ['sinə'sɔidl] *adj.* 正弦（曲线）的
prototype ['prəutətaip] *n.* 原型
waveform ['wevfɔrm] *n.* 波形
accelerometer [ək,selə'rɔmitə] *n.* 加速度计
cushioned package　缓冲包装
distribution environment　流通环境
under-packaging　欠包装
compression load　压缩载荷
drop height　跌落高度
ASTM D-3332 美国材料与试验协会标准——产品机械冲击脆值试验方法（使用冲击试验机）"Standard Test Methods for Mechanical-Shock Fragility of Products, Using Shock Machines"
ASTM D-4728 美国材料与试验协会标准——运输容器随机振动试验方法 "Standard Test Method for Random Vibration Testing of Shipping Containers"
power spectral density　功率谱密度
steady state vibration　稳态振动
ASTM D-4169 美国材料与试验协会标准——运输容器及其系统性能试验的标准规范 "Standard Practice for Performance Testing of Shipping Containers and Systems"

shock machine 冲击试验机
damage boundary curve 破损边界曲线
square wave pulse 矩形波脉冲
orthogonal axes 正交轴
resonant frequency 共振频率
vibration test machine 振动试验机
transmissibility ratio 传递率
shock cushion curve 缓冲特性曲线、最大加速度-静应力曲线
shock absorber 减震器、吸震器
flat drop 面跌落

Notes

1. *This type of evaluation allows trade-offs between damage costs and packaging costs. In most cases a certain amount of damage is acceptable because of the expense associated with trying to protect each and every unit from the highest level event.* (Para. 12) 这种评价将会在破损成本和包装成本之间寻求平衡。在大多数情况下，一定量的破损是可以接受的，因为如果要保证在最严酷的情况下都没有损坏将会产生更多的包装成本。这里用一个例子说明在选择跌落高度时要以跌落高度概率图来选择恰当的设计跌落高度，此高度并不是要保证零破损，而是要在跌落破损成本和包装成本之间折中。

2. *Because of the square wave's flat line, we can define the critical acceleration with just one set of tests. This is something that can be done in an afternoon. To define the actual shape of the damage region for other waveforms, requires performing tests along the entire length of the velocity change axis. This approach can damage hundreds of units and require weeks or even months to complete. Although it may provide more precise results for a given waveform, it says nothing about how other waveforms may cause damage. Usually the expense and effort to define the critical acceleration in this manner is not warranted because no practical benefits are gained.* (Para. 29) 因为由矩形波产生的破损边界曲线下边界为水平直线，所以在研究临界加速度时就只需要进行一组测试就可以得到了。这些测试工作将会在一个下午的时间完成。但如果要确定由其他波形产生的破损边界曲线下边界的具体形状，则需要沿着速度改变量轴进行多次测试。这将会使用上百个试件并花费数周甚至数月的时间来完成，尽管对于不同的波形这样会提供更加精确的结果，但是这与具体波形如何引起破损没有直接关系，而且通常用这种方式确定临界加速度也因不会获得实际效益，故没有必要为此花费人力和财力。"is not warranted"意指"是不必要的"。

3. *The purpose of the bare product vibration testing is to identify the natural or resonant frequencies of the critical components within the product.* (Para. 30) 对裸产品进行振动测试的目的是为了识别该产品内关键零部件的固有频率即共振频率。句中"bare product"指没有施加任何包装的产品；"identify"可理解为通过测试来"识别、辨识"。

4. *In general, the shape of the amplification/attenuation curve slopes downward as static loading increases. This results from the basic characteristics of spring-mass systems. As*

static loading increases, the amount of weight supported by a given area of cushion increases. Since the cushion/spring characteristics have not changed, the natural frequency of the system tends to decrease.（Para. 45）通常，放大/衰减曲线的形状会随着静态载荷的增大而减小。这来自于质量-弹簧系统的基本特性。随着静载荷增大，给定面积的衬垫所支撑的重量增大。既然衬垫/弹簧特性没有变化，系统的固有频率趋向减小。to slope downward 指"向下倾斜、减小、斜率为负"。

Overview Questions

1. What is the proper packaging system regarding its ability against the environment severity?
2. Why product improvement is important in the *Lansmont* Six Step Method for cushioning packaging design?
3. How to conduct a vibration test?

Lesson 13　Distribution Packaging

1　Short History Of Distribution Packaging In The USA

① Distribution packaging emerged in the 1800s as the industrial revolution blossomed and manufacturers began shipping their goods nationwide via railroad. Containers were usually made of wood, textiles, or glass. Interior packaging was mostly made from wood or forms of wood like sawdust or chips. Paper did not enter the distribution arena as protective packaging until the early 1900s, when corrugated boxes first appeared as **shipping containers**. At that time the railroads began issuing minimum packaging requirements; rules for corrugated boxes were among the first requirements instituted.

② From the end of World War I to the end of World War II, the use ratio of corrugated to wood containers went from 20/80 to 80/20. Along with **excelsior** and some paper forms, corrugated fiberboard became the predominant interior packaging material. **Pallets** became popular for industrial use following World War II, and unitizing of high-volume products for shipment accelerated in the 1950s.

③ Plastics began appearing in the early 1960s with various **foams** replacing corrugated, **rubberized fiber**, and wood-based products as interior packaging. **Steel banding**, which had been the primary unitizing material, began losing share to plastic banding and, in the early 1970s, to **stretch-wrapping**. Then the environmental challenge struck packaging in the late 1980s, and some trends from paper to plastic were reversed as new forms of paper interior packaging appeared. The interest in **returnable containers** and dunnage also accelerated.

2　Functions And Goals Of Distribution Packaging

④ The functions of distribution packaging can be summarized as follows:

2.1　Containment

⑤ The basic purpose of packaging is to contain the product. Packaging enables handling and transport of products from source to customer, supplying use value to products that would otherwise be useless to customers because their point of manufacture or production is usually remote from the customer's location.

2.2　Protection

⑥ Most products require some degree of protection from hazards in distribution. Packaging furnishes the degree of protection needed to safely transport products from source to customer.

2.3　Performance

⑦ Packaging performs in many ways in transportation, **handling**, storing, **dispensing**, and use of a product. Its performance function includes such things as orientation of the product, ease of identification, segregation of desired quantities, ease of disposal, and handling features.

2.4 Communication

⑧ A package should identify its contents and inform about package features and handling requirements. It generally provides shipping information and may include **promotional** graphics for items displayed for sale to consumers in the distribution package.

⑨ As you begin designing a package, have some goals in mind. The product, customer, distribution system, manufacturing facility, and other specifics will influence your goals, but most distribution packaging should address the following goals:

2.5 Product Protection

⑩ The primary purpose of any protective package is to ensure the integrity and safety of its contents through the entire distribution system.

2.6 Ease of Handling and Storage

⑪ All parts of the distribution system should be able to economically move and store the packaged product.

2.7 Shipping Effectiveness

⑫ Packaging and unitizing should enable the full utilization of carrier vehicles and must meet **carrier rule and regulation**.

2.8 Manufacturing Efficiency

⑬ The packing and unitizing of goods should utilize labor and facilities effectively.

2.9 Ease of Identification

⑭ Package contents and routing should be easy to see, along with any special handling requirements.

2.10 Customer Needs

⑮ The package should provide customers with ease of opening, dispensing, and disposal, as well as meeting any special handling or storage requirements the customer may have.

2.11 Environmental Responsibility

⑯ In addition to meeting regulatory requirements, the design of packaging and unitizing should minimize solid waste.

⑰ Because economy is usually of great importance in distribution packaging, you will need to **optimize** the combination of all the goals to achieve the lowest overall cost.

3 The Cost Of Packaging

⑱ Based on a number of sources, it was estimated that expenditures for all packaging materials, including **expendable** (one-way) shipping pallets, were approximately $100 billion in 1997. Of this total, about one-third was in the form of distribution packaging. The largest single segment of distribution packaging is corrugated shipping containers, at approximately 20% of total expenditures and 60% of distribution packaging costs. All industries are included in these costs, with food and beverage bearing the largest portion **due to** the volume of shipments in those industries.

⑲ All industries also share another large hidden cost, that of damage to products. *It has been estimated that although actual **freight claims** paid by carriers for damaging goods is* *approximately $2 billion, the actual cost to them and to shippers is really more than $10 billion per year.* The hidden part is in the form of claims not paid by carriers, damages **underwritten** by shippers **in lieu of** more packaging, costs of processing claims on both sides, and so forth.

20 Our goal in package design is to minimize the cost of both packaging and damage.

4 The Package Design Process

21 To develop an optimum distribution package that is both functional and cost-effective, you will need more than just assistance from your packaging suppliers. Although your experience with a product line and a supplier's experience with packaging materials are both helpful in designing packaging, both of you should consider many factors in addition to the product and the packaging.

22 Your scope of consideration should include all aspects of the distribution system, including customers, carriers, and distributors, as well as the manufacturing plant, packaging line, warehousing, and shipping. To be successful in distribution package design, take a **total-system approach.**

5 Taking A Total System Approach To Package Design

23 Once created, a package does not magically form around the product, float through shipping, travel hundreds of miles in isolation, arrive at the customer's site, and disappear. It has an influence on and is influenced by everyone and everything it encounters. Most of these **encounters** affect manufacturing and distribution costs or product integrity, with indirect impact on sales. It makes sense, then, that you should consider these events in the design process.

24 *Unfortunately, too much focus is often placed on the cost of packaging materials to the exclusion of other factors, including cost-related ones in handling, storage, and transportation.* If the package is slightly larger and/or heavier than it really has to be, costs in all three areas will be higher than necessary, perhaps producing an even greater negative effect on profits than would higher packaging costs for a smaller, lighter package.

25 For instance, although each industry and company is different, a general rule of thumb is that the total cost of transportation is between 3 and 10 times as much as packaging on average for all shipments. A small reduction in package size or weight could mean substantial savings in transportation costs, as well as in handling and storage. For example, a small kitchen appliance packaged in bulky **wraparound** die-cut may involve less material cost, but molded packaging may pack faster and require less cube, permitting more pieces per unit load, fewer handling trips, more units per storage cube, and more units per truckload. Bottom line: an overall savings.

26 An inverse relationship exists between packaging cost and maintaining product integrity with low damage rates, as shown in Figure 13.1. Other factors being equal, an increase in packaging costs provides more protection to the contents and therefore lowers the potential

for damage. Or conversely, cutting packaging costs without other improvements generally means less protection and higher damage rates. The real cost of getting the product safely to market is the sum of packaging and damage. Optimizing total cost is the true goal of packaging design. *If damages rise too high (as on the left in the graph), you will encounter an increase in both product replacement and repair costs along with the loss of customer* **goodwill** *and possible cancellation of orders.* If loss of sales and customer satisfaction are more important to your company than costs, there is not much room to move to the left of the package/damage **intersection.**

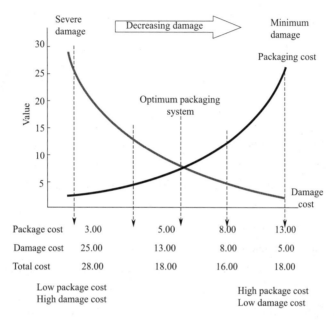

Figure 13.1 he optimum packaging system balances costs from excessive damage with the costs of overpackaging

㉗ No matter where in the company your packaging design function is located, in engineering, manufacturing, shipping, or elsewhere, try to include all factors in a total-system approach for an **optimum design.**

6　The Protective Package Concept

㉘ A simple equation is presented to explain the concept of protective packaging:

$$\text{Product} + \text{Package} = \text{Distribution Environment}$$

㉙ Figure 13.2 **depicts** the consequences of an imbalance in this equation, showing what happens when a product plus its package are not exactly what is needed to survive in the distribution process. Reading left to right, here is an explanation of the bar chart. **Severity** is the quantitative measure of the environment, which can be anyone or a combination of hazards in distribution. Here are some examples of hazards and their severity: the rough-handling hazard to a 20pound package is determined to be 30 inches of drop on any of six package surfaces; the compression (storage) hazard is determined to be 10 packages high in warehousing; the high temperature hazard is 130°F.

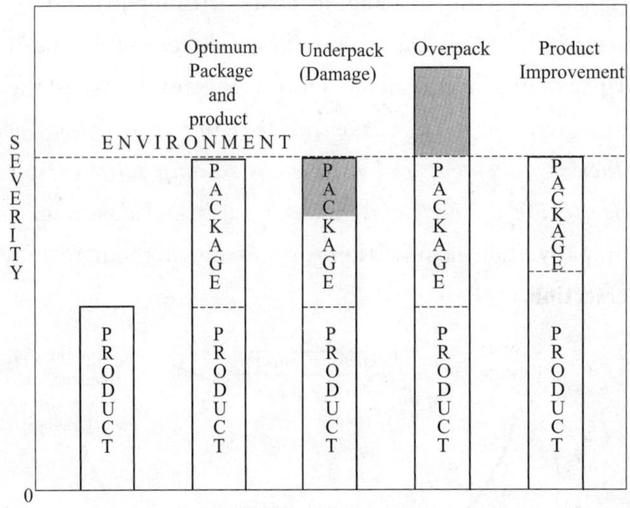

Figure 13.2 Protective package concept

50 Product represents the measured level of resistance to damage of the product. In the rough-handling example just given, the product was tested and showed the ability to survive six drops from 15 inches with no damage, although higher drops caused product impairment.

51 The third bar portrays the equation above, showing that the product's measured level of damage resistance plus the packaging's measured abilities to protect the product are exactly equal to the expected environmental hazard (s), an optimum solution. For example, a product with 15inch drop resistance is packaged in material that will dissipate the shock generated in the 30 inches of drop height the packaged product is expected to encounter in the distribution environment.

52 When the package provides less protective capacity than needed for the environment, as shown in the shaded area of the fourth bar, this underpackaging will result in damage. For example, the package will be able to absorb only 9 inches of drop height energy instead of the required 15 inches. **Overstress** results, damaging the product.

53 The next bar defines overpackaging. The package protection level is higher than the environment requires; the shaded area shows the amount wasted. *Instead of providing* 15 *inches of drop protection as needed in our example, the packaging protects to a drop height of 6 inches more than required.*

54 At times, it may be possible to improve the product as an **alternative** to more packaging. Instead of accepting a certain level of ruggedness in a product, product engineers may be able to raise the level, as shown in the bar on the far right in Figure 13.2. The result is a reduction in packaging needed. Perhaps the product had a component that malfunctioned above a 15inch drop height, but product redesign improved ruggedness to the 20inch level. Now packaging need supply only 10 inches of protection, a 33% reduction from the original.

55 The most **elusive** part of the package-plus-product equation is the distribution environ-

ment. The most difficult part of defining the environment is not so much identifying the types of hazards it contains, but determining what each hazard's expected level and probability of occurrence are. Dedicated package designers continually search for better methods of defining the environment on the right side of the equation so that they can solve for the parameters of product and package on the left side of the equation.

7 The 10-Step Process Of Distribution Packaging Design

㊱ Here is a 10-step procedure that will help you design a distribution package that provides maximum performance at least overall cost.

7.1 Identify the Physical Characteristics of the Product

㊲ Product knowledge means more than simply knowing the product's dimensions and weight. You need to be aware of surface characteristics and susceptibility to abrasion or corrosion, its ability to hold a load in compression, the effect of vibration on its internal characteristics, and particularly the product's shock and vibration fragility. Guessing about any of these factors is a sure path to potential problems.

7.2 Determine Marketing and Distribution Requirements

㊳ Package design must incorporate marketing and distribution requisites in addition to product characteristics. You will need to know, among other things: the number of units that will ship in a container, the composition and attributes of the primary package, the identity of customers and their handling and storage requirements, the package disposal criteria, total volume expected per shift/day/year, expected life cycle, the planned modes of transport, rules or regulations for packaging via those transport modes, and types of distribution channels.

7.3 Learn About the Environmental Hazards Your Packages will Encounter

㊴ Knowledge of the distribution environment is key to designing an optimum package. Major hazards to be expected in the environment are rough handling, vibration and shock in transit, compression in storage or in transit, high humidity and water, temperature extremes, atmospheric pressure, and puncturing forces. You can learn about these hazards in several ways: by observation, reading research reports, or measuring.

7.4 Consider Packaging and Unitizing Alternatives

㊵ There are many alternatives available for shipping containers, interior packaging, and unit loads. Consider and review all of them before selecting the final types for further development. Trade-off analysis techniques such as make versus buy often help. Do not limit consideration only to materials with which you have experience. Instead, periodically compare for instance paper with plastic, or wood with metal to ensure the best material for the particular project. Once you've selected the basic materials, detailed work on design can begin.

7.5 Design the Distribution Package

㊶ Once information is gathered and the basic materials are established in steps 1~4, you are ready to design the distribution package (and the unit load where appropriate). Each component of the package is analyzed for strength and other required properties and compared with technical data available from suppliers. Although you will find that some packaging materials have good design data available, most unfortunately do not. Those who are experienced in packaging frequently use their experience as the principal means for arriving at a successful solution. *Those with limited packaging experience may find that the lack of technical packaging information makes it difficult to arrive at an optimum solution.*

42   As part of the design process, be sure to consider closure methods for the shipping container. The hazards of handling, method of packing, and type of product all have a significant influence on closure selection, along with any regulatory requirements.

43   You can shorten **the trial-and-error cycle** by conducting engineering tests during package development. Setting goals for impact, vibration, and compression protection and then testing for them in the lab not only identify package shortcomings but also help to **fine-tune** the design to the optimum level of performance.

7.6 Determine Quality of Protection Through Performance-Testing

44   After you have designed the distribution package, perhaps with the aid of engineering development tests, you should then do a performance-test. This consists of **subjecting** the package (or unit load) **to** a sequence of anticipated hazards/tests in the laboratory for the purpose of a pass/fail decision. Will the shipping unit protect its contents all the way through distribution?

45   Your performance-test methods should be based on industry standards. Such standards have considerable experience and history behind their development and use, and successful completion of the test sequence almost guarantees damage-free shipments. The most widely used standard is **the International Safe Transit Association's** (ISTA's) Projects 1 and lA, in use since 1948. The American Society for Testing and Materials' (ASTM) D4169, first approved in 1982, provides a more complete set of possible hazards with corresponding test sequences, and it permits the user some flexibility in selecting test intensities. *For users with clearly defined distribution cycles different from the standard cycles in* D4169, *the ASTM standard also provides a means of developing a unique sequence of tests, resulting in performance-tests that can more precisely simulate your actual conditions.*

7.7 Redesign Package (and Unit Load) Until It Successfully Passes All Tests

46   There is an old saying: One Test Is Worth a Thousand Expert Opinions. Often performance-test results surprise even the most experienced engineers. Then it is necessary to repeat the design/test cycle, redesigning and retesting as many times as required for the packaging to "pass" the tests.

7.8 Redesign the Product if Indicated and Feasible

47   Testing occasionally reveals a product weakness that can be offset with protective packaging-but at excessive cost. If at all feasible, the product should be redesigned to correct the weakness rather than redesigning the package. This is particularly important when the cost of redesigning the product is less than the cost of extra packaging.

48   It is usually difficult for package designers to bring about product redesign when they are located organizationally in other than the product engineering group. If this is your situa-

tion, you should attempt to establish a continuing line of communication with the product engineers. Sometimes this means educating product engineers in the hazards of distribution and showing them how to correct product weaknesses.

7.9 Develop the Packaging Methods

㊾ An important part of package design is packing of the product in the shipping container and unitizing of containers. Although this may be the responsibility of someone else in your plant, you must be aware of cost factors and the appropriateness of mechanizing or automating all or part of the operations. Sometimes a trade-off in package design must be implemented to achieve over-all system economics.

7.10 Document All Work

㊿ One step frequently overlooked in the design process is documentation. This includes documenting test results, **specifications**, drawings, and methods of packing. Drawings should be in company standard formats with appropriate designations for reference in the corporate spec system. Relying on supplier sketches or drawings as reference documents is not a wise idea. They should be transferred to company format so purchasing, manufacturing, and engineering can reference them.

8 A Final Check

㉛ Here is another suggestion. For any package design project, after completing the 10-step procedure above, check your work against the list of important considerations as follows. By doing so you will significantly reduce the potential for an unpleasant surprise when shipments begin. Here is the package design project checklist:

㉜ Have you

• Considered the solid waste aspects of the package and unit load, and their alternatives, to minimize impact on the environment?

• Pondered the use of returnable/reusable containers and dunnage?

• Contemplated all cost factors in the distribution cycle: handling, storage, transportation?

• Compared the cost of this package with company/plant averages for similar products?

• Considered all possible alternatives in materials and methods?

• Used industry standards for materials and design criteria where possible?

• Performance-tested the design against accepted industry standards?

• Documented the design using the company's specification system?

• Checked damage and customer complaints on this product line?

• Satisfied all rules and regulations applying to this product for all distribution modes it is expected to encounter?

9 The Warehouse

㉝ The distribution warehouse is a central collecting point for a particular good or a particular merchandising chain. Finished goods are forwarded to and held at the warehouse until selected and assembled into a customer order. The warehouse environment is not well understood by many shippers.

54 A typical dry groceries warehouse may contain 20,000 individual stock items. A hardware chain warehouse holds upwards of 40,000 stock items. Product arrives at the central warehouse in bulk or unitized, is broken down or reunitized according to the warehouse's needs, and then is arranged for stock-picking. Stock-picking is the process of selecting individual items to **fill an order** for a particular store or destination. Central warehouses serve large customer areas; in some instances one or two warehouses may essentially serve the entire nation.

55 Product may be routed through more than one warehouse. For example, an export product may be moved from a local warehouse to dockside storage, to the cargo ship, and back to a receiving warehouse.

56 A product must fit the warehouse's material handling system. This often means palleting loose loads or repalleting loads from nonstandard pallets. Depending on the operation, anywhere from 33% to 70% of product received at a warehouse must be handled manually before an order is placed in stock. Manual handling, in addition to being costly, is also a primary source of damage from dropping.

57 In the picking aisles, stock must be clearly identifiable from every side. Multicolor graphic displays serve only to obscure vital information from the picker. A box labeled "Golden Triangle Farms" does not inform the stock-picker of the contents. Containers should be strong enough to be dragged off the pallet by one end, and stiff enough that they don't distort and release their contents when handled in less than ideal fashion. **Glue flaps** must have enough adhesive to resist **abusive handling.**

58 An assembled order may contain items as **disparate** as eight mirrors, six assorted clocks, a case of oil, four shock absorbers, a stepladder, and a Mepps #4 fishing lure. These and other items are assembled on a mixed pallet for transport to the retail outlet. Containers must be easily handled by the picker and should be readily packed onto a mixed-order pallet. Container orientation on mixed-load pallets will tend to be on a "best fit" basis, regardless of "This side up" and "Do not stack" labels. It may be possible to pack a **trapezoidal** container efficiently on your pallet, but odd shapes do not pack well in a mixed-product pallet load. Use boxes with a **rectangular cross section** wherever possible.

10 Unit Loads

10.1 Pallets

59 It is simpler to move one 1,000 kilogram load than it is to move a thousand 1 kilogram loads. Loads are most commonly unitized on pallets, a platform that can be picked up by the tines of a forklift truck. Another technique uses **slip sheets, tough** fiberboard or plastic sheets on which the load is stacked. The truck used with slip sheets has a clamp mechanism that grasps a protruding edge of the sheet and pulls the sheet and load onto a platform attached to the truck. A third method of handling a large group of assembled objects is with a clamp

truck, a mechanism that picks up loads by exerting pressure from both sides of the load.

60 Each method has its advantages and disadvantages. Slip sheets are economical, take up little space, and are light. However, the equipment is not universally available, is more expensive, and is slower to operate. Pallets are universally adaptable to a variety of handling situations and locations. However, pallets are costly, take up space, and can be difficult to dispose of. Clamp trucks use no added materials, but the geometry and character of the load must be such that it can be squeezed between the truck's clamps.

61 Most pallets are made of wood, and choice of wood species has a great impact on cost and durability. The denser and stiffer the wood, the greater the pallet's durability and usually the greater its cost. Well-made hardwood pallets are the most durable and cost-effective option of the many material choices available. Other materials are usually selected for considerations other than durability.

62 There are many possible pallet sizes and designs; however, for the sake of standardized distribution, certain sizes and designs predominate. By convention, a pallet's size is stated length first, with length defined as the top dimension along the stringer or stringer board (Figure 13.3). About a third of all pallets are nominally 40 by 48 inches, the standard set by members of **the Grocery Manufacturers of America.** This size is also very close to the international 1,000 by 1,200 mm size.

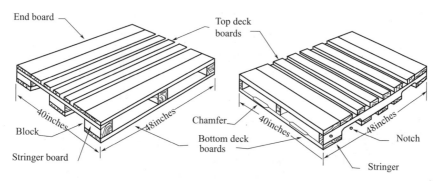

Figure 13.3 A block pallet (left) and a stringer-type pallet (right)

63 The two broad categories of pallet design are stringer and block types (Figure 13.3). A range of variations is available within each design type:

• **Reversible pallets** have similar top and bottom decks. Nonreversible designs have different top and bottom decks, with only the top deck designed to be a load-carrying platform.

• **Wing pallets** have the stringers inset so that the deck boards overhang. This allows for the pallets to be handled by slings. Pallets can be single wing or double wing, depending on whether one or both decks overhang the stringers.

• **Two-way-entry pallets** have solid stringers and can be entered only from the two ends.

• **Block-type pallets** are four-way entry, since any equipment can enter the pallet from all four directions. A partial four-way has notches cut into the stringer bottoms. A forklift's tines can enter from any direction, but a hand truck can only enter from two directions.

64 In addition to providing a product platform, the pallet is a **buffer** against the handling

environment. A forklift driver placing a pallet into position cannot see the exact placement location; he stops when he hits something.

⑥⑤ Viewed in this context, practices such as deliberate pallet perimeter overhang can only lead to problems, and warehouse operators condemn this habit. *The Food Marketing Institute holds pallet issues **responsible for** about half of all observed damage and cites poor pallet footprint as the single largest cause of shipping damage.* Of this damage, 50% is attributed to poor pallet stability and 35% is attributed to pallet overhang.

⑥⑥ Pallet maintenance programs are essential. A common and easily remedied problem is fasteners working their way out of the wood.

10.2 Unit Load Efficiency

⑥⑦ Warehouse floor space is rented by area, and the more product that can be put into that area, the better. Trucks loaded with light product should have the available volume completely filled to carry the maximum amount of product per trip. Area and cube utilization should be every packager's concern.

⑥⑧ Optimum area and cube utilization begins with the design of the primary package. Primary dimensions should be considered **in terms of** possible packing orientations in the shipping container, impact on corrugated board use in the shipping container, and palleting pattern and space utilization.

⑥⑨ "*Arrangement*" refers to packing patterns used when placing primary packages into a shipper. Traditionally, the problem was solved through intuition, experience, and a few nominal calculations. However, small cartons, packed 24 to a shipper, may have over a thousand possible orientation and palleting solutions. Computer "arrangement" programs are available that will calculate all the implications of size decisions in minutes. Typical input data for a palleting-efficiency computer program are:

- Data **pertaining to** the primary container.
- Allowed primary design changes, if required.
- Data pertaining to the proposed shipping case.
- Data pertaining to palleting requirements.

⑦⓪ Typical output data for such a program might provide the following information:

- Optimum dimensions for the primary container.
- Optimum packing orientations for selected primary containers.
- Inside and outside case dimensions for each selected case type.
- Number of units per pallet for each primary/case option.
- Area and cube utilization for each primary/case option.
- Recommended pallet patterns, including "walk-around" views.
- Dimensional details of the pallet pattern.
- Material areas used in primary, divider, and case construction.
- Relative cost factors for each construction.
- Relative compression values for corrugated board constructions.
- Proposed maximum warehouse stacking heights.

71　A thorough system analysis (including losses) can lead to substantial savings. A major business equipment manufacturer found that it had poor shipping experience because of the hundreds of different package sizes in the product line. The company designed a modular system, and all products were designed to fit one of 17 standard box sizes. Besides significant **inventory** reduction, the company gained substantial transport savings, since larger, more stable pallet loads could be built with the modular system. More-secure pallet loads resulted in further savings through reduced product damage.

10.3　Stabilizing Unit Loads

72　Unit loads often need to be **stabilized** in order to retain load geometry and order during shipping and handling. **Strapping**, usually polypropylene, is used mostly for heavier goods. Care must be taken that strapping does not cut into the corrugated container, impairing strength qualities. Cord is sometimes used as a more economical alternative, also causing cut-in problems. Corner guards should be used to prevent cut-in where strapping or cord is the necessary choice.

73　Shrink-wrapping is rarely used for load unitizing due to high installation and energy costs. Today's material of choice is stretch-wrapping. A good stretch-wrap application consists of two overlapped wraps extending 50 mm down the pallet to bind the load to the pallet. The wraps should overlap about 40% up the pallet side. Three overlapping wraps extending 50 mm past the top of the load finish the pallet. The added top wraps provide extra securement at the point in the load most likely to move.

74　Hand-wrapping a pallet with stretch material costs about $1.40. Machine-wrapping provides better material control and typically reduces the cost to about $1.00. Machines with prestretch features reduce this cost still further. More costly open netting is used where air circulation is essential.

75　Load stability can be increased through the use of high-friction printing inks and **coatings** or by the application of adhesive-like compounds. Adhesives can be designed to produce a high-tack local bond. One variation is the use of a bead of **hot-melt adhesive** formulated to have relatively poor **cohesive strength**. The bead forms a readily sheared bond between two box surfaces. However, systems that bond boxes together have caused handling problems and are not a popular load-stabilizing method with some warehouses.

76　*Caps and trays made of fiberboard or corrugated board are used to provide shape to unstable loads, to provide bottom protection against rough pallet surfaces, and, when used on top of a load, to increase the platform quality for the next pallet.* **Tier sheets** improve available compression strength and increase stability by distributing weight and encouraging layers to act as a unit.

(4867 words)

New Words and Expressions

distribution [ˌdɪstrəˈbjuʃən] *n.* 分配，流通

excelsior [ekˈselsiɔː] *n.* 细刨花

pallet [ˈpælɪt] *n.* 托盘、平台

foam [fəum] n. 泡沫材料、泡沫状物
handling ['hændliŋ] n. 装卸、搬运
dispense [di'spens] v. 分配、分发
promotional [prə'məuʃənl] adj. 促销的、增进的
optimize ['ɔptəˌmaiz] v. 使最优化
expendable [ik'spendəbəl] adj. 可消耗的、一次性使用的
underwritten [ˌʌndə'ritn] v. 给……保险、负担……费用
encounter [in'kauntə] n. 遭遇、碰撞
goodwill ['gud'wil] n. （企业的）信誉、声誉
wraparound ['ræpəˌraund] n. 环绕裹包、全裹包
intersection [ˌintə'sekʃən] n. 交叉点、交叉线
depict [di'pikt] v. 描述、描绘
severity [sə'veriti] n. 严重、严格
overstress [ˌəuvə'stres] n. 过应力
elusive [i'lusiv] adj. 难以捉摸的
alternative [ɔː'ltəːnətivz] n. 可供选择的事物、方案选项
fine-tune ['fain'tun] v. 调整、微调
specification [ˌspesəfi'keʃən] n. 技术规格、说明书
disparate ['dispərət] adj. 完全不同的、从根本上种类有区分或不同的
trapezoidal [træpi'zɔidəl] adj. 梯形的
tough [tʌf] adj. 坚韧的、牢固的
chamfer ['tʃæmfər] n. 倒棱、倒角
notch [nɑtʃ] n. 槽
buffer ['bʌfə] n. 缓冲器
inventory ['invənˌtri] n. 存货、库存
stabilize ['stebəˌlaiz] v. 变得稳定、稳固或固定
strapping ['stræpiŋ] n. 捆扎
coating ['kəutiŋ] n. 涂层、涂膜
shipping containers   运输容器
rubberized fiber   橡胶纤维
steel banding   钢带捆扎
the trial-and-error cycle   反复试验过程
returnable containers   可回收使用的容器
carrier rule and regulation   运输规则和条例
freight claims   货物索赔
in lieu of   代替
total-system approach   整体系统解决方案
optimum design   最优（佳）设计
the trial-and-error cycle   反复试验
subjecting ... to   受……管制、受到

the International Safe Transit Association (ISTA)　国际安全运输协会
fill an order　支付订货
glue flap　黏合襟片
rectangular cross section　矩形横截面
slip sheets　滑托板、薄垫板
the Grocery Manufacturers of America　美国食品杂货制造商
reversible pallet　可翻转的托盘、双面使用的托盘
wing pallet　翼式托盘
two-way-entry pallet　双向进叉托盘
block-type pallet　垫块式托盘
responsible for　为……负责、是造成……的原因
in terms of　根据、依据
pertaining to　相关的、关于
hot-melt adhesive　热熔性黏结剂
cohesive strength　黏结强度
tier sheets　层叠式薄板

Notes

1. *It has been estimated that although actual **freight** claims paid by carriers for damaging goods is approximately $2 billion, the actual cost to them and to shippers is really more than $10 billion per year.*（Para. 19）据估计，尽管实际由运输方所支付的货物损坏赔偿费用近20亿美元，但对他们和发货人来说实际的成本每年却超过100亿美元。it 是形式主语，指 that 引导的从句。

2. *Unfortunately, too much focus is often placed on the cost of packaging materials to the exclusion of other factors, including cost-related ones in handling, storage, and transportation.*（Para. 24）遗憾的是，太多的重点往往放在包装材料成本上，以至于排除了其他因素的影响，其中包括装卸、仓储、运输过程这些与成本有关的因素。too…to 指"太……以至于不能"。

3. *If damages rise too high (as on the left in the graph), you will encounter an increase in both product replacement and repair costs along with the loss of customer **goodwill** and possible cancellation of orders.*（Para. 26）如果破损率升得太高（如图左侧），将会遇到产品更换和维修成本的增加，以及客户信誉损失及可能的订单取消。along with 指"和……一起"、"除……以外（还）"。

4. *Instead of providing 15 inches of drop protection as needed in our example, the packaging protects to a drop height of 6 inches more than required.*（Para. 33）包装保护多出6英寸的跌落高度，而不是例子中要求的15英寸的跌落保护。"instead of"指"而不是……"、"替代……"。

5. *Those with limited packaging experience may find that the lack of technical packaging information makes it difficult to arrive at an optimum solution.*（Para. 41）只有有限包装经验的人可能会发现，缺乏包装技术信息，难以获得一个最佳的解决方案。that 引导的是宾语从句，it 指的是 to 引导的短语。

6. *For users with clearly defined distribution cycles different from the standard cycles in D4169, the ASTM standard also provides a means of developing a unique sequence of tests, resulting in performance-tests that can more precisely simulate your actual conditions.* (Para. 45) 对于那些确定的配送周期不同于 D4169 标准周期的用户，ASTM 标准还提供一种独特测试程序的手段，可以使性能测试更准确地模拟实际情况。ASTM D4169 即是 Practice for Performance Testing of Shipping Containers and Systems（运输器具及系统的性能测试标准）。

7. *The Food Marketing Institute pallet issues responsible for about half of all observed damage and poor pallet footprint as the single largest cause of shipping damage.* (Para. 65) 美国食品营销协会认为托盘问题约占所有观察到的损坏的一半，并将托盘与地面差作为运输损坏的最大单一原因。

8. *Caps and trays made of fiber board or corrugated board are used to provide shape to unstable loads, to provide bottom protection against rough pallet surfaces, and, when used on top of a load, to increase the platform quality for the next pallet.* (Para. 76) 用纤维板或瓦楞纸板制作的盖和浅盘用在不稳定载荷的形状固定，靠粗糙的托盘表面提供底部保护，而当盖或浅盘用在顶部时，增加第二托盘的平台质量。

Overview Questions

1. How do you ensure that your distribution packages are neither overdesigned nor underdesigned?
2. What is the size of the most commonly-used standard pallet?
3. What are the final checks of distribution packaging design?
4. ASTM and ISTA comprehensive preshipment test procedures are the most commonly-used industry methods for evaluating the effectiveness of a distribution packaging system. What are the respective merits of each method?
5. List methods of increasing the shipping stability of a palletized unit load.

Lesson 14 Test Method for Product Fragility

① Just as the weight of the product can be measured using a scale, the product ruggedness can be measured with dynamic inputs. A shock machine is used to generate a damage boundary curve, and a vibration system is used to **map out** the natural frequencies of a product.

1 Shock: Damage Boundary

② Shock damage to products results from excessive **internal stress** induced by **inertia forces**. Since inertia forces are directly proportional to acceleration ($F=ma$), **shock fragility** is characterized by the maximum tolerable acceleration level, i.e. how many G's the item can withstand.

③ When a dropped package strikes the floor, local accelerations at the container surface can reach several hundred G's. The packaging material changes the shock pulse delivered to the product so that the maximum acceleration is greatly reduced (and the **pulse duration** is many times longer). It is the package designer's goal to be sure that the **G-level** transmitted to the item by the cushion is less that the G-level which will cause the item to fail.

④ The damage boundary theory is a testing protocol which determines, in an engineering sense, which shock inputs will cause damage to a product and which will not. There are two parts of a shock which can cause damage, the acceleration level and the velocity change. The velocity change, or the area under the acceleration-time history of the shock, can be thought of as the energy contained in a shock. The higher the velocity change the higher the energy content. There is a minimum velocity change which must be achieved before damage to the product can occur. This level is called the **critical** velocity change (ΔV_c). Below the critical velocity change, no damage occurs regardless of the input acceleration level. In essence, there is not enough energy in this region of the damage boundary to cause harm to the product. Exceeding the critical velocity change, however, does not necessarily imply that damage results. If the change in velocity occurs in a manner which administers acceptable doses of acceleration to the product, the velocity change can be very large without causing damage. However, if the critical velocity and the critical acceleration (A_c) are both exceeded, damage occurs.

⑤ A typical damage boundary curve can be found in Figure 14.1. *It is a plot of the damage causing parameters of a shock pulse and defines the region where certain combinations of acceleration and velocity change fall in the clear band outside the damage region, no damage occurs.* For example, if the velocity change of the input is below that of the product's critical velocity change, then the acceleration level of the input can be in the $100G$'s, $1000G$'s, $10,000G$'s, or even infinite without causing damage. In practical terms, to achieve these very high accelerations means that the duration of the shock must be very short. If we think of it in a graphical sense, we have a certain amount of area or velocity change which we can re-arrange to produce a short pulse. If we decide to make the pulse very tall, i.e. very high in acceleration, then because of the limited amount of area that pulse can contain it must also be very short in duration. In fact, the duration is so short that the product cannot respond the acceleration level of the event, only the energy input. Because the input velocity change did not exceed the critical velocity change of the product, no damage has occurred.

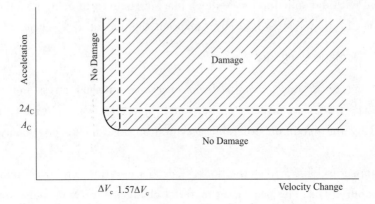

Figure 14.1 Typical damage boundary curve

⑥ When the velocity change of the input exceeds the critical velocity change of the product. However, the only way to avoid damage is to limit the input acceleration to a level below that of the product's critical acceleration. This is usually one of the functions that a cushioned package performs. It translates the high acceleration events experienced on the outside of the container to lower acceleration events experienced inside at the unit.

⑦ The low-velocity portion of the plot (at the left) is that area where damage does not occur even with very high accelerations. Here the velocity change (drop height) is so low that the item acts as its own **shock isolator.** Below the acceleration boundary portion of the plot (under the curve), damage does not occur, even for large velocity changes (drop heights). That's because the forces generated ($F=ma$) are within the strength limits of the products.

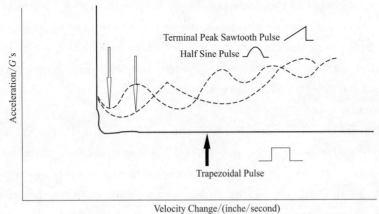

Figure 14.2 Damage boundary for pulses of same peak acceleration and same velocity change

⑧ Figure 14.2 shows that the velocity change boundary (vertical boundary line), is independent of the pulse wave shape. However, *the acceleration value (to the right of the vertical line) of the damage boundary curve for half* **sine** *and* **sawtooth pulses** *depends upon ve-*

locity change. Use of this damage boundary would require accurate prediction of drop heights and container/cushion coefficients of restitution. Since they normally cannot be predicted, a trapezoidal pulse shape is typically used.

⑨ *The damage boundary generated with use of a trapezoidal pulse encloses the damage boundaries of all the other waveforms*. This is a great advantage, since the wave shape which will be transmitted by the cushion is usually unknown. By using the trapezoidal pulse to establish the acceleration damage boundary rating, the package designer can be sure that actual shocks transmitted by the cushion will be equal to or less damaging than the test pulse. On the other hand, it should be noted that the trapezoidal wave used to determine the critical acceleration provides conservative results.

⑩ Fragility testing is the process used to establish damage boundaries of products. It is usually conducted on a **shock testing machine**. The procedure has been standardized and incorporated into several standards such as ASTM D3332, Mechanical-Shock Fragility of Products, Using Shock Machines. Use of a shock machine provides a convenient means of generating variable velocity changes and consistent, controllable acceleration levels and waveforms.

⑪ Typically, the item to be tested is fastened to the top of a shock machine table and the table is subjected to controlled velocity changes and shock pulses. The shock table is raised to a preset drop height. It is then released, free falls and impacts against the base of the machine; it rebounds from the base and is arrested by a braking system so that only one impact occurs. A **shock programmer** between the table and the base controls the type of shock pulse created on the table (and the test item mounted on it) during impact.

⑫ For trapezoidal pulses used in fragility testing, the programmer is a constant force **pneumatic cylinder**. The G-level of the trapezoidal pulse is controlled simply by adjusting the compressed gas pressure in the cylinder. The velocity change is controlled by adjusting drop height.

1.1 Conducting A Fragility Test

⑬ To determine a damage boundary requires running two sets of tests. A step velocity test is used to determine the product's critical velocity change and a step acceleration test is used to determine the critical acceleration.

① Step Velocity Test

⑭ To run the step velocity test, shock machine drop height is set at a very low level to produce a low velocity change, and the product is secured to the table surface. Either a half sine or a rectangular pulse may be used to perform this test, since the critical velocity portion is the same. A half-sine shock pulse waveform programmer is normally used for convenience. The first drop is made and the item examined to be sure damage has not occurred. Drop height is then increased to provide a higher velocity change. The second drop is made and again the specimen is examined. Additional drops are made with drop height gradually increasing until failure occurs. The velocity change and peak acceleration are recorded for each impact. Once damage occurs, the velocity boundary testing is stopped, since the minimum velocity necessary to create damage has been established as well as the velocity change portion of the damage boundary curve (Figure 14.3). The damage boundary line falls between the last drop without damage and the first drop causing damage.

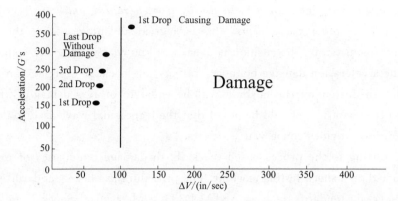

Figure 14.3 Velocity damage boundary development

⑮ In some cases, it is sufficient to determine only this vertical line of the damage boundary. If the velocity change required to damage the product will not be encountered from normal drops expected in the environment, no cushioning will be needed. However, if the product is damaged at levels which will be encountered in the environment, product improvements or cushioning for shock protection will be required. This indicates a need to establish the horizontal line of the damage boundary.

② **Step Acceleration Test**

⑯ Determining the acceleration boundary line requires that a new test specimen be attached to the shock table. The drop height is set at a level which will produce a velocity change at least 1.57 times the critical velocity. *The programmer compressed gas pressure is adjusted to produce a low G-level shock which is lower than the level which you anticipate will cause damage to the product.* Again, a first drop is made and the item is examined for damage. If none has occurred, the programmer pressure is increased to provide a higher G-level impact from the same drop height. Another drop is made and again the specimen is examined. The procedure is repeated with gradually increasing G-levels until damage occurs. This level establishes the level of the horizontal line of the damage boundary curve. The damage boundary line falls between the last drop without damage and the first drop causing damage.

⑰ You can plot the damage boundary curve by connecting the vertical velocity boundary line and the horizontal acceleration boundary line. The corner where the two lines intersect is actually rounded, not square. In most cases, this rounded corner will not be in the range of interest and a square corner can be used. If, however the corner is in the range of interest, the shape of the corner can be determined by calculation or by running an additional test in the area. Figure 14.4 shows a typical damage boundary plotted by this method.

⑱ In a rigorous testing program, damage boundary curves are generated for each orientation of the unit. To do this requires damaging 2 units per orientation, one for the step velocity test and one for the step acceleration test. Rarely are this many units available for destructive testing in the prototype stage of a product's life, the most beneficial time to do this type of work. Compromises are often made to limit the number of units which must be damaged. In certain situations it is possible to perform the testing only along the three orthogonal axes. In addition, the unit can often be repaired between tests, so that one unit and a few spare parts may be used to perform all of the testing.

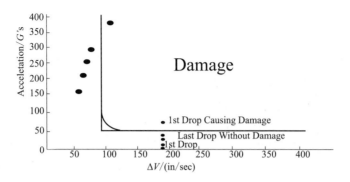

Figure 14.4 Damage boundary line development

2 Vibration: Resonance Search & Dwell

⑲ *It is generally accepted that the steady-state vibration environment is of such low acceleration amplitude that failure does not occur due to nonresonant inertial loading.* Damage is most likely to occur when some element or component of a product has a natural frequency which is excited by the environment. If this tuned excitation is of sufficient duration, component accelerations and displacements can be amplified to the failure level. The identification of those frequencies, therefore, becomes critical in designing a package system. The purpose of the bare product vibration testing is to identify the natural or resonant frequencies of the critical components within the product.

⑳ **Response** of a product or component to input vibration may be represented by a curve similar to that shown in Figure 14.5.

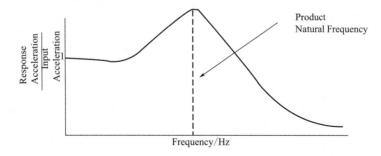

Figure 14.5 Typical resonant frequency transmissibility curve

㉑ You can see that for very low frequencies, response acceleration is the same as the input; for very high frequencies, the response is much less than the input. But in between, the response acceleration can be many times the input level. This is the frequency range where damage is most likely to occur.

㉒ To actually determine a product's vibration fragility would involve complexities which are probably not justified in terms of greatly improved results. The product test method, then, involves identifying the product and component resonant frequencies. A test method often used to accomplish this is ASTM Standard Method D3580, Vibration (Vertical Sinusoidal Motion) Test of Products.

㉓ The resonance search is run on a vibration test machine (shaker). *The item to be tested is fastened to the* **shaker table** *and subjected to a low level sinusoidal input over a broad frequency range.* As the frequency is slowly varied between lower and upper limits, the test item is observed for resonances. Sometimes, if non-critical product panels, etc., or other shielding external components are removed, resonant effects can be seen or heard directly. At other times, use of a **stroboscope** and/or various **sensors** may be necessary. The critical frequencies and components should be recorded.

㉔ In general, tests should be performed in each of the three axes, and three sets of critical frequencies recorded. If the product is mounted on a **definite skid base**, only the vertical axes need to be analyzed.

(2200 words)

New Words and Expressions

critical ['kritikl] *adj.* 临界的
sine [sain] *n.* 正弦
response [ri'spɔns] *n.* 响应、反应
stroboscope ['strəubəskəup] *n.* 频闪观测器、频闪仪
sensor ['sensə] *n.* 传感器
map out 描绘
internal stress 内应力
inertia force 惯性力
shock fragility 冲击脆值
pulse duration 脉冲持续时间
G-level G 值
shock isolator 隔振器
sawtooth pulse 锯齿波脉冲
shock testing machine 冲击试验机
shock programmer 冲击程序装置
pneumatic cylinder 气缸
step velocity test 步进式速度测试
step acceleration test 步进式加速度测试
transmissibility curve 传递率曲线
shaker table 振动台
definite skid base 确定的防滑地座

Notes

1. *It is a plot of the damage causing parameters of a shock pulse and defines the region where certain combinations of acceleration and velocity change fall in the clear band outside the damage region, no damage occurs.* (Para. 5) 这是一个由冲击脉冲引起破损的参数图,其中当某些加速度与速度改变量的组合落在破损区域之外的空白带内时,就不会有破损发生。

2. *However, the acceleration value (to the right of the vertical line) of the damage boundary curve for half sine and sawtooth pulses depends upon velocity change.* (Para. 8) 但是,对半正弦脉冲与锯齿脉冲,破损边界曲线的加速度值(垂直边界线的右边)则与速度变化有关。

3. *The damage boundary generated with use of a trapezoidal pulse encloses the damage boundaries of all the other waveforms.* (Para. 9) 用梯形脉冲产生的破损边界包含了全部其他波形的破损边界。

4. *The programmer compressed gas pressure is adjusted to produce a low G-level shock which is lower than the level which you anticipate will cause damage to the product.* (Para. 16) 调整程序装置的压缩气体压力使之能产生一个低 G 值的冲击,该 G 值低于使产品破损的期望的 G 值。

5. *It is generally accepted that the steady-state vibration environment is of such low acceleration amplitude that failure does not occur due to nonresonant inertial loading.* (Para. 19) 一般认为稳态振动环境具有如此低的加速度幅值,以至于失效不会因非共振惯性载荷而发生。

6. *The item to be tested is fastened to the shaker table and subjected to a low level sinusoidal input over a broad frequency range.* (Para. 23) 将被测试的产品固定在振动台上并使产品在一个宽泛的频率范围内受到低水平的正弦输入。

Overview Questions

1. Plot the damage boundary curve for a product from the data given below.

Test for ΔV_c			Test for A_c		
Drop No.	Height	Result	Drop No.	Gas Pressure	Result
1	2″	no damage	12	50 psi	no damage
2	3″	no damage	13	100 psi	no damage
3	4″	no damage	14	150 psi	no damage
4	5″	no damage	15	200 psi	no damage
5	6″	no damage	16	250 psi	no damage
6	7″	no damage	17	300 psi	no damage
7	8″	no damage	18	350 psi	damage
8	9″	no damage			
9	10″	damage			

2. What's the reason for shock damage to products? Why chose the maximum tolerable acceleration to characterize shock fragility?
3. What're the two parts of a shock which can cause damage? What's the meaning of velocity change (ΔV_c)? And try to describe the relationship between the velocity change and the acceleration-time curve.
4. How to avoid damage when the velocity change of the input exceeds the critical velocity change of the product?
5. What are the two sets of tests to determine a damage boundary? Try to describe their test principles and methods.

Lesson 15 Stress-Energy Method for Determining Cushion Curves

1 Current Practice

① The current industry standard practice for determining how much foam **cushion material** to use in a protective packaging application is to consult hard-copy graphs called **cushion curves**. Cushion curves are graphical representations of a foam material's ability to limit transmission of shock (called G level) to a product. G level is plotted along the vertical axis versus static loading (weight divided by bearing area) along the horizontal axis. Curves are specific to a particular material, a particular density, and a particular drop height. Simply consulting the cushion curve will visually tell how many G's will be transmitted for a given drop height, cushion thickness and static loading.

② There are at least two related limitations to the current use of cushion curves: how the cushion curves are generated and subsequently how the cushion curves can be used. **Simply put**, the current practice is static, in that only the combinations of drop height, thickness and static loading that are tested are plotted, and hence gives limited information about G levels expected. For example, it is easy to predict G level for thicknesses of one, two, three and four inch thick cushions (Figure 15.1), but what if the engineer wants to predict performance at half an inch or six inches? Although it can be estimated, this points out

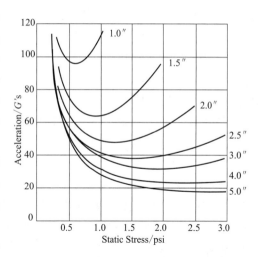

Figure 15.1 Cushion Curve for EPS Material

how limited cushion curves can be. The current method for constructing cushion curves is outlined in ASTM D1596. It is possible to overcome the limitation of selected data, but the process for collecting this information is very time consuming and resource intensive. To generate a full set of cushion curves (range of drop heights, about seven cushion thicknesses) would require somewhere on the order of 10,500 sample drops and over 175 hours of test time. Even more samples and time would be required to fill in the data for other cushion thicknesses and drop heights.

2 Something New

③ But what if there was a way to simplify the process of generating cushion curves, and at the same time give the ability to generate an unlimited number of curves with any set of variables (i. e. any drop height, any thickness, any static loading)? In fact, this is possible and has been documented by Dr. Gary Burgess of Michigan State University. The method is called the "Stress-Energy" method, or more specifically "dynamic stress versus dynamic energy". The stress-energy method is really about how much energy a cushion can absorb and the dynamic loading on the cushion during the absorption. Another way to think of this method is to realize material properties of a cushion can be described by a relationship between the specific variables of static loading, drop height, cushion thickness and G level. These are familiar terms already used in traditional cushion curves, and the stress-energy method simply uses them in a different way to achieve unlimited performance information about the cushion. Instead of conceptualizing cushion curves as lines on a graph, the concept is to reduce all combinations of drop height, static loading and thickness into a single equation that is able to generate any cushion curve you would like for that specific material.

3 Stress-Energy Method

④ **Dynamic Stress** is defined as:
$$Gs \qquad (15.1)$$
where G—fragility level (critical acceleration); s—static loading.

Dynamic Energy is defined as:
$$sh/t \qquad (15.2)$$
where s—static loading; h—drop height; t—cushion thickness.

Both have units of pounds per square inch (psi).

To illustrate, start by picking any point from a cushion curve (Figure 15.1: $s=1$, $h=36$, $t=2$). Calculating energy gives $sh/t=18$ psi, and stress is $Gs=50$ psi.

⑤ The stress-energy method says that for any calculated energy, G can be predicted. To demonstrate this, compare predicted G levels (from Gs) to actual G levels from the published cushion curve. Table 15.1 shows by picking different combinations of s, h and t (to equal 18 psi), G levels can be predicted very accurately. This exercise can be repeated for any energy level with similar results predicting G level.

Table 15.1 Predicted and actual G values

h	s	t	Gs(Predicted)	G Actual(curve)
36	0.5	1.0	96	95
36	1.5	3.0	29	30
36	2.0	4.0	22	22

⑥ The underlying relationship of these variables can be described by an equation in the form of
$$y = ae^{bx} \qquad (15.3)$$
where y—dynamic stress=Gs; x—dynamic energy=sh/t; e—constant=2.71828.
and the constants a and b are dimensionless values that describe the material properties of the cushion, derived from a curve fit operation when dynamic stress is plotted versus dynamic

energy (described later).

⑦ Once the constants a and b are found, Equation (15.3) can be rearranged and used to draw cushion curves for any combination of variables for the material:

$$G = \frac{ae^{b\frac{sh}{t}}}{s} \tag{15.4}$$

⑧ Note the form of Equation (15.3) should look familiar, and in fact is in the same form as the ideal gas model, suggesting the primary cushioning effect is related to the behavior of air. Second, there the model is limited to certain types of cushions, namely **closed-cell materials** (and corrugated materials). *This is because closed cell materials rely on displacement of air for cushioning properties, contrasted to materials that rely on mechanical means for cushioning (such as polyurethane), which will probably require a different model.*

4 Procedure To Find The Stress-Energy Equation

⑨ To find the stress-energy equation for a particular material, start by **tabulating** the independent variables s, h, and t, and the dependent value G. If **starting from scratch**, this information simply comes from following ASTM D1596. An example is shown in Table 15.2.

Table 15.2 Tabulation of Variables

Drop Height/in	G	Static Loading/psi	Cushion Thickness/in
18	30	0.4	3.0
24	33	0.5	3.0
30	55	0.5	1.5
36	37	0.6	3.0
42	65	0.35	2.0

⑩ To generate a traditional cushion curve, simply plot G versus static loading. For the stress-energy method, two more steps are required, neither requiring any more data to be collected, but rather two simple calculations with the existing data, as shown in Table 15.3.

Table 15.3 Calculated values of dynamic stress and dynamic energy

Drop height/in	G	Static Loading/psi	Cushion Thickness/in	Dynamic Stress/Gs	Dynamic Energy/sh/t
12	80	0.1	1.0	8.0	1.2
18	30	0.4	3.0	12.0	2.4
24	33	0.5	3.0	16.5	4.0
30	55	0.5	1.5	27.5	10.0
36	37	0.6	3.0	22.2	7.2
42	65	0.35	2.0	22.8	7.4

⑪ *The next step is to plot dynamic stress versus dynamic energy, and apply a simple **exponential curve** fit to the data points (LSM method by hand, or Power Trendline in Excel), as shown in Figure 15.2.*

⑫ The equation Excel displays in Figure 15.2 is now the dynamic stress-energy equation that fully describes the cushioning ability of this material. Also displayed is the R^2 value, which is an indication of how well the equation fits the data. 95% is extremely good. Figure 15.2 shows the value for a is 9.8891 and the value for b is 0.1036. We now have one equation that can be used to generate ANY cushion curve for this material.

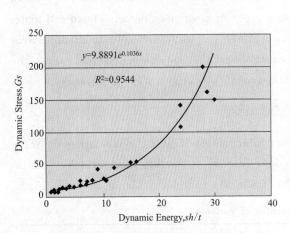

Figure 15.2 Microsoft Excel plot of dynamic stress versus dynamic energy

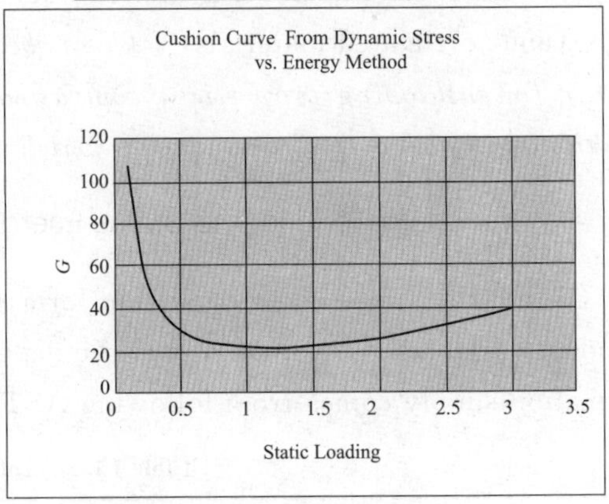

Figure 15.3 Cushion curve using Equation (15.4)

5 Using The Stress-Energy Equation To Generate Cushion Curves

⑬ A simple **spreadsheet** can be set up to use Equation (15.4) to draw any cushion curve for this material. An example is shown in Figure 15.3, a and b come from the stress-energy plot (Figure 15.2); then simply change the drop height and/or thickness and plot G versus s.

6 Test Procedure

① Step 1

⑭ Set maximum and minimum limits on the energy absorbed. Since energy $= sh/t$, the minimum energy corresponds to the smallest s, the smallest h, and the largest t that you want data for. If the intent is to eventually produce a standard set of cushion curves, then for closed-cell foams, these values are usually $s = 0.5$ psi, $h = 12$ inches, and $t = 6$ inches. These give $sh/t = 1$ in-lb/in^3. For open-cell foams, this limit will be lower because the material is not as stiff.

⑮ The maximum energy corresponds to the largest s, the largest h, and the smallest t that you want data for. If the intent is to eventually produce a standard set of cushion curves, then for closed-cell foams, these values are usually $s = 3$ psi, $h = 48$ inches and $t = 3$ inches. These give $sh/t = 48$ in-lb/in^3. For open-cell foams, this limit will be lower.

⑯ It is not necessary to set an exact range. This step is merely a guideline to establish limits within which to conduct drop tests. Machine limitations may require modifications to this range.

② Step 2

17 Divide the energy range in Step 1 into about 10 approximately evenly spaced points. If the range 1 to 48 is used, then test for energies in steps of about 5 psi. You could for example choose nine different energies equal to 5, 10, 15 ... and 45 in-lb/in³.

③ Step 3

18 For each of the energies chosen in Step 2, select five different combinations of s, h and t values that give this energy. These five combinations are in effect "replicates" for each of the energies listed in the range in Step 2. For example, five different combinations of s, h and t that give $sh/t=30$ are shown in Table 15.4.

Table 15.4 Five different combinations of s, h and t values

s/psi	h/inch	t/inch	Energy=sh/t(in-lb/in³)
1	30	1	30
1.5	40	2	30
2	30	2	30
2.5	36	3	30
3	15	1.5	30

19 Next, perform these five drops on the **cushion tester**. For the first drop, set the cushion tester up for an equivalent free fall drop height of 30 inches, select a cushion sample with an actual thickness of 1 inch, add enough weight to the platen to achieve a static stress of 1 psi, and drop the platen. Capture the shock pulse, filter it, and record the peak G for this drop. This completes the first of the five drops corresponding to an energy of 30 in-lb/in³.

20 Finish the remaining four drops the same way and summarize the experimental data in a table like the one shown in Table 15.5. The G values in the 4th column come from the drop tests. The last column of this table shows the calculated stress values corresponding to an energy of 30 in-lb/in³.

Table 15.5 The experimental data

s/psi	h/inche	t/inche	G/G's	Energy=sh/t(in-lb/in³)	Stress=Gs/psi
1	30	1	60.3	30	60.3
1.5	40	2	41.8	30	62.7
2	30	2	29.4	30	58.8
2.5	36	3	24.6	30	61.5
3	15	1.5	19.5	30	58.5

21 If the material behaves in a "normal" manner, the stress values in the last column should cluster tightly about a mean value. The **mean** in this case is 60.36 psi and the **standard deviation** is 1.78 psi, which is 2.9% of the mean.

④ Step 4

22 Repeat Step 3 for each of the energies in the range chosen in Step 2 and construct the

stress vs. energy relationship shown in Table 15.6. The stress values listed are the means for the five replicates tested for each energy. The **variations** are the standard deviations expressed as a percent of the mean.

Table 15.6 Relationship between stress and energy

Energy/(in-lb/in^3)	Stress/psi	Variation/%
5	22.47	4.5
10	27.65	5.2
15	33.08	3.1
20	42.16	1.8
25	50.92	4.3
30	60.36	2.9
35	73.48	2.6
40	90.87	3.7
45	110.23	4.4

⑤ **Step 5** (optional)

23 Fit an equation to the stress vs energy data. The relationship between stress and energy can usually be described to a high **degree of correlation** by the exponential relationship:

$$\text{stress} = a\, e^{b(\text{energy})}$$

where a, b—constants specific to foam type and density; e—2.71828.

Regression can be used to best fit this equation to the data.

7 Conclusion

24 Curve fit correlation was excellent across all densities and number of drops, almost always over 90% and in many cases over 95%. Since the stress-energy method relies heavily on energy absorption (static loading, drop height, thickness), great care needs to be taken when measuring these variables. Failure to do so will affect results. *Total test time was approximately for hours, and considering this covered four different densities and a complete cushion curve profile, this validates the stress-energy method as a quicker alternative to ASTM D1596.* In the future, it is felt far less testing than 250 drops per density would be required. It is more valuable to test a wider range and greater sampling of energy values than multiple replicates at a specific energy. Testing a variety of energy levels will account for material variability due to the application of the curve fit.

(2190 words)

New Words and Expressions

tabulate ['tæbjuleit] *vt*. 把……制成表格
spreadsheet ['spredʃi:t] *n*. 电子数据表
mean [mi:n] *n*. 平均值
variation [ˌveəri'eiʃ(ə)n] *n*. 差异、方差
cushion material 缓冲材料
cushion curve· 缓冲曲线

simply put   简单地说
dynamic stress   动态应力
dynamic energy   动态能量
closed-cell material   闭环式材料
start from scratch   从零开始
exponential curve   指数曲线
cushion tester   衬垫试验机
degree of correlation   相关度
standard deviation   标准方差

Notes

1. *Both have units of pounds per square inch (psi).* (Para. 4) 两者（动能量和动应力）的单位都是磅/平方英寸（psi）。文中单位均为英制，如 in-lb/in³ 为英寸-磅/立方英寸，pcf 为磅/立方英尺。其中，1 磅力＝4.45 牛顿，1 英寸＝2.54 厘米。

2. *This is because closed cell materials rely on displacement of air for cushioning properties, contrasted to materials that rely on mechanical means for cushioning (such as polyurethane), which will probably require a different model.* (Para. 8) 这是由于闭环式材料的缓冲特性是靠气体的移动来产生，这与靠机械方式产生缓冲（如聚氨酯材料）的那些材料（模型也许不同）形成了对照。"closed cell materials" 是指材料的结构单元为闭环式，类似地，"open-cell foams" 为"开环式泡沫"。

3. *The next step is to plot dynamic stress versus dynamic energy, and apply a simple exponential curve fit to the data points (LSM method by hand, or Power Trendline in Excel), as shown in Figure 15.2.* (Para. 11) 下一步是绘制动应力与动能量的曲线图，用简单的指数曲线拟合数据点（LSM 手绘方法，或用 Excel 趋势线法），如图 15.2 所示。"LSM" 即 Least Square Method，指最小二乘法。Excel 趋势线法，是在软件 Excel 中根据应力-能量数据建立散点图，其中能量为自变量，然后对该散点图创建趋势线，选择函数关系为幂函数，即可自动绘出趋势线并计算出 a、b 的数值及 R^2。

4. *Total test time was approximately xx hours, and considering this covered four different densities and a complete cushion curve profile, this validates the stress-energy method as a quicker alternative to ASTM D1596.* (Para. 24) 应力能量法整个测试时间大约 XX 小时，涵盖了四个不同密度和完整的缓冲曲线轮廓，证明它是比 ASTM D1596 更加快速的一种替代方法。"and considering this covered four different densities and a complete cushion curve profile" 为让步状语，意为（虽然整个测试可能要花 XX 小时，但）考虑到它是四个不同密度的缓冲曲线。ASTM1596 是由美国材料实验协会（American Society of Testing Materials）制定的关于包装材料减震性能标准试验方法。

Overview Questions

1. What is a cushion curve?
2. Which parameters can be used to describe the material properties of a cushion?
3. Try to find the a and b value with the data shown in Table 15.6 by Power Trendline in Excel.

UNIT 4

Packaging Technology and Machinery

Lesson 16 Liquid Filling

1 Introduction

① The numerous liquid and semi-liquid products that fill different types of containers have required the development of a number of different filling techniques and machines. However, most of the machines operate on common basic principles and filling methods, so it is possible to adapt similar machines and components to meet the requirements of particular products and containers.

② In this section you will become familiar with the basic types of techniques and machines used in liquid and semi-liquid filling, and in the following sections you will learn to identify the different types of techniques and machines, their appropriate application, and their operation and **maintenance** requirements.

2 Type Of Filling Machine

③ The characteristics of the product and the container being used make specific requirements upon the selection of the filling machine for any particular application. Liquid and semi-liquid products vary in **consistency** from very thin liquids such as alcohol and soda water to semi-liquids such as toothpaste, peanut butter, and **caulking compound**. Some solid products such as stick deodorants are also filled as semi-liquids by processing them at a temperature that keeps them fluid and allowing them to harden in the container.

④ The containers may be made of plastic, metal, glass, treated paperboard, or a number of other materials. They may vary in size from small **ampules** and vials used in the **pharmaceutical** industries to large **drums** used for paints and petroleum products.

Figure 16.1 The shapes of the containers

⑤ The shapes of the containers include those of bottles, jars, vials, tubes, cans, pouches, cartons, and drums (Figure 16.1).
⑥ Most liquid and semi-liquid products are filled by one of two major methods: **volumetric**, or **constant level filling**. In volumetric filling the amount of product is premeasured so that each container has the same volume of product. Constant level filling techniques fill each container to the same level, so it is frequently called the **"fill-to-a-level"** method.

3 Use Of Filling Machines

⑦ *Volumetric filling is particularly appropriate for applications in the pharmaceutical industry where it is important that each container is accurately filled with a specific volume of product.* Constant level filling is used with **see-through** bottles in which it is important that all of the bottles in a display appear to be filled to the same level, although the bottles may not be exactly the same size and the volumes may be slightly different.

4 Rotary And Straight-Line Fillers

⑧ Fillers are also classified in terms of the way the containers are moved through the filling operation. The **rotary filler** removes the containers from the conveyor onto a rotating plate which carries them in a circle through the filling machines. The filling heads rotate with the containers as they are filled, so that there is a continuous motion (Figure 16.2).

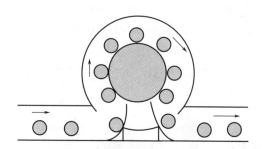

Figure 16.2 Rotary filling

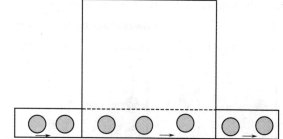

Figure 16.3 Straight-line filling

⑨ The **straight-line fillers** fill each container as it moves along the conveyor in a straight line. This may be an **intermittent motion** operation in which the conveyor is stopped until the container is filled, then it is moved just far enough to place the next container in position under the filling nozzle. It may fill only one or a number of containers in each operation depending upon the number of filler heads that are used. Some multiple head machines have the capability of following the containers along the line and filling them in a continuous motion without stopping the filling line (Figure 16.3).

5 Liquid Volumetric Filling

5.1 Introduction

⑩ Volumetric fillers deliver a premeasured volume of product to each container, and the volume of product in each container is held constant.

①　The advantages of the volumetric fillers include accuracy, **flexibility**, and reliability in addition to being relatively easy to clean. *Accurately delivering a premeasured volume of product to each container also produces cost savings by reducing the amount of product that is used in* **overfill** *in some operations to assure that each container receives at least the minimum desired amount of product.*

②　The flexibility of volumetric filling systems allows them to be adapted to fill a wide variety of products ranging from thin alcohol to thick caulking compounds. They can be used to fill rigid containers or light weight ones that may be distorted by the forces of vacuum or pressure used in some filling operations.

③　These systems are reliable and easy to maintain, because they are generally constructed from relatively simple designs.

④　*Volumetric filling systems are relatively easy to clean by flushing techniques since they normally do not have small hoses, tight passageways, or delicate sensing mechanisms that can be clogged or damaged by water or air pressure.*

⑤　Three popular types of volumetric filling methods are piston operation, **diaphragm** action, and timed flow. Each of them will be considered in some detail.

5.2　Piston Volumetric Filling

⑥　A piston filler (Figure 16.4) measures and delivers the product to the container by the action of a single piston for each filler head. On the **intake stroke** the piston draws the product from **the supply tank**, through a **valve**, and into the **cylinder** of the piston **chamber**. This cylinder may be called the **measuring or filling chamber**, since its volume remains constant for any adjustment and controls the volume of product that is delivered.

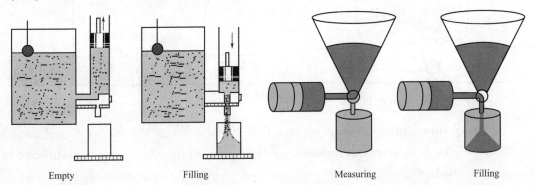

　　　　Empty　　　　　　　　Filling　　　　　　　　Measuring　　　　　　　Filling

　Figure 16.4　Piston volumetric filling with　　　Figure 16.5　Piston volumetric filling with
　　　　　　　rotary valve　　　　　　　　　　　　　　　　　　reciprocating valve

⑦　On the **downstroke** the valve moves to open the passageway, and the product is forced from the chamber through the valve and into the container.

⑧　In the illustration on the left, the rotary valve has opened so that the rising piston draws the product from the supply tank into the filling chamber (or cylinder). In the illustration on the right, the valve to the supply tank has closed, and as the piston moves downward, the product is forced from the chamber into the container.

⑨　The operations of these two machines are very similar, except one machine has a **recip-**

rocating valve (Figure 16.5) and the other has a rotating one (Figure 16.6).

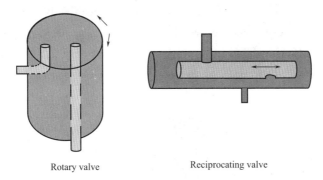

Rotary valve   Reciprocating valve

Figure 16.6 Valves

⑳ The rotating valve turns to position the channels and direct the product flow in the direction that is desired, and the reciprocating valve slides from one side to the other in order to position the slots and regulate the flow. Although mechanically these valves operate differently, they perform the same function in a similar manner.

㉑ The volume of product delivered into the container is determined by the volume of the filling chamber in which the piston is operating. This volume can be changed by adjusting the length of the piston's stroke. As the stroke is lengthened, the volume of the chamber is increased, and as the stroke is shortened, the volume of the chamber is decreased.

㉒ Figure 16.7 shows a typical way of making the volume adjustment by moving the point at which the end of the arm is attached in the slot of the wheel what provides the **cranking** motion.

㉓ The volume is usually adjustable within a 10 fold range. If the largest amount of product that can be filled on one stroke is 100 **cc**, the piston can usually be adjusted so that it can deliver as little as 10 cc on a stroke. The accuracy of the fill is easier to maintain near the maximum fill, so it is frequently recommended that the cylinder and piston be replaced with a smaller size when the desired fill approaches the lower limits of the range.

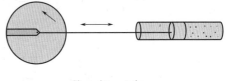

Short piston stroke

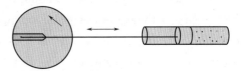

Longer piston stroke

Figure 16.7 Volume adjustment

㉔ In some applications **double filling** is used to deliver two discharges from the filler into each container. This technique works effectively, but it is considerably slower and is generally used for short runs.

㉕ Figure 16.8 shows three methods that are used to keep the product from leaking out around the piston as it moves **back and forth**. Some pistons and cylinders are machined to a fine **tolerance** which prevents **leaking**. Some pistons have rings placed in **grooves** around their **circumference**. These rings may be made of a number of different materials that are selected to work best with the characteristics of the products being filled. Metal rings are normally used when the product temperature is over 150°F, and **teflon** is a popular material for many multipurpose rings. **Cuffs** may also be applied to the ends of the pistons to create a seal, and the materials used in the cuffs should be selected specifically for the materials being filled. Each filling machinery manufacturer has information on the most appropriate uses of each type of ring or cuff, but laboratory tests may be necessary to determine the most appropriate one to provide a fast smooth operation with no leakage and long wear when a new product is introduced.

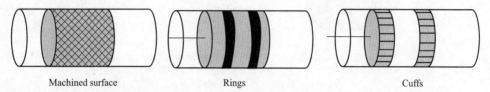

Machined surface   Rings   Cuffs

Figure 16.8　Three methods to keep product from leaking out around piston

㉖　Different types of products require different nozzle designs and sizes (Figure 16.9). The nozzle used for filling an ampule with 10cc of thin fluid obviously would not be appropriate for filling a five gallon can with paint, and there are numerous sizes and shapes available between these two extremes.

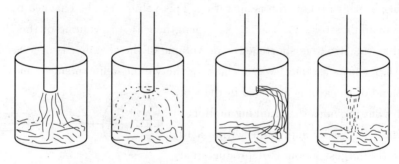

Figure 16.9　Nozzle types

㉗　The diameter of the nozzle should be selected to provide a smooth flow. If it is too small the velocity of the material will be too high as it enters the container and there may be **splashing** and foaming. If the nozzle is too large, the product may **drip** at the end of each fill. When the nozzle diameter matches the **surface tension** of the product, a film or bubble is naturally formed in the tip of the nozzle at the end of each fill, and this keeps any material left in the nozzle from dripping out. If the nozzle diameter is too large, the film cannot form and there may be a few drips at the end of each fill. *This can make it necessary to either run the process slower so that all the drips are caught in the container or to require frequent cleaning of the containers and the machine.*

㉘　Filler nozzles are designed in different sizes and shapes to prevent splashing, foaming, or excessive **aeration** of the products during filling. Some nozzles direct the flow directly downward in a solid stream, and others break it up into a **spray** or direct it to the side of the container so that it will run down the sides. **Screens** may also be mounted inside the nozzles to break up foamy products so that they will flow smoothly. Each filling machinery manufacturer has a large selection of nozzles available, and no one of them is appropriate for all products. The best way to select a nozzle for filling a new product is laboratory testing with a number of different types of nozzles to find the type and size that works most efficiently and effectively. Using in improper nozzle can cause slow operation, loss of product, uneven fills, and a messy machine.

5.3 Diaphragm Volumetric Filling

㉙ The **diaphragm** type volumetric filler uses a flexible diaphragm and **pneumatic pressure** to move premeasured amounts of product from the supply tank into a controlled volume chamber and into the container (Figure 16.10).

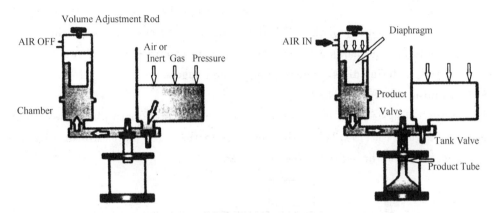

Figure 16.10 Diaphragm volumetric filler

㉚ A pneumatic pressure of up to 15 **psig** is maintained on the product in the supply tank. When the valves are open, this pressure forces the product into the volume chamber much as the piston draws the product into the chamber by its intake stroke.

㉛ When the chamber is filled, the valve from the supply tank is closed to prevent flow back into the tank. The valve directing the flow into the container is opened, and air pressure is applied to the **plunger** which presses against the top of the diaphragm to push it downward and force the product into the container.

㉜ When the stroke is completed, the air pressure on the diaphragm is released, the valves change positions, and the chamber is refilled from the supply tank.

㉝ A "no container no fill" mechanism is attached to the bottle neck guide system. The movement of a mechanical contact on the container activates valves in the pneumatic control system which directs the flow of air to operate the movement of the diaphragm.

㉞ The volume of the filler can be changed by replacing the volume chambers with different sizes or by adjusting the position of the rod stop mechanism that controls the distance the plunger can move within the volume chamber. Inserting the rod farther into the chamber shortens the distance the plunger can move and reduces the amount of product that is delivered. This system is extremely accurate, and the adjustments in the cylinder can be made within a **tenfold** range.

㉟ The diaphragm volumetric filling systems are generally used to fill small-necked bottles with relatively expensive products because of their high level of accuracy and the small loss of product in filling.

㊱ The diaphragms may be made from a variety of **elastomer** products, so a material can be selected to work well with the **viscosity** and chemical characteristics of the product being filled into the containers.

5.4 Timed Flow Volumetric Filling

㊲ The volume of product in each fill can also be regulated by controlling the amount of time the product flows at a constant rate through a standard sized tube into the container. For example, if 1cc flows through a tube in 1/10th of a second, 2cc's will flow through the same tube in 2/10ths of a second and 3cc's will flow through the tube in 3/10ths of a second. In these operations the accuracy of the fill is determined by the smoothness of the flow and the accuracy of the timing mechanisms. Three popular ways of measuring and regulating the flow time are rotating **metering discs**, rotary pumps, and **augers**.

5.5 Rotating Metering Discs

㊳ In this type of system the filling head consists of two stationary plates and a bottom rotating plate (Figure 16.11). When the openings in the three plates are **in alignment**, the product is free to pass through into the container.

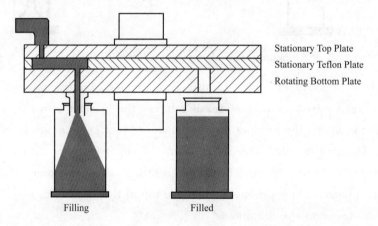

Figure 16.11 Rotating metering discs

㊴ The flow ceases when the rotating bottom plate moves out of position.

㊵ *The product flow is maintained by an external pump, and the time of fill and volume of product delivered is controlled by the size of the opening in the stationary center plate and the amount of time that is required for the opening in the rotating plate to pass under the one in the stationary plate.*

㊶ The volume of the fill can be regulated by changing the size of the opening, or slot, in the center plate or by changing the speed of the rotating plate. When a longer slot is used in the center plate, the opening in the rotating bottom plate is in contact with the product in the center slot for a longer period of time and more product is delivered to the container. In some cases, the velocity of the product coming from the pump can be regulated, but this adjustment is normally used for regulating the smoothness of the flow rather than the volume.

㊷ This filler can double fill a container by passing it under two stations, or two or more different materials can be filled into the container by supplying different products at the different filling stations.

㊸ The volumetric filler using metering discs can handle a wide range of liquid products and some semi-liquid products that may be as thick as peanut butter.

㊹ The **clearance** between the rotating metering disc and the fixed disc is critical, and it must be accurately maintained or **spillage** can occur and the volumes of the fills will not be accurate. The surfaces of the discs can be polished to remove any scratches or minor damage that may be caused by wear.

5.6 Rotary Pumps

㊺ Rotary pumps (Figure 16.12) are used for volumetric filling by regulating the operating interval of the pump to control the amount of product that is delivered.

㊻ Some of these pumps apply the force needed to move the product by squeezing it between **interlocking cam-shaped rotors** like the ones shown in Figure 16.12. Other pumps move the product with pockets in the rotating elements. The product flow is controlled by electric or air operated **clutch and brake mechanisms** that can start and stop at very precise intervals.

㊼ Electronic timing devices and counters are normally used for timing the operation and activating the clutches and brakes.

㊽ The electronic timing device is triggered by the same action that starts the pump, and the timing device pro-

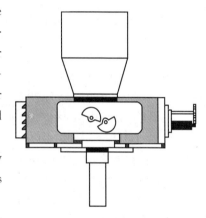

Figure 16.12 Rotary pump filler

duces a signal that operates the clutch and brake to stop the pump after a preset time interval. Irregularities in the fill may be corrected by adjusting the time interval of the pump or regulating the pump speed. These units frequently contain **plug-in type relays** that must be replaced periodically.

㊾ Some fillers measure the product flow with a **revolution counter** attached to the **drive shaft** of the pump, and the flow is regulated by adjusting the number of turns that the pump impellers make on each fill. It is possible to make adjustments as fine as 1/100 of a revolution, and volumes as small as 1cc can be delivered with accuracy. The speed of the pump rotation does not affect the volume of the product that is delivered, but variations could cause timing problems with the machine components that move the containers.

㊿ When the fillers are operating at speeds up to 150 cycles per minute the clutch and brake must operate in a period as short as 2 milliseconds. A slipping clutch or an ineffective brake can cause overfills, and a **grabbing clutch** or brake can cause underfills or irregular volumes. The **solenoid** or air valves that operate the clutch and brake mechanisms can also affect the timing sequence if they stick or fail to operate at the proper instant.

51 *Some filler heads are equipped with positive action shut-off valves to eliminate any product drip.*

5.7 Augers

⑤② Semi-liquid products that are too thick to be moved by a rotary pump may be filled by a similar filler that uses an auger in place of the pump to move the product.

⑤③ The volume of product delivered by an auger filler (Figure 16.13) is controlled by the amount of time the auger turns or the number of turns it makes on each fill, in much the same manner that is used for rotary pumps.

⑤④ Some auger fillers use an **agitator** rotating in the product **hopper** to keep the product moving smoothly and maintain an even consistency.

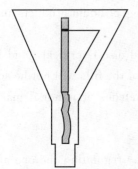

Figure 16.13 Auger filler

6 Liquid Constant Level Filling

6.1 Introduction

⑤⑤ Although there may be slight variations in the sizes or shapes of bottles and jars, it is frequently desirable to have the fill line at the same height on each container so that all of them appear to be filled exactly the same when they are on the display shelf. Some manufacturers deliberately add extra product and overfill some of their "see through" containers in order to improve their appearance and to guarantee that all of them contain at least the minimum desired amount of product.

⑤⑥ In this section you will learn the operation of five basic methods used for constant level filling of **still liquids** and semi-liquids and one method used for filling carbonated products. You will also learn ways in which each method is used and their basic machine maintenance requirements.

⑤⑦ **Molding and forming techniques** can cause minor irregularities in the sizes of shapes of rigid glass and plastic bottles and jars, and some semi-rigid containers may also change their shape and volume slightly during the filling process. *It is frequently more economical to use additional product in the oversized containers than it would be to more closely control the forming processes.* Constant level filling techniques compensate for the minor changes and variations in the containers and produce the uniform appearance that is important in the sale of many products in "see through" containers.

6.2 Pure Gravity Filling

⑤⑧ Pure gravity (Figure 16.14) is not only one of the oldest and simplest constant level filling methods, but it is also one of the most accurate ones for filling free-flowing products. Some gravity fillers can eliminate the need for overflow and recirculation of the product, and some can hold aeration of the product to a minimum.

⑤⑨ Pure gravity filling can be used for most still liquids and some semi-solids. Products with consistencies as heavy as **catsup** or **mustard** can be filled in this way, but care must be taken with the heavier products to assure that the temperature and consistency of the product is maintained at a level at which the product flows freely and evenly. Although the constant level fillers can be adjusted to automatically operate at slower speeds when the product flows more slowly, this change in rate of flow can affect the timing of the conveying system and other machines in the packaging line.

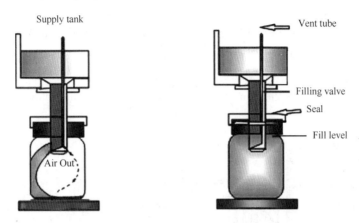

Figure 16.14 Pure gravity filling

60 The product flow is produced in the pure gravity system by locating the product supply tank high enough so that **forces of gravity** can be used to move the product through the **feed mechanisms** into the containers.

61 The flow is started and stopped by the action of a **spring loaded filling valve** that is opened by the force of making contact with the top of the container. Some machines place the container on a platform and raise it up against the valve in the filler head, and others lower the filler head down onto the top of the container.

62 When the pressure of the contact is sufficient, the spring opens the filling valve, and the product flows from the supply tank into the container. In the machine shown in the above illustration, the air in the container is forced out through the **vent tube** as the product enters the container, and when the liquid covers the port of the vent tube the air can no longer escape. The backpressure stops the flow of product into the container.

63 When the contact between the filler head and the container is broken the spring closes the filler valve and prevents dripping of the product as the head is removed from the container and positioned for the next fill.

64 Some other types of gravity fillers use timing devices to open and close the valves to control the flow of product into the containers. Instead of an air vent tube, an **overflow pipe** is located at the fill line to carry any excess product into an overflow tank from which it can be pumped back into the supply tank. This type of system does not require a tight seal between the container and the filler head.

65 The maintenance requirements of the pure gravity filling system are relatively simple. The filler components must be kept clean and well **lubricated**, and the **bearings, bushings**, and springs must be replaced when they wear excessively. The movement of the machine parts must be monitored and adjustments made to time and position them correctly for the **fill level** that is desired.

66　The consistency of the product and the free movement of all moving parts are critical for smooth operation. Underfill may be caused by **constriction** of the air vent tube or by failure of sufficient product to move into the container during the time period it is under the filler head.

6.3　Pure Vacuum Filling

67　Pure vacuum filling systems (Figure 16.15) are used primarily for filling **narrow necked** glass bottles with still liquids. It is not normally used with plastic or other non-rigid containers, because the vacuum pressure that is used to draw the product from the storage tank into the container can cause non-rigid containers to **contract** and become distorted.

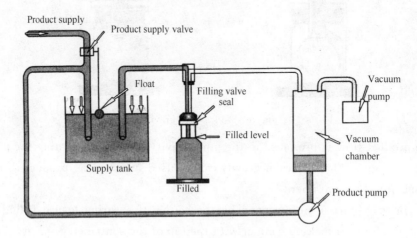

Figure 16.15　Pure vacuum filling

68　As shown in Figure 16.15, the vacuum tube within the filling **sleeve** is connected to a vacuum chamber which is maintained at a constant vacuum level by a vacuum pump.

69　When the bottle is in position, an **airtight seal** is established between the top of the bottle and the filler head. As the valves are opened a vacuum is drawn on the bottle, and this pressure differential draws the product from the supply tank into the bottle. The fill level is determined by the position of the vacuum port in the bottle, because the excess product is drawn out through the port. The length of the fill time is usually controlled by a cam operation or some other type of timing mechanism.

70　The product overflow from the container is drawn through the vacuum port into the vacuum chamber on which a vacuum is maintained by the action of the vacuum pump. The excess product is pumped back into the supply tank.

71　The pure vacuum filling system is generally faster than a pure gravity system, but the vacuum system always has some overflow and recirculation of product. The vacuum filler works only when there is a seal between the bottle and filler valve, so chipped or cracked bottles can not be filled.

72　A vacuum level of between 25 and 28 inches of **mercury** is normally maintained to provide smooth and efficient operation, but the level may vary on different machines as a result of the differences in manufacturers' designs and specifications.

73　The vacuum filler is somewhat more difficult to clean than a simple gravity system because of its complexity and small vacuum tubes that can become **plugged**. All the fittings must be kept tight for efficient operation.

74 The sealing **washers** must be replaced when they show signs of damage or wear, and the replacements should always be of the recommended type and material.

75 Care should be taken to assure that the proper type of filling nozzle is always used for the product being filled, because foaming, aeration, and splashing can be increased by rapid movement of the product.

6.4 Gravity Vacuum Filling

76 Gravity vacuum filling systems (Figure 16.16) combine the features of both the gravity and the vacuum systems. A light vacuum of from 3 to 5 inches of mercury is maintained in the system to assist in drawing the air out of the containers and to stabilize the flow of the liquid.

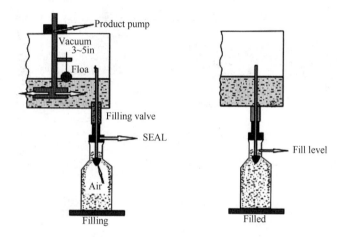

Figure 16.16 Gravity vacuum filling

77 In the filler shown in Figure 16.16 a light vacuum is maintained in the area above the product in the supply tank, and the vent tube from the container opens into this area of the supply tank. When the seal is established between the filler head and the bottle the vent is opened, and a vacuum is automatically drawn in the container. This vacuum in the container helps make the product flow into it more rapidly and smoothly.

78 As in the gravity filling systems, when the product reaches the fill level the **vent port** is closed by the rising liquid, and the flow stops.

79 The gravity vacuum system does not have the overflow or recirculation of product that is inherent in the pure vacuum system, and it produces minimum amounts of **turbulence** and aeration of the product. This type of system is particularly well suited to filling **distilled spirits** and wine because there is virtually no loss of **proof**.

80 Since the filling nozzle is spring loaded and operated only when the vacuum is present, a chipped or cracked bottle on which a seal is not formed will not be filled.

81 The vacuum gravity system must be kept clean and clear of any **obstructions** that could affect the flow through the vacuum tubes. The seals and connections are critical and should

always be kept tight and replaced when signs of wear appear.

6.5 Pure Pressure Filling

82　In pure pressure filing systems (Figure 16.17), the product is pumped from a storage tank through a filling valve and into the container. The product storage tank may be located external to the machine or as an integral part of it.

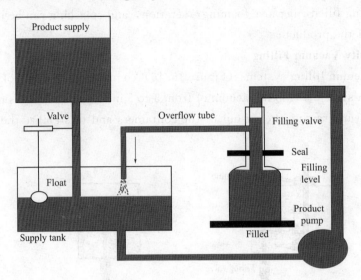

Figure 16.17　Pure pressure filling

83　A product pump draws the product from the supply tank, and applies the pressure necessary to pump it into the containers.

84　The filling valve contains an overflow tube to drain the excess product back to the supply tank and prevent overfill. The fill level of the container is determined by the location of the overflow port in this tube within the neck of the container. When the product reaches the overflow port, the fill is completed, but the flow continues through the overflow tube until the seal on the container is broken and the valve sleeve covers the ports in the filling nozzle and stops the flow.

85　Pure pressure filling is normally used for still liquids, but it can be applied to a wide variety of products with different consistencies.

86　Applying pressure to the product may tend to increase splashing or foaming, so particular care should be taken to insure that the appropriate nozzle styles and sizes are used and that the recommended pump speeds and pressure settings are used.

87　Regular cleaning and lubrication of all moving parts according to the manufacturers' recommendations is required to keep the machines in the top operating condition. Cleaning of the nozzles at the end of each run is particularly important with heavier products or materials that can build-up in the nozzles and tubes.

6.6 Level Sensing Filling

88　Level sensing (Figure 16.18) makes it possible to fill containers to a level without establishing an airtight seal between the tops of the containers and the filler valves. It is used primarily for high speed filling of small necked plastic and glass bottles. The fill level is very accurate, and there is no need for product recirculation.

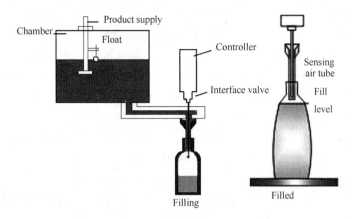

Figure 16.18 Level sensing filling

89 A pressure of up to 15 psig is applied to force the product from the supply tank through the product valve and into the bottle. Since there is no seal the displaced air escapes out of the top of the bottle around the product tube.

90 The sensing device operates by directing a stream of low pressure air at the filling station and detecting changes in the resistance to the air flow.

91 The positioning of the filler and the start of the flow can be initiated when the sensing device detects the presence of the empty bottle in the filling station. When the product has filled to the desired level, the sensing unit signals the **interface valve** to stop the flow.

92 The sensing air tube is located inside the filler tube, so it must be kept clean and free of obstruction. A small build-up of product inside the sensing air tube could cause differences in the unit's sensitivity and cause underfills.

93 The product tube is frequently designed for a particular product and bottle design, and may contain screens or other features to reduce turbulence or foaming. A new filler tube may be required when changes are made in either the product or the container.

6.7 Pressure Gravity Filling

94 **Carbonated beverages** can lose some of their carbonation during the filling process if they

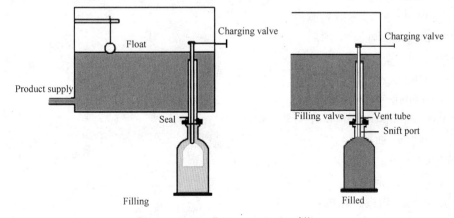

Figure 16.19 Pressure gravity filling

are handled roughly or exposed to open atmosphere. Pressure gravity filling (Figure 16. 19) is a process that was developed specifically to help hold the CO_2 in the carbonated beverages during the filling operation. This method is used almost exclusively for filling beverages such as sodas, beer, sparkling wines, and champagnes into bottles or cans.

⑤ In the pressure gravity filling operation a pressure of between 15 and 125 psig is maintained in the headspace of the product supply tank to help hold the CO_2 in solution. Otherwise the filling is basically a gravity filling process.

⑥ When the container is sealed to the filling valve a mechanical **latch** opens a **charging valve** at the top of the valve vent tube. This establishes in the container the same pressure level that is being applied to the surface of the product in the supply tank. The product is in a totally pressurized environment. This pressure does not interfere with the filling process, but it does prevent the loss of carbonation.

⑦ *After the pressure in the container is equalized with that in the supply tank, the spring loaded filler valve opens and the product flows into the container by gravity until the vent port is covered and the flow is stopped as in a normal gravity filling process.*

⑧ The filling valve also contains a mechanical **snift port** which opens the headspace of each filled container to the atmosphere in order to gently release the pressure from the filled container and prevent **gushing** when the seal is broken and the filler head is removed from the container.

(5365 words)

New Words and Expressions

maintenance ['mentənəns] *n.* 保养、维修
consistency [kən'sistənsi] *n.* 浓度、稠度
ampule ['æmpul] *n.* 安瓿瓶、一次用量的针剂
pharmaceutical [ˌfɑrmə'sutikəl] *adj.* 制药的、配药的
drum [drʌm] *n.* 鼓、圆筒
volumetric ['vɔljə'metrik] *adj.* 容量的、容积的
see-through ['siˌθru] *adj.* 透明的
flexibility [ˌfleksə'biləti] *n.* 适应性
overfill ['əuvə'fil] *n.* 过灌装、溢出
diaphragm ['daiəˌfræm] *n.* 隔膜
valve [vælv] *n.* 阀
cylinder ['siləndə] *n.* 圆筒、气缸
chamber ['tʃembə] *n.* 室、腔
downstroke ['daunˌstrəuk] *n.* 下行程
crank [kræŋk] *n.* 曲柄
tolerance ['tɑlərəns] *n.* 公差
leaking [li:kiŋ] *n.* 泄漏
groove [gruv] *n.* 沟槽、轧槽
circumference [sə'kʌmfərəns] *n.* 环状面、圆周

teflon ['teflaːn] *n.* 聚四氟乙烯
splash [splæʃ] *v.* 飞溅、喷洒
drip [drip] *v.* 滴下、漏下
aeration [erˈreiʃn] *n.* 空气混入、掺气
spray [spre] *n.* 喷雾、喷射
screen [skrin] *n.* 网
psig [ˈsig] *abbr.* pound per square inch 磅/平方英寸
plunger [ˈplʌndʒə] *n.* 柱塞
tenfold [ˈtenfəuld] *adj.* 十倍的
elastomer [iˈlæstəmə] *n.* 弹性体
viscosity [viˈskasiti] *n.* 黏性、黏度
clearance [ˈklirəns] *n.* 空隙、间隙
spillage [ˈspilidʒ] *n.* 溢出、溢出量
interlocking [intə(ː)ˈlɔkiŋ] *adj.* 联锁的、互锁的
solenoid [ˈsələˌnɔid] *n.* 电磁线圈、电磁铁
auger [ˈɔgə] *n.* 螺杆
agitator [ˈædʒiˌtetə] *n.* 搅拌器、混合器搅拌装置
hopper [ˈhapə] *n.* 加料斗
catsup [ˈkætsəp] *n.* 番茄酱
mustard [ˈmʌstəd] *n.* 芥末酱
lubricate [ˈlubriˌket] *v.* 润滑
bearing [ˈbeəriŋ] *n.* 轴承
bushing [ˈbuʃiŋ] *n.* 轴衬
constriction [kənˈstrikʃən] *n.* 压缩、狭窄
contract [ˈkanˌtrækt] *v.* 缩小、紧缩
sleeve [sliv] *n.* 套筒、套管
mercury [ˈmɜːkjəri] *n.* 汞、水银
plug [plʌg] *v.* 阻塞
washer [ˈwɔʃə] *n.* 垫圈、垫片
turbulence [ˈtəːbjələns] *n.* 湍流、涡流
proof [pruf] *n.* （酒的）标准酒精度
obstruction [əbˈstrʌkʃən] *n.* 障碍物、阻碍物
latch [lætʃ] *n.* 拨叉、碰锁
gush [gʌʃ] *v.* 喷涌
caulking compound　填隙料
constant level filling　定液位灌装
fill-to-a-level　等液位（法）
rotary filler　旋转式灌装机
straight-line filler　直列式灌装机
intermittent motion　间歇运动

intake stroke　进给冲程、吸入冲程
supply tank　供料缸
measuring chamber　计量腔（室）
filling chamber　灌装腔（室）
reciprocating valve　往复（运动）阀
rotary valve　转阀
cc (cubic centimeter)　立方厘米、毫升
double filling　双级灌装法
back and forth　来回的、往复
surface tension　表面张力
pneumatic pressure　气压
metering disc　计量盘
in alignment　成一直线、对准
cam-shaped rotor　凸轮型转子
clutch and brake mechanism　离合器和制动机构
plug-in type relay　插入式继电器
revolution counter　转速计
drive shaft　驱动轴
grabbing clutch　握式离合器
still liquids　不含气液体、静态液体
molding and forming technique　模塑成型技术
forces of gravity　重力
feed mechanism　供料机构
spring loaded filling valve　弹簧作用的灌装阀
vent tube　排气管
overflow pipe　溢流管
fill level　灌装液位
narrow necked　细颈的
airtight seal　空气密封、气密
vent port　排气孔
distilled spirit　蒸馏酒精
level sensing filling　液面感应式灌装
interface valve　接口阀
pressure gravity filling　压力重力式灌装（即等压灌装）
carbonated beverage　碳酸饮料
charging valve　充气阀
snift port　卸压口

Notes

1. *Volumetric filling is particularly appropriate for applications in the pharmaceutical industry where it is important that each container is accurately filled with a specific*

volume of product. (Para. 7) 容积式灌装特别适用于制药行业，因为每个容器准确地灌装特定量的产品很重要。

2. *Accurately delivering a premeasured volume of product to each container also produces cost savings by reducing the amount of product that is used in overfill in some operations to assure that each container receives at least the minimum desired amount of product.* (Para. 11) 精确地把预定量容积的物料装入容器中，降低了生产成本，这是由于为确保每个容器得到所需的最小容量的产品，减少了灌装操作时产生的溢流量。

3. *Volumetric filling systems are relatively easy to clean by flushing techniques since they normally do not have small hoses, tight passageways, or delicate sensing mechanisms that can be clogged or damaged by water or air pressure.* (Para. 14) 容积灌装系统易于使用冲洗技术进行清理，因为它们通常没有小软管，狭窄的通道或精致的传感装置，这些能被水或空气压力堵塞或损坏。*flushing techniques* 为"冲洗技术"。

4. *This can make it necessary to either run the process slower so that all the drips are caught in the container or to require frequent cleaning of the containers and the machine.* (Para. 27) 这就使得，要么避免漏液滴在外面导致的降低工作速度，让液滴落在容器内，要么频繁清洗灌装容器和机器。

5. *The product flow is maintained by an external pump, and the time of fill and volume of product delivered is controlled by the size of the opening in the stationary center plate and the amount of time that is required for the opening in the rotating plate to pass under the one in the stationary plate.* (Para. 40) 泵持续供给，灌装的时间和输送液体的容积由在固定中心板上的开口大小和回转底板上孔口在静止板下停留的时间长短来控制。

6. *Some filler heads are equipped with positive action shut-off valves to eliminate any product drip.* (Para. 51) 某些灌装机在灌装头上装有强制作用的截止阀，以消除物料的滴漏。*positive action* 指"强制作用"，*shut-off valve* 指"截止阀"。

7. *It is frequently more economical to use additional product in the oversized containers than it would be to more closely control the forming processes.* (Para. 57) 通常来讲，对尺寸稍大的瓶子灌装多余的物料比在成型工艺中进行精确地控制要经济得多。

8. *After the pressure in the container is equalized with that in the supply tank, the spring loaded filler valve opens and the product flows into the container by gravity until the vent port is covered and the flow is stopped as in a normal gravity filling process.* (Para. 97) 当容器中的压力跟料缸中的压力相等时，弹簧作用使灌装阀打开，物料依靠重力流入容器，直到堵住排气孔为止。流动像普通重力灌装过程一样停止。

Overview Questions

1. What is the principle of piston volumetric filling?
2. How to control the liquid level in the pure gravity filling?
3. Which factors should be considered to t select the diameter of the nozzle?
4. What are the characteristics of the diaphragm type volumetric filling?
5. Please describe the difference pressure gravity filling between gravity vacuum filling.

Lesson 17   Dry Product Filling

1   Introduction

① Some dry products are light and others are heavy; some are sticky and others are very dry; some flow freely, and some have to be moved by force. The containers may be as small as **a capsule** or as large as a one hundred pound bag or a 55 gallon drum. They can be boxes, cans, bottles, pouches, bags, drums, or other forms, and they may be made of a variety of rigid and flexible materials.

② Although the products and containers may have very different characteristics, the basic operating principles and filling techniques are very similar, and it is possible to adapt most filling machines to operate with a variety of products and containers.

③ In this section you will become familiar with the basic types of machines and techniques used for filling dry products, and in the following sections you will learn to identify the different types of machines and techniques, their appropriate applications, and their operation and maintenance requirements.

2   Type Of Dry Products

④ Some dry products, like rice and beans, are **free flowing** and easy to move. **Fine powders**, such as confectioners sugar, may be hard to contain in the bins or containers and can cause a dusty or explosive atmosphere.

⑤ **Nails, bolts,** and other hardware items may require special handling because of their weight, bulk, or shape.

⑥ Cornflakes and light bulbs must be protected from breakage during packaging, and ice cream bars must not be allowed to melt. Special care is needed to keep food packaging sanitary and drugs **sterile**.

⑦ Special machine designs are required to move the **nonfree-flowing** products. Sticky products such as cake mixes may tend to bridge across the openings of the delivery **shoots** or the containers, so the machine must include mechanisms to keep them flowing smoothly. Products like sugar and cement may tend to become **lumpy** as the result of changes in the atmosphere in which they are stored or variations in the manufacturing processes, so agitators may be used in the **bins** to keep them separated into small particles that can be moved and measured easily.

3   Type Of Dry Filling Operations

⑧ There are four basic types of dry filling machines based upon the way the amount of product being delivered is measured. The amount of dry product in each package may be measured by volume, **net weight**, **gross weight**, or **count**.

⑨ The volumetric fillers deliver a constant volume of product to each container, the net weight filler weighs the product before it is delivered into the package, the gross weight filler weighs the product in the container, and the counter places the same number of items into each container.

4 Product Delivery

⑩ Free flowing dry products are frequently delivered to the filling machine through gravity feed system in which the storage bins are situated higher than the hopper of the machine. Other products are delivered by **vibrator** systems in which vibrating pans cause the product to "walk" into the filler. A conveyor may be used to move product along a level belt or to raise it to a higher level in **buckets**. Augers are used to move the nonfree-flowing products that are sticky or have other characteristics that make them difficult to handle.

5 Dry Volumetric Filling

5.1 Introduction

⑪ Volumetric fillers load an exact volume of product into each container without consideration of the product's density or weight.

⑫ The volumetric fillers used for dry products use some of the simplest and most dependable filler designs, and they can be operated at relatively high speeds with low costs. However, delivering a constant volume does not always produce a constant weight. Variations in the density of the product can cause differences in the package weight, and variations in the size of the containers can cause differences in the height of the fill and the amount of head room in the container.

⑬ Four popular types of volumetric fillers used with dry products are the **cup or flask fillers**, **flooding or constant stream fillers**, auger fillers, and vacuum fillers.

5.2 Cup or Flask Fillers

⑭ This type of filler (Figure 17.1) gets its name from the cup or flask that is used to measure the amount of product that is delivered. The sides of the cups **telescope** so that the cup size can be adjusted to the exact volume that is needed.

⑮ The product is delivered from the filler hopper to the measuring cup by a gravity feed system. When the cup is full a brush or **scrape** moves across the top to level the fill and move the excess product into a bin from which it is returned to the filler hopper.

⑯ When the cup is filled and the container is in place, a **shutter** opens, and the product is dropped into the container.

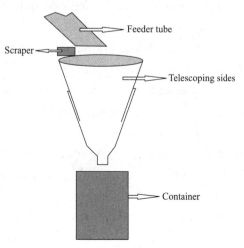

Figure 17.1 Cup or flask fillers

17　This system works particularly well with free-flowing granular products such as detergents and rice.

18　Volumetric filling is normally used for products in which the density or unit weight is relatively constant or when the volume of the product that is delivered is more important than the weight. In some industries, it is economically feasible to use volumetric filling and adjust the delivered volume so that at least the minimum desired weight is always delivered. This is usually used when the size variations in density of the product can only cause overfill. This is usually used when the size variations are relatively small or the gained speed and economy more than **offset** the cost of the extra product that is being used.

19　**Feedback** *or net weight checkweighing systems are used on some cup or flask type volumetric fillers to weigh sample loads and automatically adjust the volume of the cups to correct for variations in the delivered weight that may occur from changes in the product.*

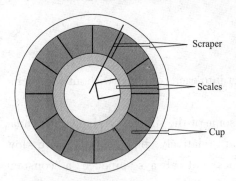

Figure 17.2　Rotary filler with feedback scale (top view)

20　For example, Figure 17.2 shows a top view of a twelve head rotary filler with a scale on one cup. This system makes corrections in the delivered volume based on the variations in the weight of the product, but it does not make direct checks on the adjustments of the seven cups that are not weighed. The accuracy of this type of system is very similar to that of a net weight system.

21　The cups of flasks normally telescope to make adjustment in the volume of product they hold and deliver, and their range of adjustment is normally within the range of a two to one ratio. The sides of the cup can be telescoped enough to double the minimum volume. Larger volume adjustments can be made by replacing the complete cup assemblies.

22　Machine adjustments may occasionally be needed to correct conditions of underfill, overfill, excessive recirculation, or spillage at the container.

23　Underfill can be caused by an incorrect cup size adjustment or by a slow or irregular feed from the hopper. If the product level in the hopper is too low, the delivery **spouts** are restricted, or the material does not flow freely, insufficient product will be delivered to fill the cups. **Misalignment** of the delivery spout and the measuring cups could also cause underfill and excessive recirculation of the product.

24　Excessive recirculation of product can also be caused by too rapid a flow of product from the hopper for the speed at which the machine is running.

25　Overfill is usually caused by incorrect adjustment of the telescoping mechanisms of the measuring cups.

26　Spillage of the product at the container is usually caused by problems of registry such as misalignment of the container and the delivery spout or incorrect timing. However, it could also be caused by an undersized container or sticky product that clings to the tube and causes a delay in delivery.

5.3 Flooding or Constant Stream Fillers

27 Flooding or constant stream fillers (Figure 17.3) work on the principle that containers passing under a constant stream of product in the same amount of time will all receive the same amount of product. *A steady stream of product flows from the hopper through a filling tube into the containers, and a system of **funnels** is used to direct the flow into the containers and prevent the loss of product as the filler moves from one container to the next.* Without the funnels, the product would spill between the containers unless they were all tightly compressed against each other.

Figure 17.3 Flooding or constant stream filler

28 Any variations in the speed of the container as it travels under the filler affects the volume of the fill. When the container travels too slowly, there is an overfill; if it travels too fast, there will be an underfill.

29 Changes in the rate of product flow from the hopper can also affect the volume of the fill. This flow can be changed by factors such as **restrictions** in the filler tube or nozzle, level of product in the hopper, and density changes in the product.

30 If the product becomes more sticky than normal it may tend to cling to the sides of the hopper and filler tubes and to flow more slowly. If it is too dry, it will flow faster. Lumps in the product can cause irregular flow or splashing.

31 Vibrator or auger feed mechanisms are frequently used to maintain a constant level of product in the hoppers and a steady stream of product into the containers. Agitators may be used to keep the product moving in the hoppers to prevent lumping or caking.

5.4 Auger Fillers

32 Auger fillers (Figure 17.4) use a rotating auger in a funnel shaped hopper to deliver a specific amount of product at a constant rate.

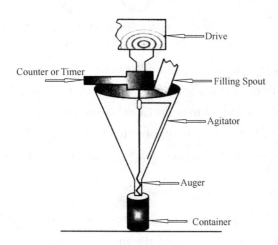

Figure 17.4 Auger filler

33 The auger is usually mounted vertically inside the funnel so that the delivery spout is directly over the container. When the container is in place, a sensing unit produces a signal that engages the electric clutch that turns the auger. When the full load is delivered, the clutch is **disengaged** and a brake is applied to stop the auger and prevent the flow of the product.

34 The volume of the fill is determined by the extent of the rotation made by the auger. Some auger fillers have timing mechanisms that allow the auger to turn for a pre-

set time on each fill, and other designs count the number of rotations that the auger makes on each fill. The clutch and brake mechanisms can be sensitive enough to control the movement of the auger to within 1/100 of a rotation or to a very small portion of a second.

35   Auger fillers can be used for a wide variety of products, but they are particularly appropriate for products that tend to bridge over the openings, such as **finely ground coffee**, cake mixes, and flour. The auger provides a steady flow and a constant volume.

36   Some auger fillers have agitators or **stirring blades** that keep the product moving in the hopper and prevent lumps or cakes from forming. The shape of the agitator is changed to meet the requirements of different types of products.

37   Different shapes of augers are used for products that flow differently. For example, the auger shown in Figure 17.5 is used to move nonfree-flowing powders, and the auger shown in Figure 17.6 is designed to provide an even flow of free-flowing powders while preventing the natural flow when the auger is not turning.

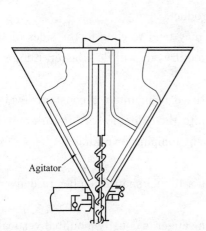

Figure 17.5   Auger for nonfree-flowing powder

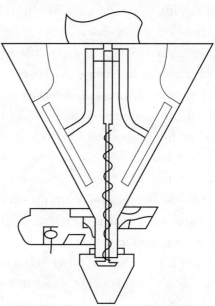

Figure 17.6   Auger for free-flowing powder

38   The electric or pneumatic sensing and control units on auger fillers are the most sensitive parts of the machines. They require relatively little maintenance, but they should be cleaned and adjusted regularly according to the machinery manufacturer's instructions.

39   Conditions of underfill or overfill on auger fillers are usually corrected by making adjustments, and other **malfunctions** are frequently corrected by identifying and replacing the faulty sensing or control components.

5.5   Vacuum Fillers

40   Vacuum fillers for dry products (Figure 17.7) are similar to the fill-to-a-level operations used with liquid products. They use the container as a measure of the fill. A vacuum is used to draw off excess product and to **tamp** the product for a tight fill or preventing product dust

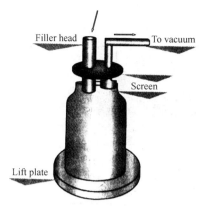

Figure 17.7 Vacuum fillers for dry products

from escaping into the air.

㊶ The **gasket** on the filler head is firmly pressed against the open top of the container by raising the container with a lift plate or lowering the filler head. When a seal has been established between the container and the filler head, a vacuum is drawn on the container and against the exposed surface of the product in the filler tube. The air in the incoming powder expands as it enters the vacuum and this action distributes the product evenly over the surface of the fill. In some operations the vacuum is released and reapplied two or more times during the fill to allow the product to settle and be tamped into the container more tightly.

㊷ The operation of the filler is usually controlled by a **cam** action or other timing device, and the excess product is drawn into the vacuum chamber and returned to the hopper.

㊸ Thin walled cans and other fragile containers may be damaged by the vacuum pressure, but these containers can be filled by the vacuum technique by placing the containers in a **shroud** in which the pressure inside and outside the containers is equalized during filling.

㊹ The optimum vacuum pressure will be different for different types of products. A vacuum that is too high can cause dusting in some products, and one that is too low may fail to provide the tamping action that is needed. The vacuum pressure should be monitored and maintained at the level established for the product by the engineering department or machinery manufacturer.

6 Dry Filling By Weight

6.1 Introduction

㊺ *For some products the weight of the product in each package is more important to the user than the volume, and the cost of the product may make it uneconomical to deliberately overfill some of the containers in order to guarantee that they all contain a minimum weight.*

㊻ There are two basic techniques for filling by weight, and in this section you are going to learn the basic principles of operation and the operation and maintenance requirements for net weight and gross weight filling systems.

6.2 Net Weight Filling

㊼ Net weight is the most accurate of the filling techniques. Each product load is weighed separately in the filling machine before it is loaded into the container. This technique can easily be used for filing bags and other flexible containers that must be supported during the filling operation, since the holding apparatus does not interfere with the weighing.

㊽ In the typical net weight filler (Figure 17.8), the product is fed into a filling chamber that is attached to a scale. The scale weighs the product as it is delivered and **triggers** a signaling apparatus to stop the fill when the desired weight is reached.

㊾ The weighed product is dropped from the filling chamber into the container.

㊿ The accuracy of the net weight fill can be affected by the amount of product in the stream when the shutter closes to stop the fill. If the product is flowing in a large stream, much more of it enters the container after the scale signals the shutter to close than enters when the product is being slowly **dribbled** into the container. The scale can be adjusted to compensate for flow during this reaction period, but any variations in the flow of the stream can have an effect upon the accuracy.

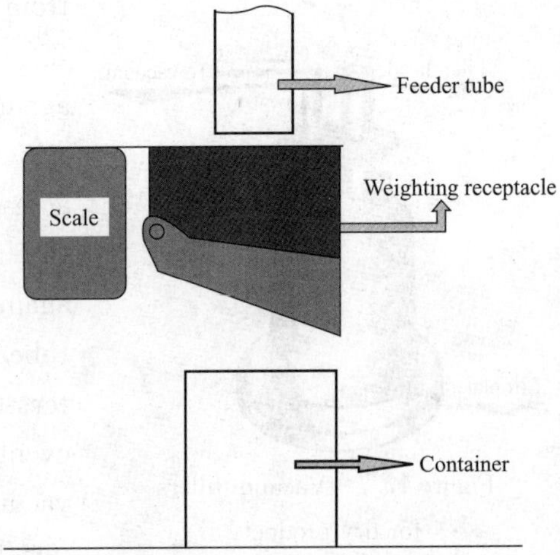

Figure 17.8 Net weight filler

(51) Some net weight fillers use a two-stage operation (Figure 17.9) to increase their accuracy and speed. In the first stage the bulk of the fill is dropped rapidly into the chamber by a large filling spout, then the chamber is moved under a smaller spout that dribbles in amount of product that is needed to complete the fill.

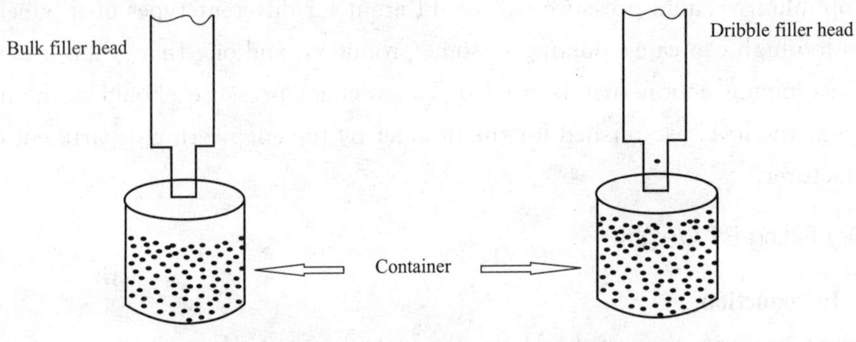

Figure 17.9 Bulk-dribble filler

(52) This type of system allows the container to be filled more rapidly while keeping the amount of product in the stream at a minimum when the shutter is closing.

(53) Maintenance of net weight fillers generally involves thorough and frequent cleaning, lubrication, and adjustment. Malfunctioning sensing units are usually plug-in types that can be easily replaced.

(54) Consistent overfills and underfills can usually be corrected by adjusting the sensing or control units.

(55) *Irregular fills may be caused by malfunctioning sensing units, but they are more likely to be the result of an obstruction in the flow path or something interfering with the operation of the shutter or causing an irregular flow of product from the hopper.*

6.3 Gross Weight Filling

(56) In gross weight filling (Figure 17.10), the product is weighed inside the container as it

is filled, and the scale operates a signaling device that operates the filler.

57 *Allowances for the weight of the container are made in the computation of the net weight of the product, but any variations in the weight of the container will cause irregularities in the weight of the product it contains.* However, this type of system can exceed the accuracy requirements for many products while using the average run of containers, and the accuracy can be improved by applying more rigid standards to the selection of the containers that are used.

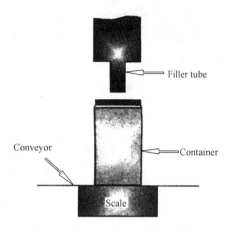

Figure 17.10　Gross weight filler

58　Filling by gross weight is faster than by net weight, because it requires one less step. It is also preferred for handling more fragile products such as cornflakes and potato chips that can be broken or crushed by handling, since they can be moved gently into the package without dropping them from one container to another.

59　Some products, such as brown sugar, may need to be settled into the package during the filling process in order to obtain a tight pack.

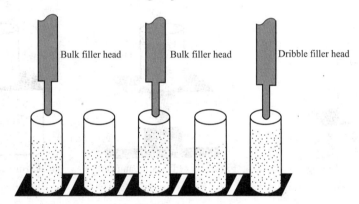

Figure 17.11　Vibrating fill

60　Product settling can be accomplished during gross weight filling by partially filling each package at two or more filling stations and vibrating the packages between each fill (Figure 17.11). The speed of the operation can be increased by bulk filling at all but the last filler and dribbling the amount that is needed to complete the fill with the package on the scale.

61　Most gross weight filler heads can not fill more than 15 or 20 packages a minute, so multiple head fillers are used on the larger machines to produce high speed operations.

62　Machine maintenance normally consists of cleaning, lubricating, and adjustment along with monitoring the machine operation for signs of worn or broken parts.

6.4　Types of Scales

63　The four types of scales commonly found on filling machines are the **balance beam scale, spring balance scale, air balance scale, and liquid displacement scale.**

64 The balance beam (Figure 17.12) uses a standard weight on one end of a balance beam and the product to be weighed on the other. When the product weight balances the standard weight a signal is produced to stop the filling operation.

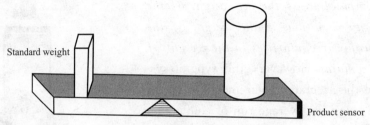

Figure 17.12 Balance beam scale

65 The spring balance scale (Figure 17.13) uses a spring under the weighing platform, and when the spring is compressed to a preset level the stop-filling signal is produced.

66 The air balance scale (Figure 17.14) operates when the weight of the product **depresses** a plate against a stream of air. When the back pressure reaches a preset level a signal is produced to stop the fill.

67 Liquid displacement scale (Figure 17.15) uses a plate that displaces liquid in a chamber as a weight is applied, and a signal is produced when the liquid reaches a given level.

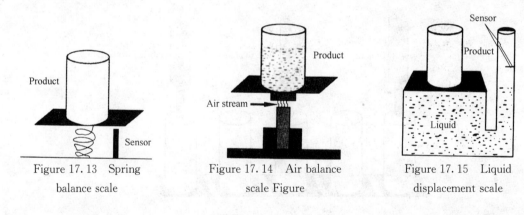

Figure 17.13 Spring balance scale 　　Figure 17.14 Air balance scale Figure 　　Figure 17.15 Liquid displacement scale

7 Filling By Count

7.1 Introduction

68 **Nuts**, bolts, capsules, cookies, and a wide variety of other products are marketed by number rather than volume or weight. There may be 10 screws in a pouch, a dozen eggs in a carton, and a hundred tablets in a bottle.

69 Four popular techniques are frequently used for filling products by count, and in this selection you will learn the basic uses, principles of operation, and operation and maintenance requirements of board or disc counters, **slat** counters, column counters, and electronic counters.

7.2 Board or Disc Counters

70 The board and disc counters (Figure 17.16) are mechanized variations of the **hand**

paddle technique. *The product is dropped or brushed into a group of holes in a disc or board and moved to a position from which it is dropped into the container, the number of items in each fill is determined by the number of holes in the area of the board or disc that is used.* This technique is widely used in the pharmaceutical industry for filling containers with tablets.

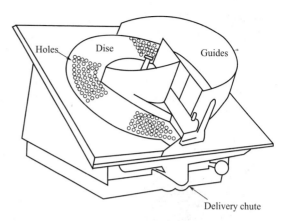

Figure 17.16 Disc counter

71 In the disc filler shown in the illustration, the product is held in a hopper like area formed by the **guides** near the bottom of the inclined disc. As the disc turns upward in a counter-clockwise direction, one product item falls into each hole in the disc and the excess falls or is brushed back into the holding area. As the disc rotates, the counted product slides over the stationary plate under the disc until it reaches the opening through which the product falls into a chute and into the container.

72 During the descending portion of the movement, the disc is usually visible so that an inspector can easily check to make sure that all the holes are filled and the product is not broken or damaged.

73 The holes in the disc can be arranged to feed one or several containers at a time, and the movement of the containers can be timed to allow double fill in which the product in more than one set of holes is used to fill larger containers. For example two fills of 50 items each can be used to fill a bottle with 100 tablets.

74 Change-over to another size of product or a different number of items in each fill is usually accomplished by removing the disc and replacing it with one that has a different hole size or **configuration.**

75 Variations of this technique include machines that use horizontally moving boards or trays instead of rotating discs. The product is flooded, vibrated, or wiped into the holes, and delivery is made by moving the tray so that the product falls out of the tray, through a chute, and into the container.

76 Other variations have a shutter under the product area, and the shutter opens to deliver the product.

7.3 Slat Counters

77 The **slat counter** (Figure 17.17) is a chain driven conveyor made up of slats which have pockets or **cavities** to hold product items.

78 The slats in Figure 17.17 each have ten cavities or pockets that are aligned so that rows of product can be directed into specific containers.

79 As the slats pass under the feeder, one product item is dropped or vibrated into each cavity, and any excess is brushed back to fill the following slats. While the product is being

carried downward over the face of the filler, it is clearly visible for inspection.

(80) Some slat counters operate with an intermittent motion. Each time the machine moves, the number of slats needed to fill the container are advanced past the chute, then the slat drive stops, the filled container is moved away, and an empty container is placed for the next fill. The operation is then repeated.

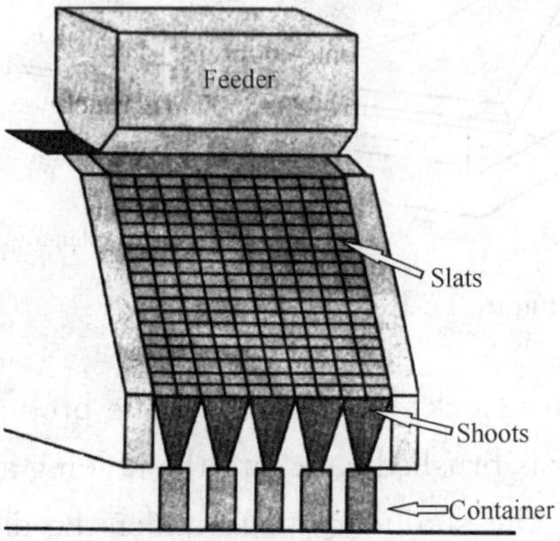

Figure 17. 17 Slat counter

(81) In the filler shown in Figure 17. 17, the product carried in 2 rows of cavities is delivered to each container, so 5 containers can be filled at a time with the 10 cavity slats that are shown. The number of product items delivered to each container is controlled by the number of rows and the number of slats that are used for each fill. For example, when a 100 count is needed, 2 rows can be fed into each container from each of 50 slats, or 4 rows from 25 slats. Partially blank slats can be inserted when the desired count is not a multiple of the number of rows being used.

(82) A mechanical or electrical counter may be mounted behind the slats to count the slats that pass and produce signals that stop the movement when the preset number is reached. Some machines count banks of slats by use of a finger mounted on the back of the last slat of each group to activate a triggering device when the total group has passed.

(83) Vibrators may be used to slightly vibrate the slats and help make the product fall into the cavities. The locations of these vibrators can be adjusted to compensate for different sizes, weight, and shapes of the products.

(84) **Thumpers** can be mounted behind the slats to assure that the product drops out at the proper point. The thumper strikes each slat giving it a slightly forward motion as it passes.

(85) Continuous motion is obtained on some slat counter type filling machines by using larger delivery chutes or bins equipped with shutters or gates. The load can be counted and dropped into the closed bin while the containers are being moved into place, and the load can be dropped into the container all at once. Blank slats with no cavities may be inserted between the banks of slats to provide extra time for positioning the containers when it is needed.

7.4 Column Counters

(86) Products that have a constant thickness and flat surfaces, such as cookies, tablets, or washers, can be mounted by measuring the height of stacks of the product. A six inch stack, or column, of 1/4 inch cookies will contain 24 cookies, and a 12 inch stack will contain 48. This count will be accurate as long as the thickness of the cookies remains constant.

(87) Single track machines with only one column may be used, or machines with two or more tracks may be used to gain extra speed.

(88) When the product reaches the desired height, a trip release, sensing device, or signal from the operator releases the product and allows it to drop into the container.

⑧⑨ Change-over is relatively simple and does not require many change parts or much time, but the versatility of the system is limited to products that have a constant size.

⑨⓪ Horizontal models of this type of counter filler may collect the product in tracks in a tray and push them into the containers rather than dropping them.

7.5 Electronic Counters

⑨① Electronic counters use electric-eyes or other sensing devices to detect the product item passing a given point. They produce signals to operate the counters and related machinery components when a pre-determined number of items has been counted.

⑨② The electronic counters can count objects of a wide variety of sizes and shapes with only minor change-over procedures. It may be necessary to adjust the height of the beam, change the product handling features on the conveyor, or the size or shape of the gates, funnels, and feed tubes that lead into the container, but these changes are relatively minor.

7.6 Maintenance

⑨③ The primary maintenance requirements for counting type fillers are cleaning, lubricating, and minor adjustments. Cleaning can be of prime importance particularly with products that can produce dust, because the dust can **clog** the holes or cavities that carry the product and cause a low count.

⑨④ Timing of the product delivery mechanisms and container flow can affect the accuracy of the fill, so the proper adjustments must be maintained. For example, a loose tray on a vibrator could cause an uneven flow of product into the counter and cause underfills.

⑨⑤ Repairs will generally consist of making adjustments and replacing worn or defective parts.

(4565 words)

New Words and Expressions

capsule ['kæpsjuːl] *n.* 胶囊
nail [neɪl] *n.* 钉子、钉状物
bolt [bɒlt] *n.* 螺栓、螺钉
sterile ['sterəl] *adj.* 无菌的
shoot [ʃuːt] *n.* 滑道、滑槽
lumpy ['lʌmpi] *adj.* 结块的
bin [bɪn] *n.* 储料箱
count [kaʊnt] *n.* 计数
vibrator ['vaɪˌbreɪtə] *n.* 振动器
bucket ['bʌkɪt] *n.* 水桶、料斗
telescope ['telɪˌskəʊp] *v.* 伸缩
scrape [skreɪp] *n.* 刮、擦
shutter ['ʃʌtə] *n.* 底门
offset ['ɒfˌset] *v.* 抵消、补偿
feedback ['fiːdˌbæk] *n.* 反馈
spout [spaʊt] *n.* 管口、(容器的) 嘴

misalignment [ˌmisəˈlainmənt] *n.* 未对准
funnel [ˈfʌnəl] *n.* 漏斗
restriction [riˈstrikʃən] *n.* 约束、限制
disengage [ˌdisinˈgeidʒ] *v.* 释放、脱离
malfunction [mælˈfʌŋkʃən] *n.* 故障、失效
tamp [tæmp] *v.* 夯实
gasket [ˈgæskit] *n.* 垫圈、垫片
cam [kæm] *n.* 凸轮
shroud [ʃraud] *n.* 护罩、保护罩
trigger [ˈtrigə] *v.* 引发、触发
receptacle [riˈseptəkəl] *n.* 容器、放置物品的地方
dribble [ˈdribəl] *n.* 涓滴、细滴
depress [diˈpres] *v.* 压下、推下
nut [nʌt] *n.* 螺母、螺帽
slat [slæt] *n.* 板条、狭板
guide [gaidz] *n.* 导板、护罩
configuration [kənˌfigjəˈreʃən] *n.* 结构、构造
cavity [ˈkæviti] *n.* 腔、洞
thumper [ˈθʌmpə] *n.* 重击物、重锤
clog [klɔg] *v.* 阻碍、堵塞
free flowing 自由流动
fine powder 细粉末
nonfree-flowing 非自由流动
net weight 净重
gross weight 毛重
cup or flask filler 量杯式充填机
flooding or constant stream filler 溢流或等流量式充填机
finely ground coffee 细磨咖啡
stirring blades 搅拌叶片
bulk-dribble filler 粗加料-细加料双级充填机
balance beam scale 杠杆秤
spring balance scale 弹簧秤
air balance scale 空气秤
liquid displacement scale 排量秤
hand paddle 手划桨技术
slat counter 板条式计数装置
column counter 高度式计数装置

Notes

1. **Feedback** *or net weight checkweighing systems are used on some cup or flask type volumetric fillers to weigh sample loads and automatically adjust the volume of the cups to*

correct for variations in the delivered weight that may occur from changes in the product. (Para. 19) 有的量杯式容积充填机带有反馈系统或净重检验系统对充填量进行抽样检查,自动调节量杯的容量,以纠正因物料变化而引起的重量误差。

2. *A steady stream of product flows from the hopper through a filling tube into the containers, and a system of **funnels** is used to direct the flow into the containers and prevent the loss of product as the filler moves from one container to the next.* (Para. 27) 当充填机从一个容器移到下一个时,连续稳定的物流从料斗通过充填管进入容器,漏斗直接把物料引入容器,防止物料漏损。

3. *For some products the weight of the product in each package is more important to the user than the volume, and the cost of the product may make it uneconomical to deliberately overfill some of the containers in order to guarantee that they all contain a minimum weight.* (Para. 45) 对于某些物料来说,用户注重的是每个包装中产品的重量而不是体积,而且,如果为了保证每个容器都有最小的重量而有意使某些容器过度充填将会很不经济。

4. *Irregular fills may be caused by malfunctioning sensing units, but they are more likely to be the result of an obstruction in the flow path or something interfering with the operation of the shutter or causing an irregular flow of product from the hopper.* (Para. 55) 若是充填量忽多忽少,则可能是传感元件失灵引起的,但更可能是由于流动路径或底门操作受阻导致料斗中物料流动不畅而造成的。

5. *Allowances for the weight of the container are made in the computation of the net weight of the product, but any variations in the weight of the container will cause irregularities in the weight of the product it contains.* (Para. 57) 在计算物料净重时,扣除了容器的重量,但是容器重量的变化会影响容器内的物重。make allowance for 指"扣除"。

6. *The product is dropped or brushed into a group of holes in a disc or board and moved to a position from which it is dropped into the container, the number of items in each fill is determined by the number of holes in the area of the board or disc that is used.* (Para. 70) 将物品落入或扫入带有计数孔的盘中,然后再转到充填位置,让物品落到容器中,每次充填物品的数量由盘在充填区域中的孔数决定。

Overview Questions

1. What are characteristics of the volumetric fillers used for dry products?
2. What can cause underfill?
3. Discuss the difference between gross weight filler and net weight filler.
4. Please describe the principle of the board or disc counters.
5. What are the primary maintenance requirements for counting type fillers?

Lesson 18　Applied Packaging (I)

1　Carded Display Packages

① *Carded display packages offer maximum product visibility and self-service convenience, while at the same time providing reasonable product protection against contamination and shipping damage.* The **backing card** discourages theft of smaller articles, provides convenient space for product identification or instructions, and is useful as a means of retail display. Although staples, ties, or other physical means can be used to attach the product to a backing card, blister and **skin packaging** add additional protection by covering the product and are more adaptable to automated production.

② A blister package uses a preformed plastic shape that holds the product and is attached to a backing card by heat sealing. Skin packaging places the product on a backing card and uses a vacuum to draw a plastic film into close conformity with the object. A heat-sealable coating bonds the film or blister to the backing card (Figure 18.1). The inks used to decorate the cards must withstand the hot temperatures involved with either process.

③ The majority of carded packages are displayed on **pegboards**. A significant problem with suspended display packs is inadequate strength in the pegboard, or "butterfly", hole area. The backing card should be able to hold several times the product weight and be capable of being readily removed and replaced on a pegboard with no special care. Cards narrower than 50 millimetres (2in) might better be displayed by other means. Where permitted by the manufacturing process, card corners should have a radius to reduce curling or ply separation.

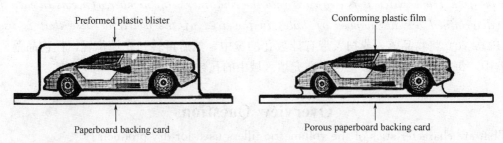

Figure 18.1　A paperboard-backed blister pack (left) and a skin pack (right)

④ Most retailers have standard dimensions for spacing the **hangers** for carded display packages. The length and width of the backing card should be selected to provide maximum use of available display space and should require minimum hanger relocation.

⑤ Package depth is controlled by product geometry and placement. Heavy blister and skin packages with a **center of gravity** significantly in front of the backing card will hang on an angle facing downward from the viewer's eyes. Such designs tend to twist the pegboard hole against the hanging peg, often tearing the backing card.

2 Blister Packages

⑥ Blister packaging is composed of a rigid, preformed thermoformed plastic shape, usually attached to a paperboard backing card. The plastic shell is usually adhesive bonded but sometimes is otherwise attached to the backing card.

⑦ The most common blister package is the blister-on-card type shown in Figure 18.1. Perforating the backing card provides a convenient opening feature. Sandwich, or foldover, cards effectively increase the thickness of the backing card (Figure 18.2). Sliding designs provide repeated or easy access to a product. Sliding designs do not require heat-sealable coatings.

⑧ Double blisters and clamshells (Figure18.2) are used where it is an advantage to have product visibility from all sides, where the thermoform is shaped to hold an irregularly shaped product, or where it is necessary to keep a design's center of gravity close to the package midpoint. Clamshell designs can be used as a **hinged** storage container. Information is usually provided on a paper or card inserted into the blister along with the product.

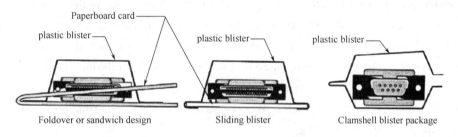

Figure 18.2 Examples of foldover, slide, and clamshell blister design variations

⑨ Plastic blisters are produced by thermoforming: heating a plastic sheet to a temperature at which the sheet can be shaped to a mold with the desired configuration. The key properties of blister material are cost, moldability, impact resistance, scuff resistance, low-temperature performance, and clarity. While most thermoplastics can be thermoformed, blister packs are usually made from one of the following:
- Poly (vinyl chloride)(PVC).
- **Poly (ethylene terephthalate) copolymer (PETG)**.
- Polystyrene (PS).

⑩ **Cellulose acetates, propionates,** and **butyrates** offer excellent thermoformability and clarity and at one time were the material of choice, but higher cost has almost eliminated their use.

⑪ The greater proportion of blisters and clamshells are thermoformed from PVC. PVC performance will vary depending on formulation, and PVC characteristics should be verified for the application. PETG has recently become more cost competitive and offers superior stiffness. In some applications, PETG is considered to be the environmentally more friendly choice. **Styrenics** have excellent clarity but low impact resistance unless an impact grade is used. The caliper of blister material will vary depending on the material, blister geometry, and product nature. A thickness of 0.12 to 0.18 millimetre (0.005 to 0.007 in) is used for the majority of applications.

02　Paperboard is selected according to the weight of the product being packaged, and must be suitable for the intended graphic presentation. Paperboard for quality blister packaging should be about 500 micrometres (0.020 in) thick and can go up to 800 micrometres (0.030 in) for heavier or larger objects. Suitable paperboard thickness can be achieved by doubling a lighter sheet rather than by using a single heavy stock. This technique is particularly useful for reinforcing pegboard holes. Lighter boards would be used only for small items, for items that will not be displayed on pegboards, and for designs that have other means for developing structural integrity. *Most paper-backed blister packs are flat and do not require that the paperboard have good folding properties.*

03　The board surface must be receptive to the printing process and must have enough internal bond strength to resist ply separation under use conditions. Clay-coated newsback or its equivalent is a good choice for most hardware applications. **Double-white-lined boardstock** would be used for applications where the back of the sheet will be printed or decorated. **Solid bleached boardstock** is used where an overall high-quality appearance is necessary.

04　After printing, the card is coated with a heat-sealable top coating selected to be compatible with the blister material being used. Most PVC blisters are attached with a PVC-based heat-sealing material. **Acrylic** and **ethylene-vinyl** acetate based formulations are also used. The blister is attached by the application of heat applied from either the blister side or the paperboard side. A properly produced blister package will have a fiber-tearing bond between the blister and the paper-board backing card.

05　In the packaging operation, the product is most commonly dropped into the open top of the blister, the paperboard **substrate** card is placed over the blister, and heat is applied to form the seal.

(1042 words)

New Words and Expressions

pegboard ['pegbɔːd] *n.* 钉板
hanger ['hæŋə] *n.* 挂架
hinge [hindʒ] *v.* 装铰链
propionate ['prəupiəneit] *n.* 丙酸盐、丙酸酯
butyrate ['bjuːtireit] *n.* 丁酸盐
styrenic ['staiəˌrinik] *n.* 苯乙烯
bleach [bliːtʃ] *v.* 漂白、变白
acrylic [əˈkrilik] *adj.* 丙烯酸的
substrate ['sʌbstreit] *n.* 底层、基材
carded display package　衬卡展示包装
backing card　底层衬板
skin packaging　贴体包装
center of gravity　重心

poly（ethylene terephthalate）copolymer（PETG）　　聚对苯二甲酸乙二醇酯-1，4-环己烷二甲醇酯

cellulose acetate　　醋酸纤维素、纤维素乙酸酯

double-white-lined boardstock　　双面贴白色面纸的木浆纸板

solid bleached boardstock　　硬质漂白木浆纸板

ethylene-vinyl　　乙烯-醋酸乙烯

Notes

1. *Carded display packages offer maximum product visibility and self-service convenience, while at the same time providing reasonable product protection against contamination and shipping damage.* （Para. 1）衬卡展示包装提供最大的产品可视性和自助销售的便利性，同时也提供合理的产品免受污染和运输破损的保护性。

2. *Most paper-backed blister packs are flat and do not require that the paperboard have good folding properties.* （Para. 12）大多数纸基泡罩包装都是平整的，不需要纸板有多么好的折叠性。have 前省略了 should，是 require 的要求。

Overview Questions

1. What difference between the blister packaging and skin packaging?
2. What materials can be thermoformed into the blister packs?
3. How to select the paperboard for the blister packaging?

Lesson 19　Applied Packaging (II)

1　Carded Skin Packaging

1　A carded skin package is made by first placing the product on a flat paperboard sheet. A plastic film mounted in a frame above the substrate card is heated to softening and then draped over the product (Figure 19.1). A vacuum applied through the substrate card draws the film down to conform **intimately** around the product. A heat-activated adhesive bonds the film firmly to the underlying board wherever contact is made.

Figure 19.1　Examples of skin packages

2　An alternative to blister packaging, skin packages are more economical since no special tooling or mold is required; the product becomes the mold. Plastic film is used rather than thicker sheet stock, a factor that increases in importance with larger parts. The process is readily adapted to small or large **production runs.** Unlike blister packaging, skin packaging immobilizes or secures the product to the backing sheet. Skin packaging can be designed to hold several parts securely and in such a manner that each part can be inspected individually.

3　Skin-packaging films are usually polyethylene or **ionomer** (for example, Du Pont's Surlyn). Ionomers have good clarity, are abrasion resistant, are exceptionally tough, and have rapid cycle times. This makes ionomers the material of choice for retail display applications, even though there is a price **premium.**

4　Polyethylenes are more economical but are not as clear and are easily abraded or scuffed. Polyethylene requires more heat (longer cycle time) and has a higher shrinkage factor than other films. The high shrinkage factor can curl board edges. The advantage of polyethylene's material economy is largely lost to longer cycle times. Industrial applications where clarity and appearance are not critical issues use polyethylene.

5　Since a vacuum needs to be drawn through the board to create the conforming skin, paperboards used in skin packaging must be porous. **Clay-coated paperboard** is rarely used since the clay seals the board surface. If a clay-coated or other non-porous board is selected for appearance reasons, the board must be perforated to ensure that air can be withdrawn from the skin enclosure. Perforating a high-quality board tends to defeat the appearance objective, and a weakness of skin packaging from a presentation point of view is the difficulty of creating a high-quality graphic image. In some instances the product is large enough or the geometry is such that perforations can be concealed behind the product.

⑥ Paperboards need to be stiff enough to provide a good display card and not curl or delaminate when the skin is applied. Thicknesses of 450 to 635 micrometres (18 to 25 points) are the most popular. Corrugated board is successfully used in skin packaging for larger items.

⑦ The heat-seal material must not seal off the board surface to the extent that a vacuum cannot be quickly drawn. These materials are usually formulated from **ethylene-vinyl acetate.**

2 Pharmaceutical Packaging

2.1 Drug Properties

⑧ Most drugs are complex organic compounds. The chemical nature of many of them is such that their exact composition can be easily altered, resulting in significant changes in the drug's **potency.** *Proper packaging should not contribute any influences stemming from the nature of the materials used that would encourage chemical changes and should protect the drug from outside influences that might promote changes.* At the extreme, some preparations are inherently chemically unstable, and almost any outside influence, including heat and light, can cause detrimental changes. In addition to the usual protective packaging, these products typically require refrigeration.

⑨ Many pharmaceuticals, particularly those containing **ester** or **amide** functional groups, are prone to hydrolysis. Hydrolysis of these groups breaks the molecules down into acid and alcohol or acid and amide species. For these products, an ultimate moisture barrier such as would be provided by a blister package (Figure 19.2) made from polychlortrifluoroethylene (Aclar) is necessary. Glass bottles also provide a superior barrier, but repeated opening and closing of the bottle introduces moisture into the headspace, continuously degrading the product over time.

Figure 19.2 Blister package for tablets

⑩ Many drugs (for example, some vitamins, **steroids,** and **antibiotics**) are adversely affected by oxidation reactions. Often only small amounts of oxygen are needed since the oxygen acts as an initiator that sets off a chain reaction. Superior oxygen barriers are required for these products. In some instances packages are flushed with **inert gases.**

⑪ Light is an energy source. When an organic molecule is exposed to light, the molecule's energy level is increased. Shorter light wavelengths have greater energy levels. Ultraviolet (UV) light is more energetic than visible light and can rupture **chemical bonds** to form **free radicals.** The free radicals can encourage a chain reaction, continuing degradation. Oxidation reactions are often photochemically initiated where both UV light and oxygen are present.

② Of the furnace glass colors, only **amber glass** offers a significant filter to UV light. Opacity of a plastic material to visible light does not necessarily mean it is opaque to UV light. Specific UV barrier pigments must be incorporated into plastics used for **light-sensitive preparations.** Similarly, a plastic can be clear and still offer some UV barrier. Poly (**ethylene naphthalate**) is a clear plastic said to have a good UV barrier.

③ *This concern for knowing the exact nature of packaging material and guarding against any possible sources of contamination impacts heavily on the materials that can be used-and particularly, on recycled plastics.* Even new plastics can contain very low molecular weight prepolymer fragments, which can leach out into a contained product. *Suppliers of polymers sometimes designate particular resins as being pharmaceutical grade, meaning that some care has been taken to minimize low-molecular-weight components and that no additives (**antistatic agents**, **process aid**, and so on) that might affect the drug have been used.*

④ Recycled plastics by definition have an unknown history, and their use is restricted to secondary packaging roles if used at all. Even **in-plant** regrind is not allowed for many applications since there is a possibility of some molecular breakdown during the harsh conditions of regrinding, remelting, and extrusion.

2.2 Packaging Emphases

⑤ Food and Drug Administration defines pharmaceuticals or drugs as articles intended for use in the diagnosis, cure, medication, treatment or prevention of disease in man or other animals and includes nonfood particles intended to affect the structure of any function of the body of man or other animals. Pharmaceutical packaging manufacture is under heavy regulation by FDA and is strictly monitored.

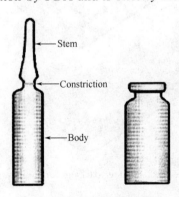

Figure 19.3 An ampoule and a vial

⑥ Pharmaceutical packaging uses all types of packaging materials that are available, from glass sealed items such as ampules for injectable drugs to folding cartons for containing **over the counter drugs. Prescription drugs** require little or no sales appeal having packaging emphasis on protection and identification. Prescription drugs are commonly put into glass containers (Figure 19.3) or plastic containers with good reclosure systems. Over the counter drugs must have sales appeal. Graphics plays an important role in over the counter drugs' packaging, uses many folding cartons to contain the primary protection package in glass, plastics, tubes and others. Over the counter drugs must compete with all the other available non-prescription pharmaceutical products. Cost of packaging of pharmaceuticals is secondary to performance. The processing and packaging operation of pharmaceuticals is mostly short production runs, little automation and very labor intensive for quality control and inspections. Food and beverages packaging machines do a lot of the inspecting for check weighing, fill levels, missing parts and metal detection. The pharmaceutical companies rely on extensive human visual inspection.

⑰ The pharmaceutical packaging emphasis is first on moisture control because most pharmaceuticals are hygroscopic, meaning that moisture is readily adsorbed resulting in caking, **efflorescence**, or inactivation of the product. Second, emphasis is on oxygen control because many drugs will oxidize resulting in reduced potency. Third, emphasis is on **volatility** where many of the solvents and oils have volatile organic backgrounds that may require extra headspace for one-way **venting**. Fourth, light protection is essential for some drugs due to degradation from exposure to ultraviolet light. Fifth, heat-since pharmaceuticals are chemical products, heat can cause accelerated toxin formation and deterioration of the product. Sixth, sterility-many items require sterility from initial packaging until use. **Surgical** tools, **dressings**, injectables, and some medications require special bacteria preprocessing in packaging rooms with specially protective packaging. Sterility can be accomplished in many ways using chemicals, heat or radiation.

⑱ Packaging for pharmaceuticals must be childproof for the protection of children. This is tested by giving potential packages to children of certain age groups. The test requires that at least 80% of these packages be intact after a specified time period. Also when testing for childproof packaging, the packages must be capable of opening by older people who might have handicaps. Emphasis is also placed on tamper-evident packaging to discourage tampering of the product. Packages containing drugs, especially over the counter drugs, are being manufactured to have either tamper-resistant or tamper-evident packaging systems. It is accepted within the packaging field that tamper proofing is not possible. Lastly, there is now an emphasis on unit-dose packaging which has reduced costs and increased accuracy in health care administration.

2.3 Regulations

⑲ *It is essential that drugs be taken in the prescribed amount and at the prescribed time periods for them to achieve their desired effects.* Furthermore, there needs to be a high degree of confidence that the drug, as taken by the patient, is actually the one that was prescribed and that it has not lost or changed in its potency. Any departure from these conditions could result in an ineffective treatment or, since many drugs are potentially harmful or even toxic when improperly administered, could have more serious consequences.

⑳ *It should come as no surprise that the packaging and sale of drugs is a highly regulated business.* The primary responsibility for drug regulations falls under the **jurisdiction** of the Food and Drug Administration (FDA) in the United States and the Health Protection Branch (HPB) in Canada. The spirit and intent of regulations in both countries are similar, although regulation details vary.

㉑ For regulatory purposes, packaging is regarded as part of a drug, so an application for approval of a drug also requires a full and detailed disclosure of the intended packaging. Package information often calls for explicit data on the composition and manufacture of a material.

㉒ A manufacturer who wants to change an existing approved package format will usually need to submit a supplementary application to request approval of the change.

2.4 Manufacturing Practice

㉓ The manufacture of packaged pharmaceutical products is also the subject of **scrutiny**. *Producers must validate their production process, meaning "... establishing documented evidence which provides a high degree of assurance that a specific process will consistently produce a product meeting its predetermined specifications and quality attributes"* (*FDA definition*).

㉔ In essence it must be proved and documented that the manufacturing process does not in any way change the drug and packaging from what has been approved.

㉕ This intense focus has led to many manufacturing procedures unique to the drug packaging industry. For example:

Gang printing of packaging (the printing of several designs on one press sheet) for different drug products or for the same product in different strengths is prohibited, with the possible exception of items that are absolutely differentiated by size, shape, or color.

㉖ Almost all pharmaceutical labeling is done from pressure-sensitive roll stock. Automatic counting of labels is easily done by machine, and the possibility of a loose label slipping into the wrong pile is eliminated.

㉗ On-line 100% verification of packages or labels is a common feature of pharmaceutical production. Typically these are optical character recognition systems that inspect a simple bar code on each label or package to ensure that it matches the product being packed.

(1931 words)

New Words and Expressions

intimately ['intimitli] *adv.* 密切地
ionomer ['aiənəmə] *n.* 离聚物、离子交联聚合物（同杜邦公司的 Surlyn）
premium ['primjəm] *n.* 额外费用、奖金
potency ['pəutənsi] *n.* 效力、效能
ester ['estə] *n.* 酯
amide ['æmaid] *n.* 氨基化合物
steroid ['stiərɔid] *n.* 类固醇
antibiotic [ˌæntibai'ɔtik] *n.* 抗生素、抗生学
in-plant ['inplɑːnt] *adj.* 内部的、厂内
efflorescence [ˌeflɔː'resəns] *n.* 开花、风化
volatility [ˌvɔlə'tiliti] *n.* 挥发性
venting ['ventiŋ] *n.* 通风、泄去
surgical ['səːdʒikəl] *adj.* 外科的、手术上的
dressing ['dresiŋ] *n.* 敷料剂
jurisdiction [ˌdʒuəris'dikʃən] *n.* 司法权、权限
production run　生产批量
clay-coated paperboard　瓷土涂布纸板
ethylene-vinyl acetate　乙烯-醋酸乙烯酯

inert gases　惰性气体
chemical bond　化学键
free radical　自由基
amber glass　琥珀玻璃
light-sensitive preparation　光敏制剂
ethylene naphthalate　萘二甲酸乙二醇酯
antistatic agent　抗静电剂、抗静电物
process aid　加工助剂
over the counter drug（OTC）　非处方药
prescription drug　处方药
Manufacturing Practice 药品生产质量管理规范

Notes

1. *Proper packaging should not contribute any influences **stemming from** the nature of the materials used that would encourage chemical changes and should protect the drug from outside influences that might promote changes.* （Para. 8）适当的包装不应造成任何来自所使用材料性质方面的影响，这些材料会促使一些化学变化，包装应保护药品免受可能促进这些变化的外界影响。这里，contribute 意指"有助于、促进"。stem from 意指"来源、由……造成"。encourage 意指"促进、助长"。promote 意指"促进、推动"。

2. *This concern for knowing the exact nature of packaging material and guarding against any possible sources of contamination impacts heavily on the materials that can be used-and particularly, on recycled plastics.* （Para. 13）知晓包装材料准确性质和预防任何可能的污染源方面的担心对可使用的材料，尤其是回收利用的塑料材料有严重影响。concern for 意指"为……担心"。guard against 意指"避免、预防"。impact ... on（upon）意指"对……影响"。

3. *Suppliers of polymers sometimes designate particular resins as being pharmaceutical grade, meaning that some care has been taken to minimize low-molecular-weight components and that no additives (**antistatic agents, process aid**, and so on) that might affect the drug have been used.* （Para. 13）高分子材料供货商有时标明作为药用级的特殊树脂，这就意指注意了将低分子量成分减至最低程度并且未使用那些影响药效的添加剂（抗静电剂、加工助剂等）。pharmaceutical grade 意指"药用级、药品级"，即指纯度能够被用于人类药物的化学品，每种化学品的药用级都不同。药品级的纯度一般高于食品级和医用级。

4. *It is essential that drugs be taken in the prescribed amount and at the prescribed time periods for them to achieve their desired effects.* （Para. 19）为取得期望的效果，药品必须按规定量且在一定的时间服用。be taken＝should be taken，是 essential 要求的虚拟语气。

5. *It should come as no surprise that the packaging and sale of drugs is a highly regulated business.* （Para. 20）药品包装及销售是一个被高度管制的行业就不足为奇。It 是形式主语；come as no surprise 意指"不足为奇、不感到意外"。

6. *Producers must validate their production process, meaning "... establishing documented*

evidence which provides a high degree of assurance that a specific process will consistently produce a product meeting its predetermined specifications and quality attributes" (FDA definition). (Para. 23) 生产商必须使他们的生产过程合法化，这意指"……备有提供高度保证的书面证据，确保特定的生产过程一直按照预先规定的产品规格和质量属性在生产药物"（FDA 定义）。这里，that 引导了 assurance 的同位语从句。

7. *Gang printing of packaging (the printing of several designs on one press sheet) for different drug products or for the same product in different strengths is prohibited, with the possible exception of items that are absolutely differentiated by size, shape, or color.* (Para. 25) 对于不同的药品或者不同浓度的相同药品，包装的组合施印（即把几个不同的设计图样编排在一起，在同一纸面上印刷）禁止使用，但对于大小、形状或颜色完全不同的药品，可能有例外。在这里，gang printing 指"组合印刷"技术；strengths 指"浓度、药力"；items 与 drug products 同义，指"药品"。

Overview Questions

1. What is the carded skin packaging?
2. What materials can be used for the skin-packaging films?
3. What packs are needed according to the drug properties?
4. Where should the pharmaceutical packaging emphasis be put?

UNIT 5

Packaging Development

Lesson 20 Packaging Development Process

1 Managing The Packaging Project

① It would be convenient to be able to draw a tidy flowchart of the steps needed to develop a package, accompanied by another block diagram detailing who was responsible for each activity. However, this is just not possible. There are about as many ways of developing and managing the packaging project as there are companies.

② Packaging is an extraordinarily complex endeavor that must be viewed as a part of a larger system, within which every activity has some impact or demand on package. Thus, purchasing, receiving, warehousing and materials handling, production, marketing, **shipping**, distribution, and sales each has its own particular demands on the package. Quite often, these demands are not mutually compatible:

- Purchasing wants a good price and reliable suppliers.
- Product development wants a package that contains, protects, and preserves.
- Production wants trouble-free operation on existing equipment.
- The warehouse staff want to stack three pallets high.
- Shipping wants packages that will withstand every shipping hazard.
- Marketing wants a unique seven-sided package printed in 11 colors.
- Sales wants a package that will impress their customers into providing more shelf space.
- Legal wants protection against all real and imaginable possibilities.

③ To complete the picture, add the needs of the retailer and the final consumer.

④ No part of the product production system can be altered without affecting the parts. The purchase of a faster packaging machine may require tighter packaging specifications. A small change in package size may have a detrimental impact on palleting and transport efficiencies. A change from corrugated distribution boxes to shrink-wrap may require an increase in the compression strength of plastic bottle. The most economical package to purchase may be the most difficult to fill. A unique design having superior shelf impact may require extensive retooling. The packaging challenge is to meet all the individual system requirements, including the company's **long-term strategy** and **profit objectives.**

⑤ How the packaging process is managed is primarily a function of the company's philosophy and its view of packaging's role in the enterprise. Company size and packaging cost relative to total cost are other important factors. In a small company, packaging responsibility may be a part of some other officer's job, typically a **purchasing agent** or the **production supervisor.** Time constraints or lack of in‑depth knowledge usually forces a heavy reliance on the **package supplier.**

⑥ As company size or expenditures on package material increase, a full-time packaging specialist may be appointed to oversee packaging activities. This person may report to the production manager, to marketing, or to other departments, depending on company structure. The person may have total responsibility for organizing, coordinating, and directing/implementing the packaging process, or he or she may work under the direction of a committee or a senior manager. Companies with professional packaging staff and facilities are less dependent on suppliers.

⑦ Packaging responsibilities at very large companies are typically divided amongst various departments. Package development, quality control and testing, graphic design, and package purchasing are typical divisions. The management and interrelationships of the individual departments vary.

⑧ If there is a single universal statement to be made, it is that product and package must be developed in a parallel, closely coordinated fashion, with all parties contributing, rather than sequentially, as is often done.

2　Project Scope

⑨ Package development can take from a month to several years, depending on the scope of the project. The simplest situation is one where changes are made to an established line to change the demographic/psychographic appeal, add a seasonal or holiday note or announce a promotional offering. These changes are invariably restricted to copy or graphics. Whatever the change, the product parentage remains obvious, and because the change affects only the graphics, most of the work centers on the marketing function. With few exceptions, new suppliers do not have to be located, production line changes are not needed and material compatibility or shelf-life studies are not necessary (Table 20.1, Situation 1).

Table 20.1　Package development projects vary in complexity depending on the project's scope

	Product	Physical Design	Materials	Graphic Design
Situation 1	same	same	same	change
Situation 2	same	similar	same	similar
Situation 3	similar	same	same	similar
Situation 4	same	different	same	different
Situation 5	same	different	different	different
Situation 6	all new	unknown	unknown	unknown

⑩ Repositioning is primarily a graphic change, assuming that the package's physical structure does not need significant alteration. The usual challenge is to reposition the product to appeal to a new demographic/psychographic market while still retaining the identity or equity of the original product. Obviously, this is a more risky undertaking.

¶1 A second situation is somewhat more complex and might involve changing the package's physical design (Table 20.1, Situation 2). The package size might have changed, the new graphics may include a **hologram** or the design may have been altered to accept a promotional/bonus compact disk. Now, the changes affect more than just the marketing and graphics functions. Production needs to check if such a package can run on existing machinery, and if not, what changes are required.

¶2 Formulation changes, changes in the package's physical design or changes to different packaging materials (Table 20.1, Situations 3, 4 and 5) might trigger the need for product/package shelf-life tests, material compatibility studies or material specification changes. The impact on production, warehousing, shipping container size, distribution, customer acceptance, and other parts of the system will need to be determined.

¶3 Situation 6 in Table 20.1 illustrates the most complex packaging projects — a truly all-new product. No consumer is going to the supermarket with this item on a shopping list. For such a product, there is no history of consumer attitude toward the product, nor is there any existing marketplace experience (competitors) on which to base package design directions. At the beginning of such a project, there is, at best, only a general feel for the demographic/psychographic audience and a limited idea of what the package material should be and what form the package should take.

¶4 Packaging form is sometimes dictated by technical necessity, similar existing product lines or the consumer's preconception of what the packaged product should look like. For example, there is no technical reason why a breakfast cereal or a detergent could not be offered in a round, **spiral-wound** paperboard container. However, many consumers would not recognize the product because it was not in a familiar container. For example, Canadian smokers expect cigarettes to come in rigid paperboard **slide boxes**; U.S. consumers prefer a soft pack. Using either nation's packaging system in the other country would be going against established consumer perceptions.

3 The Package Development Process

3.1 An Overview of the Package Development Process

¶5 The several figures in this section provide a generalized model for the package development process. *It is important to understand that the model does not have an orderly information flow from left to right or top to bottom. Rather, information flows continuously among the respective bodies; and it passes or is coordinated through "package design function" several times during its course.* (In this text, "package design function" does not imply major technical and graphic activity by a specific design department. Rather, it implies important milestones at which some person or body must give serious thought to some aspect of the final package.)

¶6 It is obvious that the package design function, whatever its form, acts as a central

clearing point and a vital communication channel through which ideas are assembled and evaluated and a consensus established.

3.2　Generating Ideas

17　Ideas for change can come from many sources inside or outside the company (Figure 20.1). Some companies have specific product development departments. Such departments continuously scan the field for new product ideas and work on new product concepts. Some projects might be internally motivated. For example, company growth may be stagnant, and the management may have been **brainstorming** for a major new market offering. The following are some ways in which companies generate new product ideas:

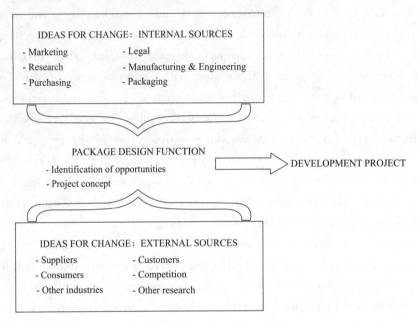

Figure 20.1　Ideas for change can come from many sources

• Many concepts, particularly in the fashion industries, are developed in the marketing department or consumer research department. Perhaps some color trend can be applied to cosmetic offerings. A competitor's success or consumer perceptions of some issue can motivate new product ideas.

• Company R&D laboratories generate new technologies that can be used to create products with market possibilities. This is an important source of ideas for technology areas. Alternatively, technologies and product ideas can be licensed or purchased from outside sources.

• Suppliers are often good sources of ideas. Most suppliers are actively involved in getting that little edge on the competition, and their efforts often lead to **cost-cutting** innovations or new possibilities.

• Ideas can often be generated by observing techniques used in other industries or other countries. For example, can some technique used in the cosmetics industry be used to enhance a food product? Is a business product finding increased home use?

18 Whatever the origin, management must judge project ideas in view of manufacturing expertise and limitations, corporate goals, and financial capabilities. In a dynamic field such as packaging, there are literally thousands of options available at any point in time.

19 Any package change or development must have clearly stated objectives. Change simply for the sake of change is not a valid reason for altering a package. The following are some examples of specific, quantified objectives for new or revised packaging:

- To successfully launch a new product, with success being identified by specific sales targets.
- To revitalize a dormant brand and raise sales to a specified level.
- To increase sales by providing a new convenience or utility to the consumer.
- To respond to environmental concerns, whether voluntarily or mandated.
- To respond to newly identified customer needs.
- To reposition an existing product in response to changing market conditions.
- To reduce costs by changing to more efficient packaging or processing.
- To maintain market share by responding to a competitor's initiative.

20 The objective of all business, of course, is to create profitable sales. Every proposed option must be tested against the prime objective. New product developments are notoriously risky and expensive undertakings. It is said that of 100 concepts developed to a working stage, only ten are actually offered to the consumer. Of these ten, only one or two will still be available two years later.

21 The most effective way of decreasing risk is to do your homework first. Marketing analysis, product mapping studies, focus group sessions, exhaustive development trials and market tests are costly, but their cost is minimal compared with the cost of a failed product launch.

22 A new product/package launch represents a team effort, with each team member wanting something different from the package. What management, purchasing, marketing, sales, production, distribution, retail and customer want from a package varies dramatically. All must be consulted and actively involved in screening out the insignificant many from the significant few. The significant few are the real business opportunities.

3.3 The Package Design Brief

23 The ideas generated in the project initiation phase are not enough to form the basis of a full development program. Perhaps the most important first task is to expand and quantify the project objectives. The next task is to generate a comprehensive catalog of all possible information related to the new product launch.

24 Among others, the various departments and groups listed under "Information Development" in Figure 20.2 and further expanded in Figures 20.3 through 20.6 will contribute information according to their needs and expertise. Compiling the required information for the brief is an excellent example of the kind of interdisciplinary, interdepartmental role a packaging professional may be expected to fill.

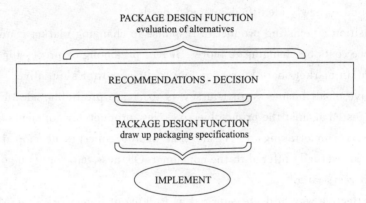

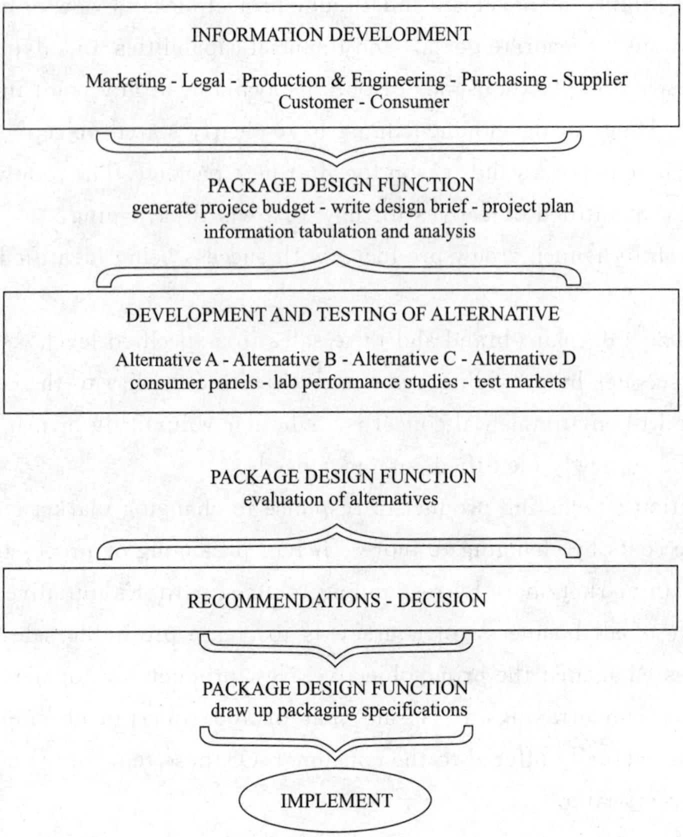

Figure 20.2 The general package design process

PRODUCT SPECS	PACKAGE SPECS	INFORMATION	PERFORMANCE
protection required	material requirements	ingredient lists	trials and tests
form required	testing requirements	instructions	field evaluations
handling characteristics	storage data	legal claims	hazards
hazards	field conditions	government	

Figure 20.3 Package design: technical responsibilities

EQUIPMENT	PACKAGE	STAFF	SCHEDULE
available capacity	production trials	quality control	supplier
new machines	size specification	machine operators	plant
product performance	storage specification	handling staff	seasonal
package performance	receiving specification	occupational hazards	
production methods	pallet patterns		
layout	rejection criteria		
required modifications	shipping tests		
production costs	plant hazards		
	inspection		

Figure 20.4 Package design: manufacturing and engineering responsibilities

CONCEPT	MARKET FACTORS	INFORMATION	MARKET TESTING
Product 　form 　arrangement 　features Package 　type 　unit size or sizes 　artwork and copy 　features	identification buying habits product exposure market practice competition seasonal factors distribution positioning targeted customer	use instructions storage instructions safety recipes pricing	customer consumer

Figure 20.5 Package design: marketing responsibilities

LEGAL	PURCHASING	TRAFFIC
brand names trademarks patents net weight ingredients tamper evidence child resistance DIN number government	company-supplier liaison materials availability package type alternatives graphic art alternatives cost estimates prototype samples supplier identification specification agreements quality control standards government approvals delivery schedules	freight classifications rates gross weights distribution costs customer practice shipping hazards coding handling and warehousing

Figure 20.6 Package design: legal, purchasing and traffic responsibilities

25　The information is compiled in a comprehensive document, usually called the "**packaging design brief**". The brief summarizes what the proposed package design is supposed to achieve:

- In what marketplace.
- With what product.
- By what means.
- Against what competition.
- Targeted to what group.
- In conjunction with what other activities.

26　The objective is to know as many facts about the proposed launch as possible. All of the information is entered into the packaging design brief, ensuring that all needs are met and all compromises are acceptable.

27　The design brief can be compared to a musical score in that it ensures that all participants are playing the same tune. The brief is not a static document. As a project evolves, information on the initial brief may change and new information may be added. The important thing is that everyone involved with a project should clearly understand the objectives and the means by which they will be achieved. It is vital that input be sought from everyone involved in the launch. Finding out after several months of intensive packaging design work that the hot new material has not been FDA approved is costly. Input from suppliers should be sought as early as possible in any project. They are aware of new technologies **on the horizon**, as well as knowing the arts and tricks that will keep costs in line and quality at its peak.

㉘ Not all the facts will be readily available at this early stage. Depending on the nature of the launch, some consumer focus-group studies may be conducted in order to better quantify market potential or product parameters. The group responsible for the package design function assembles and coordinates this information for presentation to management.

3.4 The Development Timetable

㉙ Assuming favorable management response to the project concept, a timetable is developed that lists activities, milestones and critical decision points. Gantt charts, or critical path charts, are popular project scheduling and tracking methods. The simplified Gantt chart in Figure 20.7 shows only the most general details. Gantt charts clearly identify those activities that can start only after the completion of another, and they allow staff and resource allocation planning over the project time period.

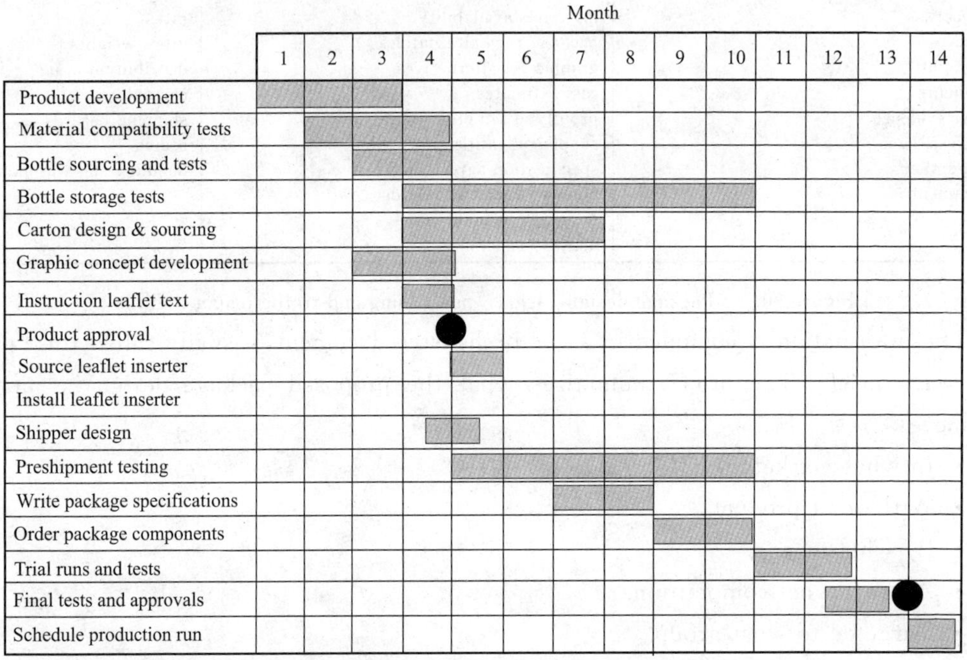

Activity	1	2	3	4	5	6	7	8	9	10	11	12	13	14
Product development	▨	▨	▨											
Material compatibility tests		▨	▨	▨										
Bottle sourcing and tests			▨	▨										
Bottle storage tests					▨	▨	▨	▨	▨	▨				
Carton design & sourcing					▨	▨	▨							
Graphic concept development				▨	▨									
Instruction leaflet text					▨									
Product approval					●									
Source leaflet inserter					▨	▨								
Install leaflet inserter														
Shipper design				▨										
Preshipment testing					▨	▨	▨	▨	▨					
Write package specifications							▨	▨						
Order package components									▨	▨				
Trial runs and tests											▨	▨		
Final tests and approvals												▨	●	
Schedule production run														▨

Figure 20.7 Example of a Gantt chart

3.5 Development and Testing of Alternatives

㉚ Based on the information provided in the package design brief, the package design group will generate ideas for how the product might be packaged. Some organizations conduct wide-open brainstorming sessions, at which all ideas, regardless of their apparent practicality, are considered. From these, the choices are narrowed to those most likely to succeed. Others use a strictly logical and practical step-by-step scientific approach. Both approaches have their merits.

㉛ Eventually, the possibilities are narrowed to a few options, all of which, on the surface, appear good. **Laboratory evaluations** are used to detect the not-so-obvious flaws or problems in the approaches. **Product compatibility** tests, shelf life studies, and **simulated** shipping tests are three of the most important types of laboratory evaluation. *This part of the program may also include pilot product runs and consumer test panels or markets.* Laboratory evaluation data are often useful when writing the package specifications.

52 If the data are positive, the project may proceed to the last phrases: recommendations, final decisions, drawing up the specifications, and implementing the design. **As often as not**, however, new information may send a project back to an earlier position. A whole new set of package alternatives may need to be explored, developed, and tested.

4 Specifications

53 The last step in the development process prior to actual production is to negotiate a supply of packages and packaging components of adequate quality from suitable suppliers. "Quality" is often defined as "conforming to requirements". In well-run companies, "requirements" are documented in the company's specifications. This is essential for communicating corporate expectations. Without such written definitions, your suppliers and your own staff will adopt the quality standards that are most convenient to them.

54 When dealing with suppliers, the production of a quality package rests finally on good communication, as formalized in the incoming packaging material specification. This specification:

- Communicates your exact needs to the supplier.
- Provides your supplier with a basis for judging production.
- Provides your staff with a basis for accepting packages and components.
- Allows for supplier bids on a fair and identical basis.
- Serves as the contractual **benchmark** in cases of dispute.
- Serves as a benchmark for package improvement.

55 Writing a specification requires an understanding of what performance factors are critical, for example, to good machinability and to the production of a completed package at the desired aesthetic level. Great care should be taken to establish the correct **tolerance level** for every critical package performance factor. Too broad a tolerance can cause machine or aesthetic problems. On the other hand, establishing unreasonably tight tolerances may limit the number of potential suppliers and can significantly increase costs.

56 A complete product specification system is not a single document, but rather three groups of documents (Figure 20.8). The documents describe all the materials and activities that will result in the efficient production of a product possessing the characteristics that have been identified as representing the appropriate quality.

57 Corporate policy standards and specifications are those documents that govern the entire specification process. For example, they will identify:

- A consistent corporate specification format.
- Who is responsible for writing specifications.
- How the specifications are written.
- How they are issued, who gets copies and where they are kept.

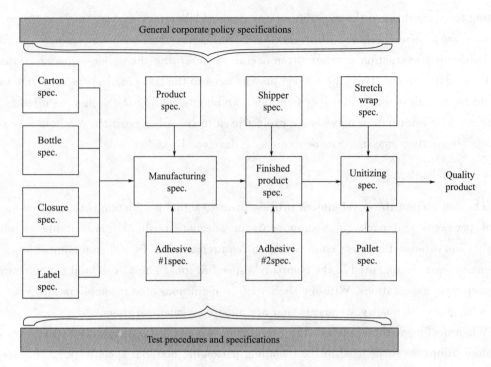

Figure 20.8 A product manufacturing specification is a set of documents that defines a quality product and all of the materials and activities that will result in a quality product

- How a specification is revised and how a specification is withdrawn.
- Implementation.
- Courses of action when events occur that are outside of the specifications.

38 Material and manufacturing specifications identify the critical properties of all raw materials, packages and components that are used in the manufacturing process, as well as the production sequence that will lead to the desired quality level. It ensures that every manufacturing step receives quality material from a previous step and forwards a quality item to the next production step.

39 The final group of standards and specifications will detail the sampling protocol and test methods used to quantify the attributes identified as being critical in the material and manufacturing specifications. For example, an adhesive material specification may call for an adhesive of a particular viscosity, **open time** and **peel strength.** These values would be determined using established test methods. This group of documents should also specify the instrument calibration methods that will be followed to ensure measurement accuracy. Such thoroughness is essential for International Organization for Standardization (ISO) certification or any other certification or accreditation.

40 A good specification is a joint document drawn up in close consultation with the supplier. The supplier must have **production facilities** that are capable of producing product within the tolerance range required by the purchaser. For example, a purchaser requires that a plastic bottle have a weight tolerance of $\pm 2\%$, but the supplier's machine, because of wear,

inadequate controls, untrained staff or other reasons, has an actual production tolerance of ±4%. There will always be a percentage of production that is out of specification. In such an instance, the purchaser must consider lowering the specification, or the supplier must improve the manufacturing process. Or, the purchaser must seek an alternative supplier.

㊶ Evidence that a supplier is actively engaged in a statistical process control program is usually a good sign that the company is working toward establishing better levels of production control.

㊷ As well as spelling out the critical values, the method of measuring these values must also be agreed to. In the majority of instances, standard methods such as those spelled out by ASTM International should be referenced.

㊸ Causes for rejection of a shipment should be clearly spelled out. In many instances, defects are classified into three categories:

A Critical defects. Those defects that prevent the packaging component from fulfilling its purpose. For example, an incorrect dimension.

B Major defects. Those defects that will likely seriously reduce the performance of the packaging component under stressed conditions, although it may perform adequately under most conditions. For example, a corrugated case with compression strength slightly below specified levels.

C Minor defects. These are mostly aesthetic defects that don't substantially reduce package function.

100 A specification may, for example, allow up to 2% of level C defects, 1% of level B defects, and 0.01% of level A defects.

(3416 words)

New Words and Expressions

shipping ['ʃipiŋ] *v.* 运输
hologram ['hɔlgræm] *n.* 全息图、激光防伪标
brainstorming ['brein,stɔːmiŋ] *n.* 头脑风暴、集体研讨
alternative [ɔl'tənətiv] *n.* 方案
benchmark ['bentʃmaːk] *n.* 基准、参照
long-term strategy 长期计划
profit objective 利润目标
purchasing agent 采购部
production supervisor 生产主管
package supplier 包装供应商
spiral-wound 螺旋缠绕（纸罐），指通过螺旋缠绕的方式生产的圆筒容器
slide boxes 滑盖盒（箱）
clearing point 中央结算点、中枢点
cost-cutting 削减成本
on the horizon 即将出现的
packaging design brief 包装设计纲要

laboratory evaluation　实验室评估
product compatibility　产品兼容性
simulated shipping　仿真运输过程
as often as not　多半、往往
tolerance level　误差等级
open time　间隔时间（涂胶后至压合前允许的最长时间）
peel strength　剥离强度
production facility　生产设施
critical defect　严重缺陷

Notes

1. *It is important to understand that the model does not have an orderly information flow from left to right or top to bottom. Rather, information flows continuously among the respective bodies; and it passes or is coordinated through "package design function" several times during its course.*（Para 15）重要的是要理解该模型没有从左至右或从上至下有序的信息，而是，这些信息在各部门之间持续不断的流动，在整个过程中，信息会经过"包装设计职责"多次或者通过"包装设计职责"来协调多次。本句主要是针对图20.2 的说明，重点是想说明包装开发过程是一个复杂的过程，其模型与普通流程图不同，其中任意一个环节的完成可能要与其他相关环节进行多次的信息交换。句中"package design function"不是指单一部门的包装技术或外观装潢设计活动，而是指重要的里程碑式的过程节点，在这些节点上相关人员或部门必须完成最终包装的某些功能。

2. *This part of the program may also include pilot product runs and consumer test panels or markets.*（Para. 31）这部分内容也包括产品小批试生产和消费者体验意见反馈即市场。"pilot product runs"指"产品试生产"，"panel"指"小组会议"。

Overview Questions

1. Package design is a very complex process. What are the factors that contribute to this complexity?
2. What is a package design brief? Why is it important?
3. What are the reasons for maintaining packaging specifications?
4. Why is it important for a specification to be jointly drawn up in consultation with the supplier?

Glossary[1]

A

a case in point 佐证、恰当的例子	(9)
a gob of （玻璃）球坯、滴料	(1)
abiotic [ˌeibaiˈɔtik] *adj.* 非生物的、无生命的	(2)
abrasion [əˈbreiʒən] *n.* 磨损、磨耗	(2)
abrasion resistance 耐磨性	(8)
abrasive [əˈbreisiv] *adj.* 研磨的	(10)
absolute value 绝对值	(11)
accelerometer [əkˌseləˈrɔmitə] *n.* 加速度计	(12)
acetal [ˈæsətæl] *n.* 乙缩醛	(7)
acetic [əˈsitik] *adj.* 醋的、乙酸的	(2)
acidity [əˈsidəti] *n.* 酸度、pH 值	(2)
acronym [ˈækrɔnim] *n.* 首字母缩略词	(4)
acrylic [əˈkrilik] *adj.* 丙烯酸的	(18)
aeration [erˈreiʃn] *n.* 空气混入、掺气	(16)
aerobic [eəˈrəubik] *adj.* 需氧的	(2)
aerosol packaging 喷雾包装	(5)
aggravate [ˈægrəvet] *v.* 加重	(2)
agitator [ˈædʒiˌtetə] *n.* 搅拌器、混合器搅拌装置	(16)
air balance scale 空气秤	(17)
air bubble sheet 气泡垫	(10)
airtight seal 空气密封、气密	(16)
aliphatic [ˌæliˈfætik] *adj.* 脂肪质的	(2)
alternatives [ɔːlˈtəːnətivz] *n.* 可供选择的事物、方案选项	(13)
amber glass 琥珀玻璃	(19)
ambient temperature 室温	(2)
amide [ˈæmaid] *n.* 氨基化合物	(19)
amphora [ˈæmfərə] *n.* 双耳细颈椭圆土罐	(1)
amplitude [ˈæmplitjuːd] *n.* 振幅	(10)
ampoul [ˈʌmpuːl] *n.* 安瓿	(6)
anaerobic [ˌæneiəˈrəubik] *adj.* 厌氧的	(2)
anisotropic material 各向异性材料	(3)
anneal [əˈniːl] *v.* 退火	(5)
antibiotic [ˌæntibaiˈɔtik] *n.* 抗生素、抗生学	(19)
antioxidant [ˌæntiˈɔksidənt] *n.* 抗氧化剂、防老化剂	(2)
antiskid [ˌæntiˈskid] *adj.* 防滑的	(4)

---

[1] 词汇索引号与课文对应，如"(3)"表示"第 3 课"。

antistatic agent　抗静电剂、抗静电物　(19)

anvil ['ænvil] *n*. 铁砧　(5)

apparent compression strength　表观（视）抗压强度　(10)

apparent strength　表观（视）强度　(3)

aromatic ['ærəmætik] *adj*. 芳香族的　(2)

arsenic ['ɑːs(ə)nik] *n*. 砒霜、三氧化二砷　(9)

art nouveau　（流行于 19 世纪末的）新艺术　(1)

arthritic [ɑːrθritik] *adj*. 关节炎的、关节炎患者　(9)

arts and crafts　工艺　(1)

as a rule of thumb　根据经验　(2)

as often as not　多半、往往　(20)

aseptic [ə'septik] *adj*. 无菌的　(1)

aspect ratio　长宽比　(10)

assembled box　组装盒　(4)

ASTM　美国试验与材料协会　(1)

ASTM D-3332 美国材料与试验协会标准——产品机械冲击脆值试验方法（使用冲击试验机）"Standard Test Methods for Mechanical-Shock Fragility of Products, Using Shock Machines"　(12)

ASTM D-4169 美国材料与试验协会标准——运输容器及其系统性能试验的标准规范"Standard Practice for Performance Testing of Shipping Containers and Systems"　(12)

ASTM D-4728 美国材料与试验协会标准——运输容器随机振动试验方法"Standard Test Method for Random Vibration Testing of Shipping Containers"　(12)

at first thought　乍一想　(2)

attenuate [ə'tenjueit] *v*. 削弱、衰减　(10)

auger ['ɔgə] *n*. 螺杆　(16)

authorized ['ɔːθəraizd] *adj*. 经授权的、经认可的　(4)

B

back and forth　来回的、往复　(16)

backing card　底层衬板　(18)

bacteriostat [bæk'tiriəstæt] *n*. 抑菌剂　(2)

bagasse [bə'gæs] *n*. 甘蔗渣　(3)

bag-in-box　盒中袋　(2)

balance beam scale　杠杆秤　(17)

bare foil　原铝箔　(8)

barring [bɑːriŋ] *prep*. 除……之外　(4)

basis weight　基重、基本重量　(3)

batch process　批量生产　(8)

bauxite ['bɔksait] *n*. 铝土矿　(8)

be characterized by　具有……的特征　(8)

be credited with　被认为　(3)

be inversely proportional to　与……成反比例　(8)
be likened to　与……相比　(7)
Beach puncture　戳穿试验（即 puncture test）　(4)
beading ['bi:diŋ] *n.* 珠状凸缘　(5)
bearing ['bɛəriŋ] *n.* 轴承　(16)
bearing area　承载面积　(10)
beefcake ['bi:fkeik] *n.* 牛肉蛋糕、健美男子　(1)
benchmark ['bentʃmɑ:k] *n.* 基准、参照　(20)
benchtop ['bentʃtɔp] *n.* 台式　(10)
bending chipboard　耐折纸板　(3)
bending stiffness　弯曲刚度　(10)
benzoic acid　苯甲酸　(2)
beryllium [bə'riljəm] *n.* 铍　(7)
biaxial orientation　双向拉伸　(7)
biaxially oriented nylon (BON) films　双向拉伸尼龙薄膜　(8)
bill of lading　提货单　(4)
billow forming　波浪成型　(7)
bin [bin] *n.* 储料箱　(17)
binary fission　（细胞的）二分体　(2)
bird swing　瓶内粘丝　(6)
blank [blæŋk] *n.* 盒坯、箱坯　(3)
blank mold　初模　(6)
bleach [bli:tʃ] *v.* 漂白、变白　(18)
bleached stock　漂白浆　(3)
blemish ['blemiʃ] *n.* 瑕疵、污点　(3)
blister ['blistə] *n.* 泡罩、气泡　(9)
block-type pallets　垫块式托盘　(13)
blow mold　吹模　(6)
blow-and-blow　吹-吹法　(6)
body hook　身钩　(5)
boil away　煮干　(8)
boil-in-bag　蒸煮袋　(1)
bolt [bəult] *n.* 螺栓、螺钉　(17)
BOPP　双向拉伸的聚丙烯　(7)
bottoming out　触底、从低点回升　(11)
bowed [bəud] *adj.* 弯曲成弓形的、弯如弓的　(10)
box compression strength　纸箱抗压强度　(4)
box maker's stamp　纸箱制造商证章　(4)
box manufacturer's certificate (BMC)　纸箱制造证章　(4)
boxboard ['bɔksbɔ:d] *n.*（纸盒、纸箱用）硬纸板（或木板）　(3)

bracing ['breisiŋ] *n.* 支撑、支柱 (10)
brainstorming ['brein,stɔ:miŋ]*n.* 头脑风暴、集体研讨 (20)
brand mark　印记、商标 (1)
brand name　商标、品牌 (1)
branding iron　烙铁、火印 (1)
bucket ['bʌkit] *n.* 水桶、料斗 (17)
buffer ['bʌfə] *n.* 缓冲器 (13)
bulge [bʌldʒ] *n.* 突出量、凸出部分 (10)
bulk packaging　散装包装、裸装的、大包装 (1)
bulk-dribble filler　粗加料-细加料双级充填机 (17)
bulkhead ['bʌlkhed] *n.* 隔壁、防水壁 (10)
bumper sticker　（汽车上的）保险杠贴 (1)
bushing ['buʃiŋ] *n.* 轴衬 (16)
butyrate ['bju:tireit]*n.* 丁酸盐 (18)
by convention　按照惯例 (3)

C

cake mix　做蛋糕用的配料 (1)
calcium carbonate　碳酸钙 (3)
calendaring ['kæləndəriŋ] *n.* 压光、压延成型 (3)
caliper ['kælipə] *n.* （材料）厚度 (3)
cam [kæm] *n.* 凸轮 (17)
cam-shaped rotor　凸轮型转子 (16)
canister ['kænistə] *n.* （茶叶、烟等）罐 (5)
cannula ['kænjulə] *n.* 套管 (6)
cap [kæp] *n.* 盖帽 (9)
capper ['kæpə] *n.* 压盖机、封口机 (1)
capsule ['kæpsju:l]*n.* 胶囊 (17)
carbohydrate [kɔ:bəhaidreit]*n.* 碳水化合物、糖类 (2)
carbonated beverage　碳酸饮料 (16)
carboy ['kɑrbɔi] *n.* 用藤罩保护的大玻璃瓶 (6)
carcase ['kɑ:kəs]*n.* 肉牛酮体 (2)
cardboard ['kɑ:dbɔ:d]*n.* 纸板（总称）、尤指中厚度纸板 (3)
carded display package　衬卡展示包装 (18)
carrier rule　运送规则 (2)
carrier rule and regulation　运输规则和条例 (13)
carton blank　盒坯 (3)
cartonboard　（折叠纸盒用）纸板 (3)
cartoner ['kɑ:tənə]*n.* 装盒机 (1)
cascade [kæ'skeid] *n.* 小瀑布、瀑布状物 (1)
case packer　装箱机 (1)

cask [kæsk] *n.* （尤指盛酒精饮料的）桶 (1)
casting process　流延（平挤）过程（工艺） (1)
catsup [ˈkætsəp] *n.* 番茄酱 (16)
caulking compound　填隙料 (16)
caulking tube　堵缝管 (1)
caustic solution　苛性碱溶液 (8)
cavity [ˈkæviti] *n.* 腔、洞 (17)
cavity mold 阴模 (7)
cc　立方厘米、毫升（cubic centimeter） (16)
cell telephone　手机 (10)
cellophane [ˈseləˌfein] *n.* 玻璃纸 (1)
cellulose [ˈseljəˌləʊs] *n.* 纤维素 (1)
cellulose acetate　醋酸纤维素、纤维素乙酸酯 (18)
cellulose fiber　纤维素纤维、木质素纤维 (3)
cellulosic [ˌseljuˈləusik] *adj.* 纤维质的 (10)
center of gravity　重心 (18)
cereal [ˈsiriəl] *n.* 谷类、谷物 (2)
chamber [ˈtʃembə] *n.* 室、腔 (16)
chamfer [ˈtʃæmfər] *n.* 倒棱、倒角 (13)
charging valve　充气阀 (16)
cheesecake [ˈtʃiːzˌkeik] *n.* 乳酪蛋糕、半裸体的女人照片 (1)
chemical bond　化学键 (19)
chemical pulp　化学浆 (3)
childproof　防止孩童的 (9)
child-resistance closure　儿童安全盖 (1)
chipboard [ˈtʃipbɔːd] *n.* 粗纸板、灰纸板 (3)
chlorodifluoromethane [ˌkləurəˌdifluərəˈmeθein] *n.* 氯二氟甲烷 (5)
chlorofluorocarbon (CFC) [ˌklɔːrəuˈfluərəukɑːrbən] *n.* 含氯氟烃 (5)
chlorophyll [ˈklɔːrəfil] *n.* 叶绿素 (2)
choke neck　瓶颈阻塞 (6)
chrome oxides　氧化铬、氧化铬绿 (6)
chromium [ˈkrəmiəm] *n.* 铬 (8)
chuck [tʃʌk] *n.* 卡盘 (5)
chute [ʃuːt] *n.* 瀑布、斜道 (10)
circumference [səˈkʌmfərəns] *n.* 环状面、圆周 (16)
city complex　城市综合体 (1)
clamp truck　钳式卡车、夹抱车 (10)
clay coated　瓷土涂布 (3)
clay-coated newsback (CCNB)　瓷土涂布新闻纸 (3)
clay-coated paperboard　瓷土涂布纸板 (19)

clearance ['klirəns] *n.* 空隙、间隙 (16)
clearing point　中央结算点、中枢点 (20)
clinch [klintʃ] *v.* 钉 (5)
clog [klɔg] *v.* 阻碍、堵塞 (17)
closed-cell material　闭环式材料 (15)
clostridium botulinum　肉毒杆菌 (2)
closure ['kləuʒə] *n.* 封闭物 (9)
clutch and brake mechanism　离合器和制动机构 (16)
coating ['kəutiŋ] *n.* 涂层、涂膜 (13)
cobalt 60　钴 60 (2)
cobalt oxides　氧化钴 (6)
cobb size test　科布施胶度试验 (4)
coefficient of friction (CoF)　摩擦系数 (4)
cohesive strength　黏结强度 (13)
collapsible tube　（金属）软管 (5)
collapsing frame　倒人字夹板 (7)
cologne [kə'ləun] *n.* 古龙香水 (2)
colorant ['kʌlərənt] *n.* 着色剂 (6)
column counter　高度式计数装置 (17)
combined board　（瓦楞纸）复合板、瓦楞纸板 (4)
commercial sterility　商业无菌 (2)
common carrier　公共承运商 (4)
compression load　压缩载荷 (12)
compression strength　抗压强度 (10)
conducive [kən'djuːsiv] *adj.* 有益的、有助于……的 (2)
configuration [kən,figju'reiʃən] *n.* 结构、构造 (17)
consistency [kən'sistənsi] *n.* 浓度、稠度 (16)
constant level filling　定液位灌装 (16)
constriction [kən'strikʃən] *n.* 压缩、狭窄 (16)
consumer commodity　消费品 (1)
consumer package　销售包装（件） (2)
Consumer Product Safety Commission　（美国）消费者产品安全协会 (9)
contact lens　隐形眼镜 (9)
containerboard [kən'teinəbɔːd] *n.* 箱板纸 (3)
content [kən'tent] *n.* 内装物 (9)
contentious [kən'tenʃəs] *adj.* 有异议的、诉讼的 (2)
contract ['kan,trækt] *v.* 缩小、紧缩 (16)
controlled atmosphere packaging (CAP)　可控气氛包装 (2)
converter [kən'vəːtə] *n.* 加工机械 (1)
cooling jacket　冷却套管 (7)

cork [kɔːks] n. 软木塞 (9)
corrugated board 瓦楞纸板 (3)
corrugated box 瓦楞纸箱 (4)
cost-cutting 削减成本 (20)
couch roll 伏辊 (3)
count [kaʊnt] n. 计数 (17)
CPET 结晶聚酯类 (7)
crank [kræŋk] n. 曲柄 (16)
creasing score 压痕线 (3)
creep [kriːp] v. 蠕变 (3)
creosote [ˈkriət] n. 木馏油 (2)
crepe [kreip] v. 起绉、绉纸 (3)
critical [ˈkritikl] adj. 临界的 (14)
critical acceleration 临界加速度 (10)
critical defect 严重缺陷 (20)
critical value 临界值 (2)
cross direction 纸张横向 (3)
cross section 横截面 (6)
crown [kraʊn] n. 王冠 (9)
CTFE (Aclar) 三氟氯乙烯、三氟氯乙烯均聚物 (7)
cullet [ˈkʌlit] n. 碎玻璃 (6)
culprit [ˈkʌlprit] n. 问题、麻烦 (4)
cup or flask filler 量杯式充填机 (17)
curb [kəːb] n. 路边 (10)
curing process 熟化工艺 (8)
curling [ˈkəːliŋ] n. 卷曲 (3)
cushion [ˈkuʃən] n. v. 衬垫、缓冲 (12)
cushion curve 缓冲曲线 (15)
cushion material 缓冲材料 (15)
cushion pad 缓冲垫 (4)
cushion tester 衬垫试验机 (15)
cushioned package 缓冲包装 (12)
cushioning material 缓冲材料 (10)
custom business 定制业务、定制生意 (1)
cyanide [ˈsaiənaid] n. 氰化物 (9)
cylinder [ˈsiləndə] n. 圆筒、气缸 (16)
cylinder machine 圆网纸机 (3)

D

damage boundary curve 破损边界曲线 (12)
damping [ˈdæmpiŋ] n. 阻尼 (11)

date coder　日期打码机 (1)

dead-fold　折叠充分、残余褶皱 (5)

dead-load　静载 (10)

debris ['debri:] *n*. 碎片、碎屑 (9)

decolorizer [di:'kʌləraizə] *n*. 脱色剂 (6)

deeper draw　深冲罐 (5)

definite skid base　确定的防滑地座 (14)

deflection [di'flekʃən] *n*. 偏差、挠曲 (4)

deflector [di'flektə] *n*. 导流板 (3)

deformation [,di:fɔ:'meiʃən] *n*. 变形 (2)

degradation [,degrə'deiʃən] *n*. 降解 (2)

degree of correlation　相关度 (15)

dehydration [,di:hai'dreiʃən] *n*. 脱水 (2)

delamination [di:,læmi'neiʃən] *n*. 层离、分层 (3)

demographics [,demə'græfiks] *n*. 人口统计学 (1)

demolition waste　工地废渣料 (1)

dent [dent] *v*. 削弱、使产生凹痕 (5)

deodorants [di'əud(ə)r(ə)nt] *n*. 除臭剂 (5)

depict [di'pikt] *v*. 描述、描绘 (13)

depletion [di'pli:ʃən] *n*. 消耗、损耗 (5)

depress [di'pres] *v*. 压下、推下 (17)

desiccant ['desikənt] *n*. 干燥剂 (2)

deteriorate [di'tiəriəreit] *v*. 恶化、变坏 (2)

dewater [di:'wɔ:tə] *v*. 脱水 (3)

dexterity [dek'steriti] *n*. 灵活、灵巧 (9)

diamond-like coating　类金刚石涂层 (8)

diaphragm ['daiəfræm] *n*. 隔膜 (16)

dictate [dik'teit] *v*. 命令、口述 (2)

die-cutting　模切 (3)

dimethyl [dai'meθil] *n*. 乙烷、二甲基 (5)

discrete [dis'kri:t] *adj*. 不连续的、离散的 (10)

disengage [,disin'geidʒ] *v*. 释放、脱离 (17)

disparate ['dispərət] *adj*. 完全不同的、从根本上种类有区分或不同的 (13)

dispense [di'spens] *v*. 分配、分发 (13)

dispensing spout　喷洒口 (5)

distilled spirit　蒸馏酒精 (16)

distort [di'stɔ:t] *v*. 扭曲、变形 (8)

distribution [,distrə'bjuʃən] *n*. 分配、流通 (13)

distribution environment　流通环境 (12)

distribution package (shipper)　运输包装（件）（同 transport packaging） (2)

divider [di'vaidə] n. 分割物、间隔物 (10)
dosage unit 剂量单位 (9)
double face 双面瓦楞纸板（三层） (4)
double filling 双级灌装法 (16)
double wall 双瓦楞纸板（五层） (4)
double white-lined (DWL) paperboard 双面白色贴面纸板 (3)
double-backer 双面机 (4)
double-seamed 二重卷边的 (5)
double-tight can 双重密封罐 (5)
double-white-lined boardstock 双面贴白色面纸的木浆纸板 (18)
downstroke ['daʊnˌstrəʊk] n. 下行程 (16)
drape [dreip] v. 用布帘覆盖、使呈褶皱状 (4)
drape over 披上 (10)
draw-and-iron 变薄拉深罐 (5)
draw-and-redraw 深冲拉拔、二次拉深 (5)
dressing ['dresiŋz] n. 敷料剂 (19)
dribble ['dribəl] n. 涓滴、细滴 (17)
drip [drip] v. 滴下、漏下 (16)
drive shaft 驱动轴 (16)
drive train 传动系统 (10)
drop height 跌落高度 (12)
drum [drʌm] n. 鼓、圆筒 (16)
dry bonding 干法复合 (8)
dry product 干燥物品、固体类物料 (1)
dryer ['draiə] n. 干燥机 (3)
ductile ['dʌktail] adj. 柔软的 (5)
dunnage ['dʌnidʒ] n. （在回收再用包装容器内不作为缓冲垫的）填充物、衬板（如填补空隙、定位、隔开等） (10)
durable ['djʊərəbl] adj. 持久的、耐用的 (10)
duration [djʊə'reiʃən] n. 持续时间 (11)
dwell time 停留时间 (8)
dynamic compressive strength 动态抗压强度 (3)
dynamic cushioning curve 动态缓冲曲线 (10)
dynamic energy 动态能量 (15)
dynamic stress 动态应力 (15)

E

edge crush test 边压强度 (4)
edgewise compression test (ECT) 边压测试 (4)
EEA 乙烯-丙烯酸乙酯、经常分组作为酸共聚物 (7)
efflorescence [ˌeflɔː'resəns] n. 开花、风化 (19)

elastomer [i'læstəmə] *n.* 弹性体 (16)
electrode [i'lektrəud] *n.* 电极 (5)
electrostatic shielding 静电屏蔽 (8)
elusive [i'lusiv] *adj.* 难以捉摸的 (13)
embossed [im'bɔsd] *adj.* 有浮雕图案的 (5)
emerald ['emərəld] *adj.* 翠绿色的 (6)
emulsion [i'mʌlʃən] *n.* 乳剂、乳浊液 (8)
enamel [i'næm(ə)l] *n.* 瓷釉 (5)
encounter [in'kauntə] *n.* 遭遇、碰撞 (13)
end hook 盖（底）钩 (5)
end panel 端板 (4)
environmentally acceptable packaging 环境可接受包装 (1)
enzyme ['enzaim] *n.* 酶 (2)
EPS 发泡聚苯乙烯 (7)
equilibrium [,i:kwi'libriəm] *n.* 均衡、平衡 (2)
equilibrium relative humidity(E. R. H.) 相对平衡湿度 (2)
equivalent shock frequency 等效冲击频率 (11)
essential oil 香精油、精油 (2)
ester ['estə] *n.* 酯 (19)
ethylene ['eθili:n] *n.* 乙烯 (2)
ethylene naphthalate 萘二甲酸乙二醇酯 (19)
ethylene-vinyl 乙烯-醋酸乙烯 (18)
ethylene-vinyl acetate 乙烯-醋酸乙烯酯 (19)
European Federation of Manufacturers of Corrugated Board（FEFCO） 欧洲瓦楞纸板制造商联合会 (4)
European Solid Fiberboard Case Manufacturer's Association（ASSCO）欧洲硬纸板箱制造商协会 (4)
EVA（EVAC）乙烯-乙酸乙烯酯 (7)
EVOH（EVAL）乙烯-乙烯醇 (7)
excelsior [ek'selsi,ɔ:] *n.* 细刨花 (13)
expanded polystyrene 发泡聚苯乙烯 (10)
expendable [ik'spendəbəl] *adj.* 可消耗的、一次性使用的 (13)
exponential curve 指数曲线 (15)
exponentially [,ekspəu'nenʃəli] *adv.* 以指数方式 (2)
eye-mark 眼标记、定位标 (8)

F

feed hopper 进料斗 (7)
feed mechanism 供料机构 (16)
feedback ['fid,bæk] *n.* 反馈 (17)
feedstock ['fi:dstɔk] *n.* 原料、给料 (1)

felted ['feltid] adj. 黏制的 (3)
fill an order　支付订货 (13)
fill level　灌装液位 (16)
filler ['filə] n.　充填机（对干料而言）、灌装机（对液体类物料而言） (1)
fill-in pad　填充衬垫 (4)
filling chamber　灌装腔（室） (16)
fill-to-a-level　等液位（法） (16)
fine powder　细粉末 (17)
finely ground coffee　细磨咖啡 (17)
fine-tune ['fain'tun] v. 调整、微调 (13)
finger ['fiŋgə] n. 导向板 (4)
finish ['finiʃ] n. 瓶口 (9)
finish surface　瓶口表面 (10)
fin-style vertical seal 鳍形立式封口 (8)
fire-bearer　炉箅托架、火炉子 (1)
flanged [flændʒd] adj. 带凸缘的，法兰的 (5)
flap [flæp] n. 摇翼、摇盖 (4)
flask [flæsk] n. 烧瓶、长颈瓶 (6)
flat crush test　平压强度 (4)
flat drop　面跌落 (12)
flat-tasting　淡味 (8)
flexibility [ˌflɛksəˈbiləti] n. 适应性 (16)
flexography　[flekˈsɔgrəfi] n. 苯胺印刷（术）、柔性印刷 (1)
flexural stiffness　弯曲刚度 (10)
flit insecticide　杀虫剂 (5)
flooding or constant stream filler　溢流或等流量式充填机 (17)
fluorocarbon [ˌfluərə(u)ˈkɑːb(ə)n] n. 碳氟化合物 (5)
flute profile　楞型 (4)
fluted ['fluːtid] adj. 有凹槽的 (4)
foam [fəum] n. 泡沫材料、泡沫状物 (13)
foam-in-place　现场发泡 (10)
foil/fiber composite can　铝箔/纤维复合罐 (1)
fold endurance　耐折度 (3)
folder box　折叠型纸箱 (4)
folding carton　折叠纸盒 (3)
Food and Drug Administration（FDA）　美国食品和药物管理局 (9)
food board　包装食品用纸板 (3)
footprint ['futprint] n. 脚印、占地面积 (8)
forces of gravity　重力 (16)
forcing (input) frequency　激励频率 (10)

forklift ['fɔːklift] *n.* 叉车 (11)

forming collar　翻领式成型装置 (8)

four sided seal pouch　四边封袋 (8)

Fourdrinier machine　长网纸机 (3)

fragility [frə'dʒiliti] *n.* 易碎性、脆值 (4)

fragility test　脆值试验 (14)

free flowing　自由流动 (17)

free radical　自由基 (19)

freezer burn　冻斑、冷冻食品表面干燥变硬 (2)

freight claims　货物索赔 (13)

freight classification　货物分类 (4)

funnel ['fʌnəl] *n.* 漏斗 (17)

furnish ['fəːniʃ] *n.* 纸浆配料、浆料 (3)

G

gasket ['gæskit] *n.* 垫圈、垫片 (17)

gas-transmissionrate　气体透过率 (2)

gate [geit] *n.* 浇口 (7)

gauge [gedʒ] *n.*（厚度）计量单位 (3)

gearreducer　减速器 (7)

gel [dʒel] *v.* 胶化 (4)

glass bead　玻璃珠 (1)

glass beaker　玻璃烧杯 (1)

glass blowpipe　玻璃器皿吹管 (1)

glassine paper　玻璃纸 (3)

G-level　G 值 (14)

glossy ['glɔsi] *adj.* 光滑的、有光泽的 (3)

glue applicator　上胶器 (8)

glue flap　黏合襟片 (13)

glue tab　糊头 (4)

glycol ['glaikəul] *n.* 乙二醇 (10)

goodwill ['gud'wil] *n.*（企业的）信誉、声誉 (13)

grabbing clutch　握式离合器 (16)

grammage ['græmidʒ] *n.* 克重 (3)

granular material　颗粒料 (2)

graphicdesign　图形设计、平面设计 (1)

gravure coating　凹版涂布 (8)

gravy preparation　肉汁配制品 (1)

greaseproof paper　防油纸 (3)

groove [gruv] *n.* 沟槽、轧槽 (16)

gross weight　毛重 (17)

ground meat 碎肉	(2)
groundwood pulp 磨木浆	(3)
guide [gaidz] n. 导板、护罩	(17)
Gurley porosity 格利孔隙度	(4)
gush [gʌʃ] v. 喷涌	(16)
gusseted pouch 有三角褶的袋子、折角袋	(8)

H

halfsine 半正弦	(11)
halogenated dimethyl ether 卤代烷二甲醚	(5)
halogenated hydrocarbon 卤代烃	(5)
hand paddle 手划桨技术	(17)
handling ['hændliŋ] n. 装卸、搬运	(13)
hanger ['hæŋə] n. 挂架	(18)
havoc ['hævək] n. 破坏、损坏	(3)
HCFC (hydrochlorofluorocarbon) 氢氯氟碳化合物	(5)
headbox ['hed͵bɔks] n. 流浆箱	(3)
headspace ['hedspeis] n. 顶部空间	(6)
heat-shrink 热收缩性	(2)
heel tap 斜底、瓶底厚薄不均	(6)
hermetic seal （真空）密封	(5)
hexagonal [heks'ægnəl] adj. 六角形的、六边形的	(10)
hierarchy ['haiəraːki] n. 层次（级）、等级	(1)
high-barrier packaging 高阻隔性包装	(2)
high-volume production 大量生产	(10)
hinge [hindʒ] v. 装铰链	(18)
hinged lid 铰链盖	(5)
HIPS 高抗冲聚苯乙烯	(7)
hologram ['hɔlgræm] n. 全息图、激光防伪标	(20)
homogenous [hə'mɔdʒinəs] adj. 同质的	(3)
hoop [huːp] n. 加强环、箍	(5)
hopper ['hɔpə] n. 加料斗	(16)
horizontal form-fill-seal (HFFS) machine 卧式成型-充填-封口机	(8)
hormone ['hɔːməun] n. 激素、荷尔蒙	(2)
hottack 热黏性	(8)
hot-melt adhesive 热熔性黏结剂	(13)
hot-melt bonding 热熔复合	(8)
humidor ['hjuːmidɔː] n. 雪茄盒	(5)
hydraulic cylinder 液压缸	(7)
hydraulic pressure 液压	(7)

hydrocarbon [ˌhaidrə(u)'kɑːb(ə)n] n. 碳氢化合物 (5)

hygroexpensive 湿润膨胀 (3)

hygroscopic [ˌhaigrə'skɔpik] adj. 易潮湿的 (2)

hysteresis [ˌhistə'riːsis] n. 滞后 (3)

I

impact strength 冲击强度 (3)

impact velocity 冲击速度 (11)

impact-extrusion 冲挤 (5)

impregnation [ˌimpreg'neiʃən] n. 注入、浸渗 (10)

in alignment 成一直线、对准 (16)

in lieu of 代替 (13)

in terms of 根据、依据 (13)

in the final analysis 总之、归根结底 (9)

incineration [inˌsinə'reiʃən] n. 焚化、焚烧 (1)

indented [in'dentid] adj. 锯齿状的、犬牙交错的 (10)

individual packaging 单独包装、小包装 (1)

industrial package 工业包装（件） (2)

industrial waste 工业废料 (1)

inert gases 惰性气体 (19)

inertia force 惯性力 (14)

infrared radiation 红外线辐射 (2)

ingest [in'dʒest] v. 摄取、吞咽 (9)

ingot [ˈiŋgət] n. 铸块、锭 (8)

inner face 里面纸 (3)

inorganic [ˈinərˈgænik] adj. 无机的、无生物的 (6)

in-plant ['inpl'ɑːnt] adj. 内部的、厂内 (19)

insert [in'səːt] n. 插入物、添加物 (10)

inside dimension 内尺寸 (4)

inside liner 里纸 (4)

instant coffee 速溶咖啡 (2)

intake stroke 进给冲程、吸入冲程 (16)

integrity [in'tegrəti] n. 完整性 (2)

interface valve 接口阀 (16)

interior form 内附件 (4)

interlocking [intə(ː)'lɔkiŋ] adj. 联锁的、互锁的 (16)

intermingle [intə'miŋg(ə)l] v. 混合、混杂 (3)

intermittent motion 间歇运动 (16)

internal stress 内应力 (14)

International Corrugated Case Association（ICCA） 国际瓦楞纸箱协会 (4)

International Fibreboard Case Code 国际纤维板箱代码 (4)

intersection [ˌintəˈsekʃən] n. 交叉点、交叉线 (13)
intimately [ˈintimitli] adv. 密切地 (19)
inventory [ˈinvənˌtɔri] n. 存货、库存 (13)
ionization [ˌaiəniˈzeʃən] n. 离子化 (2)
ionizing energy 游离能、电离能量 (2)
ionomer [ˈaiənəmə] n. 离聚物、离子交联聚合物（同杜邦公司的 Surlyn) (19)
IoPP 包装专业技术人员协会 (1)
irreversible [ˌiriˈvəːsəbl] adj. 不可逆的 (2)
isocyanine [aisəuˈsaiənin] n. 异花青 (10)
ISTA 国际安全运输协会 (1)

J

jurisdiction [ˌdʒuərisˈdikʃən] n. 司法权、权限 (19)

K

keg [keg] n. 小桶 (1)
key-opening can 卷开罐（带有开罐钥匙的金属罐） (5)
kinetic energy 动能 (11)
knocked-down 拆散压扁（纸盒纸箱和其他容器压扁后储存和运输） (3)
kraft [krɑːft] n. 牛皮纸 (3)

L

label [ˈleibəl] n. 标签 (1)
label paper 标签纸 (3)
laboratory evaluation 实验室评估 (20)
lacquer [ˈlækə] v. 涂漆、使表面光泽 (5)
lactic [ˈlæktik] adj. 乳化的 (2)
laminate [ˈlæməˌnei] n. 层压（材料）、叠压 (1)
landfill [ˈlændˌfil] n. 垃圾掩埋法（场） (1)
lap seal 搭接封口、叠封 (8)
laptop computer 笔记本电脑 (10)
lard [lɑːd] n. 猪油 (3)
latch [lætʃ] n. 拨叉、碰锁 (16)
LDPE 低密度聚乙烯 (7)
leaking [ˈliːkiŋ] n. 泄漏 (16)
leeway [ˈliːwei] n. 灵活性、回旋余地 (2)
lehr [liz] n. 玻璃韧化炉 (6)
lend itself to 有助于 (3)
level sensing filling 液面感应式灌装 (16)
lid [lid] n. 盖子 (9)
light-sensitive preparation 光敏制剂 (19)
lignin [ˈlignin] n. 木质素 (3)
lined chipboard 贴面灰纸板 (3)

liner ['lainə] *n.* 内衬 (3)
liner facing （瓦楞纸外层的）面纸 (3)
liner materials 衬料、里衬材料 (9)
linerboard ['lainəbɔːd] *n.* 瓦楞纸板面纸 (3)
liquid displacement scale 排量秤 (17)
liquid phase 液相 (5)
lithographed can blank 彩印罐坯 (5)
LLDPE 线性低密度聚乙烯 (7)
load-bearing ability 承载能力 (10)
longitudinal seal 纵封 (8)
long-term strategy 长期计划 (20)
lubricate ['lubriˌket] *v.* 润滑 (16)
lug [lʌg] *n.* 支托、耳状物 (6)
lumpy ['lʌmpi] *adj.* 结块的 (17)

M

machine direction 纸张纵向、机器方向 (3)
magnesium oxide 二氧化镁 (3)
maintenance ['mentənəns] *n.* 保养、维修 (16)
malfunction [mælˈfʌŋkʃən] *n.* 故障、失效 (17)
malleable ['mæliəbl] *adj.* 有延展性的、易适应的 (8)
mandrel ['mændrəl] *n.* 心轴 (7)
manganese [mæŋgəˈniz] *n.* 锰 (8)
manhandle ['mænˌhændl] *v.* 粗暴地对待、野蛮装卸（同 mishandle） (11)
manual handling 人工搬运 (10)
manufacturer's joint 制造接头 (4)
Manufacturing Practice 药品生产质量管理规范 (19)
map out 描绘 (14)
marginal ['mɑːdʒin(ə)l] *adj.* 少量的 (4)
marmalade ['mɔːməled] *n.* （橘子或柠檬等水果制成的）果酱 (2)
mass production 大规模生产、批量生产 (1)
mastitis tip 针型口 (5)
matte [mæt] *adj.* 无光泽的 (3)
matted ['mætid] *adj.* 无光泽的 (3)
mean [miːn] *n.* 平均值 (15)
measuring chamber 计量腔（室） (16)
mechanical pulp 机械浆 (3)
medium ['miːdiəm] *n.* 瓦楞芯纸、瓦楞原纸 (3)
melamine ['meləmin] *n.* 三聚氰胺 (7)
melt index 熔融指数 (7)
membrane ['membrein] *n.* 薄膜、隔膜 (5)

mercury [ˈmɜːkjəri] n. 汞、水银 (16)
mesophyll [ˈmesəʊˌfil] n. 叶肉 (2)
metalized paper 敷金属纸 (8)
metering disc 计量盘 (16)
microflute 微瓦 (4)
microorganism [ˌmaɪkrəʊˈɔːɡənɪzəm] n. 微生物 (2)
microscopic [ˌmaɪkrəˈskɒpɪk] adj. 微观的 (2)
misalignment [ˌmɪsəˈlaɪnmənt] n. 未对准 (17)
modified atmosphere packaging (MAP) 气调包装 (2)
modified natural resin 改性天然树脂 (7)
modulus of elasticity 弹性模量 (11)
moisture content 湿量、水分（含量） (3)
moisture-barrier packaging 防潮包装 (2)
molded pulp 纸浆模塑 (10)
molding and forming technique 模塑成型技术 (16)
molecular weight 分子量 (7)
monolithic [mɒnəˈlɪθɪk] adj. 整体的 (6)
monomer [ˈmɒnəmə] n. 单体 (7)
monostearate [mɒnəˈstɪəˌreɪt] n. 单硬脂酸盐 (6)
mottled [ˈmɒtld] adj. 杂色的 (3)
mPE 茂金属聚乙烯 (7)
Mullen burst test 纸箱耐破度测试 (4)
multicellular [ˌmʌltɪˈseljʊlə] adj. 多细胞的 (2)
multiple draw 多级拉深罐 (5)
municipal solid waste (MSW) 城市固体垃圾 (1)
mustard [ˈmʌstəd] n. 芥末酱 (16)
myoglobin [ˌmaɪəʊˈɡləʊbɪn] n. 肌红蛋白 (2)

N

nail [neɪl] n. 钉子、钉状物 (17)
Nanocomposite [ˌnænə(ʊ)ˈkɒmpəzɪt] n. 纳米复合材料 (8)
nanocomposite film 纳米复合薄膜 (8)
narrow necked 细颈的 (16)
narrow-mouthed 窄口的 (7)
nasal tip 鼻型口 (5)
National Motor Freight Classification (NMFC) 国家汽车货物分类 (4)
natural biological life 自然生物寿命（周期） (1)
natural frequency 固有频率 (10)
natural kraft paper 天然牛皮纸 (3)
net weight 净重 (17)
newsboard [ˈnjuːzbɔːd] n. 旧报纸做的纸板 (3)

newsprint ['njuːzprint] n. 新闻纸 (3)
niche [niʃ] adj. 有利可图市场的 (8)
nick [nik] n. 裂口、刻痕 (8)
nip roll 压送辊 (7)
nitrous oxide 一氧化氮 (5)
nominal cleanliness 名义洁净度 (2)
None 无离聚物、沙林（杜邦商品名） (7)
nonfree-flowing 非自由流动 (17)
notch [nɑtʃ] n. 槽 (13)
nozzle ['nɔzl] n. 管口 (5)
nuisance ['njuːs(ə)ns] n. 讨厌的东西（人、行为） (9)
nut [nʌt] n. 螺母、螺帽 (17)

O

objectionable [əbˈdʒekʃənəbl] adj. 引起反对的、讨厌的 (10)
oblong ['ɔblɔŋ] adj. 椭圆形的、长方形的 (5)
obstruction [əbˈstrʌkʃən] n. 障碍物、阻碍物 (16)
octagonal [ɔkˈtægnə] adj. 八角形的、八边形的 (10)
octave ['ɔkteiv] n. 倍频 (10)
off flavor 异味、败味 (2)
offset ['ɔˌfset] v. 抵消、补偿 (17)
offset letterpress 凸版印刷 (5)
off-white 灰白色 (3)
ointment ['ɔintm(ə)nt] n. 药膏、油膏 (5)
oleic acid 十八烯酸、油酸 (6)
on the horizon 即将出现的 (20)
opacifying agent 遮光剂 (6)
opaque [əuˈpeik] adj. 不透明的、迟钝的 (2)
open time 间隔时间（涂胶后至压合前允许的最长时间） (20)
open-celled foam 开孔泡沫 (10)
ophthalmic [ɔfˈθælmik] adj. 眼的、眼科的 (5)
OPP 拉伸聚丙烯 (7)
optimize ['ɑptəˌmaiz] v. 使最优化 (13)
optimum design 最优（佳）设计 (13)
orthogonal axes 正交轴 (12)
outer face 外面纸 (3)
outside dimension 外尺寸 (4)
over the counter drug (OTC) 非处方药 (19)
overfill ['əuvəfil] n. 过灌装、溢出 (16)
overflow pipe 溢流管 (16)
overhang [ˌəuvəˈhæŋ] n. 伸出、悬空 (10)

overlap [ˌəuvəˈlæp] n. 重叠 (4)
overpackaging [ˈəuvəˌpækədʒiŋ] n. 过度包装 (1)
overpress [ˈəuvəpres] n. 飞刺（玻璃制品的缺陷） (6)
overstress [ˈəuvəˈstres] n. 过应力 (13)
oxygen scavenger 去氧剂 (2)
oxygen-deficient 缺氧的 (2)
oxymyoglobin [ˌɔːksimaiəˈgləubin] n. 氧合肌红蛋白 (2)
oysterboard 灰纸板 (4)
ozone [ˈəuzəun] n. 臭氧 (5)

P

PA（NY） 聚酰胺、尼龙 (7)
package insert 插页、药品说明书 (7)
package supplier 包装供应商 (20)
packaging design brief 包装设计纲要 (20)
pail [peil] n. 桶、提桶 (10)
pallet [ˈpælit] n. 托盘、平台 (13)
palletize [ˈpælətaiz] v. 托盘化 (4)
PAN（AN） 聚丙烯腈、巴雷斯 (7)
paperboard [ˈpeipəbɔːd] n. 纸板、卡纸 (3)
papermaking machine 造纸机 (3)
paper-mill 造纸厂 (3)
papyrus [pəˈpairəs] n. 纸莎草、草纸 (3)
papyrus reed 纸莎草芦苇 (1)
parchment paper 羊皮纸 (3)
parison [ˈpærisən] n. 型坯 (6)
parting line 分型线、模缝线 (7)
partition [pɑːˈtiʃ(ə)n] n. 隔离物、隔板（同 divider） (3)
pasta [ˈpɑːstə] n. 意大利面 (2)
pasteurization [ˌpæstəraiˈzeiʃən] n. 加热杀菌法、巴斯德杀菌法 (2)
pathogen [ˈpæθədʒən] n. 病原菌、致病菌 (2)
pathogenic [ˌpæθəˈdʒenik] adj. 致病的、病原的 (2)
PC 聚碳酸酯 (7)
PE 聚乙烯 (7)
peak acceleration 峰值加速度 (11)
peel strength 剥离强度 (20)
pegboard [ˈpegbɔːd] n. 钉板 (18)
PEN 聚萘二甲酸乙二醇酯 (7)
penalty [ˈpen(ə)lti] n. 罚款、处罚 (4)
pendulum arm 摇臂 (4)

penetrate [ˈpenitreit] v. 渗透、穿透 (2)

perforate [ˈpəːfəreit] v. 打孔 (2)

perimeter [pəˈrimitə] n. 周长、边长 (10)

permeability [ˌpəːmiəˈbiliti] n. 渗透性 (2)

permeate [ˈpəmiet] v. 渗透、渗入 (8)

persona [pəːˈsəunə] n. 人、角色 (2)

personal care products 个人护理用品 (2)

pertaining to 相关的、关于 (13)

PET 聚（对苯二甲酸乙二醇酯）、聚酯纤维 (7)

PETG 聚（对苯二甲酸乙二醇酯）二醇 (7)

petrochemical industry 石油化工业 (7)

pharmaceutical [ˌfɑrməˈsutikəl] adj. 制药的、配药的 (16)

pharmaceutical packaging 药品包装 (1)

phase out 淘汰、停止 (5)

phenol [ˈfinɔl] n. 苯酚 (7)

phenol formaldehyde plastic (Bakelite) 苯酚甲醛塑料 (1)

pigment [ˈpigm(ə)nt] n. 色素、颜料 (3)

pin adhesion 点黏合性测试 (4)

pinch-off 交错断裂 (7)

pinholing [ˈpinˌhəuliŋ] n. 针眼、针孔 (8)

piston [ˈpist(ə)n] n. 活塞 (5)

place a premium on 重视于、鼓励 (8)

plank [plæŋk] n. 厚木板、支架 (10)

plastic tub 塑料管 (2)

plasticizer [ˈplæstisaizə] n. 可塑剂 (3)

platen [ˈplæt(ə)n] n. 压盘、压板 (4)

pliable [ˈplaiəbl] adj. 易曲折的、柔软的 (4)

plug [plʌg] v. 阻塞 (16)

plug gauge 测孔规 (6)

plug mold 阳模 (7)

plug-in type relay 插入式继电器 (16)

plunger [ˈplʌndʒə] n. 柱塞 (16)

ply separation 分离层测试 (4)

PMMI 包装机械制造者协会 (1)

pneumatic cylinder 气缸 (14)

pneumatic pressure 气压 (16)

point-of-purchase 购货点 (7)

polarity [pəˈlærəti] n. 极性 (7)

poly (ethylene terephthalate) copolymer (PETG) 聚对苯二甲酸乙二醇酯-1,4-环己烷二

甲醇酯 (18)
poly（vinylidene chloride）（PVDC） 聚偏二氯乙烯 (2)
polyethylene（PE）[ˌpɒliˈeθiliːn] n. 聚乙烯 (1)
polyolefin [ˌpɒliˈəuləfin] n. 聚烯烃 (2)
polyolefin plastic 聚烯烃塑料 (1)
polypropylene（PP）[ˌpɒliˈprəupiliːn] n. 聚丙烯 (1)
polysulfone [pɒliːˈsʌlfəun] n. 聚砜 (7)
polyurethane [ˌpɒliˈjuərəθein] n. 聚氨酯 (10)
popcorn [ˈpɒpkɔːn] n. 爆米花 (10)
postconsumer recycled（PCR） 消费后再循环的 (1)
potato chips 油炸薯片 (2)
potency [ˈpəutənsi] n. 效力、效能 (19)
potential energy 势能 (11)
pothole [ˈpɒthəul] n. 壶穴 (10)
pouch paper 纸袋纸 (3)
power spectral density 功率谱密度 (12)
power tool 电动工具 (2)
PP 聚丙烯 (7)
precondition [priːkənˈdiʃ(ə)n] v. 预处理 (4)
preform [priˈfɔrm] n. 粗加工的成品 (7)
premium [ˈpriːmjəm] n. 额外费用、奖金 (19)
prepared food 预加工食品 (1)
prescription drug 处方药 (19)
press-and-blow 压-吹法 (6)
pressure gravity filling 压力重力式灌装（即等压灌装） (16)
primary package 初级包装、一级包装、内包装 (2)
pristine [ˈpristin] adj. 原始的 (6)
process aid 加工助剂 (19)
product clearance 产品许可证 (2)
product compatibility 产品兼容性 (20)
production facility 生产设施 (20)
production run 生产批量 (19)
production supervisor 生产主管 (20)
profile extrusion 仿型（靠模）挤出 (7)
profit objective 利润目标 (20)
promotional [prəˈmouʃənl] adj. 促销的、增进的 (13)
proof [pruːf] n.（酒的）标准酒精度 (16)
propagate [ˈprɒpəgeit] v. 传播、繁殖 (2)
propellant [prəˈpelənts] n. 推进剂 (5)
propionate [ˈprəupiəneit] n. 丙酸盐、丙酸酯 (18)

propionic [prəu'piɔnik] adj. 丙酸的 (2)

proprietary name　专利商品名 (3)

prototype ['prəutətaip] n. 原型 (12)

PS　聚苯乙烯 (7)

psig ['sig] abbr. pound per square inch　磅/平方英寸 (16)

psychographics [ˌsaikə'græfiks] n. 消费心理学 (2)

psychrophilic [ˌpsaikrəu'filik] adj. 好寒性的 (2)

PTFE　聚四氟乙烯、特氟龙 (7)

pulpy fruit　软果 (1)

pulse duration　脉冲持续时间 (14)

punch [pʌn(t)ʃ] n. 冲头 (5)

puncture resistance　抗戳穿性 (8)

puncture test　戳穿试验 (4)

purchasing agent　采购部 (20)

put a premium on　重视、助长 (8)

PVAC（PVA）聚醋酸乙烯酯、聚乙酸乙烯酯 (7)

PVAL（PVOH）　聚乙烯醇 (7)

PVC　聚氯乙烯 (7)

PVDC　聚偏二氯乙烯 (7)

pyramidal [pi'ræmid(ə)l] adj. 椎体的 (4)

Pyrex ['paireks] n. 派热克斯玻璃（一种耐热玻璃） (6)

Q

quantify ['kwɔntifai] v. 量化、定量 (2)

R

radioactive isotope　放射性同位素 (2)

rag [ræg] n. 破布 (3)

rail shunting　铁路调车 (10)

railroad's Freight Classification Committee　铁路货物分类委员会 (4)

ram-screw-type machine　柱塞螺杆型机器 (7)

rancid taste　（油脂变质后的）哈喇味、腐臭味 (2)

realm [relm] n. 领域 (2)

ream [riːm] n. 令（纸张计数单位） (3)

rebound velocity　回弹速度 (11)

receptacle [ri'septəkəl] n. 容器、放置物品的地方 (17)

recess [ri'ses] v. 使凹进 (10)

reciprocating valve　往复（运动）阀 (16)

reclaim [ri'kleim] v. 回收再利用 (8)

reclosability [rikləuzbiliti] n. 重新闭合 (9)

rectangular [rek'tæŋgjələ] adj. 矩形的 (10)

rectangular cross section 矩形横截面 (13)
recycle pulp 回收浆 (3)
refillable aluminum can 可再装铝罐 (1)
reflectance [ri'flekt(ə)ns] *n.* 反射率 (3)
regenerated cellulose 再生纤维素 (7)
register point 定位点 (8)
regular slotted container (RSC) 普通开槽箱（即0201箱） (4)
reincarnation [,ri:inkɑ:'neiʃən] *n.* 再生 (1)
relative humidity 相对湿度 (2)
resealability [risilə'biliti] *n.* 重新密封 (9)
reservoir ['rezəvɔr] *n.* 蓄水池 (7)
residential waste 住宅垃圾、生活垃圾 (1)
residual stress 残留应力 (7)
resiliency [ri'ziliənsi] *n.* 弹性、回弹 (10)
resistance to curling 耐折度 (8)
resonance ['rezənəns] *n.* 共振、共鸣 (10)
resonant frequency 共振频率 (12)
response [ri'spɔns] *n.* 响应、反应 (14)
responsible for 为……负责、是造成……的原因 (13)
restriction [ri'strikʃən] *n.* 约束、限制 (17)
retard ['ritɑ:d] *v.* 延迟、使减速 (2)
retort [ri'tɔ:t] *n.* 蒸煮锅 (2)
retortable [ri'tɔ:təbl] *adj.* 耐蒸煮的 (6)
retortable pouch 蒸煮袋、软罐头 (2)
returnable containers 可回收使用的容器 (13)
reversible pallet 可翻转的托盘、双面使用的托盘 (13)
revolution counter 转速计 (16)
rewind roll 收卷辊 (8)
rigid box (Bliss box) 固定型纸箱 (4)
rigidity [ri'dʒidəti] *n.* 硬度、刚性 (4)
ring crush 环压强度 (10)
ring pull-top can 易拉罐 (5)
rodent ['rəudənt] *n.* 啮齿动物 (10)
rolled aluminum 压延铝 (8)
roll-fed 滚筒供料的 (8)
rotary filler 旋转式灌装机 (16)
rotary valve 转阀 (16)
rough handling 野蛮装卸（同abusive handling） (10)
rubber diaphragm 橡皮膜 (4)
rubberized fiber 橡胶纤维 (13)

rugged ['rʌgid] *adj.* 结实的、粗糙的 (10)
ruggedness ['rʌgidnis] *n.* 强度、坚固性 (12)
rule of thumb　经验法则 (4)
runner ['rʌnə] *n.* 浇道 (7)
rupture ['rʌptʃə] *v.* 破裂 (4)

S

sack [sæk] *n.* 麻袋、包；*v.* 掠夺 (1)
sanitary ['sæni,təti] *adj.* 清洁的、卫生的 (5)
saran　[sə'ræn] *n.* 聚偏二氯乙烯，PVDC (2)
sawtooth pulse　锯齿波脉冲 (14)
score [skɔː] *v.* 压痕 (3)
scrape [skrep] *n.* 刮、擦 (17)
screen [skrin] *n.* 网 (16)
screw-on closure　旋盖 (2)
scuff [skʌf] *v.* 刮伤、划伤 (10)
scuff resistance　耐磨损性 (8)
seal integrity　密封完整性 (8)
sealability ['siːlə'biliti] *n.* 密封性 (8)
sealant ['siːlənt] *n.* 密封剂 (5)
seamed [siːmd] *v.* 接口、卷边 (5)
secondary package　二次包装（件）、中包装 (2)
see-through ['siːθru] *adj.* 透明的 (16)
self-erecting box　能自动装配的纸箱 (4)
sensor ['sensə] *n.* 传感器 (14)
sensory active agents　感官活性剂 (2)
septum ['septəm] *n.* 隔膜 (6)
settling ['setliŋ] *n.* 沉淀物 (10)
setup box　自立纸箱 (3)
setup carton　固定（自立）纸盒 (3)
severity [sə'vεriti] *n.* 严重、严格 (13)
shaker table　振动台 (14)
shallow drawn can　浅冲拔罐 (5)
shallow-profile can　浅罐 (5)
sheen [ʃin] *n.* 光辉、光泽 (8)
sheet material　片材、板材 (1)
sheet-fed　片材供料的 (8)
shelf life　货架寿命、保存期限 (2)
shipping ['ʃipiŋ] *v.* 运输 (20)
shipping containers　运输容器 (13)
shock absorber　减震器、吸震器 (12)

thwart [θɔːt] *v.* 阻挠 (9)

tier sheets　层叠式薄板 (13)

tin-plated　马口铁、镀锡钢板 (5)

tinted ['tintid] *adj.* 着色的 (8)

tissue paper　棉纸、薄纸 (3)

titanium dioxide　二氧化钛 (3)

titanium tetrachloride　四氯化钛 (6)

tolerance ['tɔlərəns] *n.* 公差 (16)

tolerance level　误差等级 (20)

tong [tɔŋ] *n.* 钳子 (6)

tooling cost　加工成本 (7)

top [tɔp] *n.* 顶盖 (9)

total-system approach　整体系统解决方案 (13)

tough [tʌf] *adj.* 坚韧的、牢固的 (13)

tout [taut] *v.* 兜售、招揽顾客 (2)

traction device　牵引装置 (8)

trade journal　行业杂志 (1)

trade-off　折中、权衡 (2)

trailer-on-flat-car（TOFC）　平板拖车 (10)

transmissibility curve　传递率曲线 (14)

transmissibility ratio　传递率 (12)

trapezoidal [træpi'zɔidəl] *adj.* 梯形的 (13)

trigger ['trigə] *v.* 引发、触发 (17)

trim [trim] *v.* 修剪 (5)

triple wall　三瓦楞纸板（七层） (4)

triple-tight can　三重密封罐 (5)

trough [trɔf] *n.* 水槽 (8)

tumbler ['tʌmblə] *n.* 平底玻璃杯 (6)

turbulence ['təːbjələns] *n.* 湍流、涡流 (16)

TV dinner　（食前加温即可的）冷冻快餐 (1)

twin-wire machine　双网（夹网）纸机 (3)

two-piece can　两片罐 (5)

two-way-entry pallet　双向进叉托盘 (13)

typography [tai'pɔgrəfi] *n.* 印刷样式、排印 (1)

U

undamped ['ʌn'dæmpt] *adj.* 无阻尼的 (10)

under-packaging ['ʌndə'pækədʒiŋ] *n.* 包装不足、欠包装 (12)

underwritten [ˌʌndə'ritn] *v.* 给……保险、负担……费用 (13)

unicellular [juːni'seljələ] *adj.* 单细胞的 (2)

Uniform Freight Classification（UFC）　统一货物分类 (4)

unit load 单元化装载（集装） (2)
unit package 单元化包装（件）、组合包装 (1)
unit-dose 单元剂量 (8)
unlined [ʌn'laind] adj. 无贴面纸的 (4)
unsupported foil 无载体铝箔 (8)
unwind roll 放卷辊 (8)
urea [ju'riə] n. 尿素 (7)
urethane ['juərəθein] n. 氨基甲酸乙酯、尿烷、聚氨酯 (10)

V

vacuum metallizing 真空镀敷金属 (8)
vacuum packaging 真空包装 (2)
vacuum-cup 真空吸盘 (4)
valve [vælv] n. 阀 (16)
vapor phase 气相 (5)
variable speed motor 变速电动机 (7)
variation [vɛəri'eiʃ(ə)n] n. 差异、方差 (15)
varnish ['vɑ:niʃ] n. 清漆、凡立水 (10)
vat [væt] n. 大桶、制剂桶 (3)
velocity change 速度改变（量） (11)
vent [vent] v. 排放、发泄 (2)
vent port 排气孔 (16)
vent tube 排气管 (16)
ventilated [,vɛntl'et] adj. 通风的 (2)
venting [,ventiŋ] n. 通风、泄去 (19)
vermin ['və:min] n. 害虫、寄生虫 (10)
vertical form-fill-seal (VFFS) machine 立式成型-充填-封口机 (8)
verti-former 竖式造纸机 (3)
vial ['vaiəl] n. 小瓶、药水瓶 (6)
vibration magnification 振动放大（因子） (11)
vibration test machine 振动试验机 (12)
vibration-isolation cushioning system 隔振缓冲系统 (10)
vibrator ['vai,bretə] n. 振动器 (17)
vinyl chloride 氯乙烯 (5)
virgin ['və:dʒin] adj. 没有处理过的 (4)
viscoelastic [,viskəui'læstik] adj. 黏弹性的 (3)
viscosity [vi'skasiti] n. 黏性、黏度 (16)
vitreous ['vitriəs] adj. 玻璃的、玻璃状的 (6)
volatile ['vɔlətl] n. 挥发物 (2)
volatility [,vɔlə'tiliti] n. 挥发性 (19)
volumetric [,vɔljə'metrik] adj. 容量的、容积的 (16)

Resources

[1] Soroka W. Fundamentals of Packaging Technology. USA: Institute of Packaging Professionals, 2002.
[2] 陈满儒. 包装工程概论（双语）. 北京: 化学工业出版社, 2005.
[3] 陈满儒, 黄涛. 最新英汉包装词典. 北京: 印刷工业出版社, 2007.
[4] 秦秀白, 刘洊波. 新世纪大学英语系列教材·综合教程. 上海: 上海外语教育出版社, 2011.
[5] Soroka W. Glossary of Flexible Packaging Terms. USA: Flexible Packaging Association, 2003.
[6] Coles R, et al. Food Packaging Technology. UK: Blackwell Publishing, 2003.
[7] Ideas and Innovation. USA: Paperboard Packaging Council, 2000.
[8] The Wiley Encyclopedia of Packaging Technology. USA: John Wiley & Sons, Inc., 2007.
[9] ASTM D 996-95EI Terminology of Packaging and Distribution Environments.
[10] ASTM D 3332-98 Test Method for Mechanical-Shock Fragility of Products, Using Shock Machines.
[11] ASTM D 999-96 Test Methods for Vibration Testing of Shipping Containers.
[12] ASTM D 3580-95EI Test Method of Vibration (Vertical Sinusoidal Motion) Test of Products.
[13] ASTM D 1596-97 Test Method for Dynamic Shock Cushioning Characteristics of Packaging Material.
[14] http://www.lansmont.com/SixStep/Summary.html.
[15] Daum M. A Simplified Progress for Determining Cushion Curves: The Stress-Energy Method.
[16] Brandenburg R K, et al. Fundamentals of Packaging Dynamics. 4th Edition. USA: L. A. B., 1991.
[17] Fibre Box Handbook. USA: Fibre Box Association, 1999.
[18] MIL-HDBK-304C-1997 Package Cushioning Design.

W

wadding [ˈwɔdiŋ] *n.* 软填料、纤维填料 (10)
warpage [ˈwɔːpeidʒ] *n.* 翘曲、弯曲 (4)
washer [ˈwɔʃə] *n.* 垫圈、垫片 (16)
water-vapor transmission rate (WVTR)　水蒸气传输速率 (8)
waveform [ˈwevfɔrm] *n.* 波形 (12)
wavy [ˈweivi] *adj.* 波浪状的 (4)
weather-resistant　耐候的 (4)
web [web] *n.* 卷筒材料 (8)
web-fed　卷筒材供料的 (8)
wet bonding　湿法复合 (8)
wicker basket　柳条篮 (2)
wide-mouthed　广口的 (7)
wing pallets　翼式托盘 (13)
wood distillation　木材蒸馏法 (2)
wood pulp　木浆 (1)
woodcut [ˈwudkʌt] *n.* 木刻画 (1)
wooden barrel　木桶 (1)
workhorse [ˈwəːkhɔːs] *adj.* 负重的 (4)
wraparound [ˈræpəˌraund] *n.* 环绕裹包、全裹包 (13)

Y

yield point　屈服点 (8)

slotted box 开槽型纸箱	(4)
slug [slʌg] *n.* 金属块	(5)
slurry ['slʌri] *n.* 混合液、浆料	(3)
snap-on lid 按扣盖、搭锁盖（可咯嗒一声盖住的）	(2)
snift port 卸压口	(16)
soda ['səudə] *n.* 苏打、碳酸水	(6)
sodium ['səudiəm] *n.* 钠	(6)
soldered ['səuldəd] *v.* 焊接	(5)
solenoid ['sələnɔid] *n.* 电磁线圈、电磁铁	(16)
solid bleached boardstock 硬质漂白木浆纸板	(18)
solid bleached sulfate (SBS) 漂白硫酸盐硬纸板	(3)
solid fibreboard 硬纸板	(4)
solid unbleached sulfate (SUS) 非漂白硫酸盐硬纸板	(3)
solubility [,sɔlju'biləti] *n.* 溶解度	(2)
sophisticated [sə'fistikeitid] *adj.* 先进的	(10)
sorbic ['sɔ:bik] *adj.* 山梨酸的	(2)
sound packaging 良好包装、完好包装	(1)
specialty paper 特种纸	(3)
specification [,spesəfi'keʃən] *n.* 技术规格、说明书	(13)
spillage ['spilidʒ] *n.* 溢出、溢出量	(16)
spiral-wound 螺旋缠绕（纸罐），指通过螺旋缠绕的方式生产的圆筒容器	(20)
splash [splæʃ] *v.* 飞溅、喷洒	(16)
spoilage ['spɔilidʒ] *n.* 损坏、变质	(2)
spore [spɔ:] *n.* 孢子	(2)
spot label 点标	(6)
spout [spaut] *n.* 管口、（容器的）嘴	(17)
spray [sprei] *n.* 喷雾、喷射	(16)
spreadsheet ['spredʃi:t] *n.* 电子数据表	(15)
spring balance scale 弹簧秤	(17)
spring loaded filling valve 弹簧作用、灌装阀	(16)
sprout [spraut] *n.* 芽、萌芽、苗芽	(2)
square up 将纸箱撑开成直角	(4)
square wave pulse 矩形波脉冲	(12)
stabilize ['stebə,laiz] *v.* 变得稳定、稳固或固定	(13)
stacking strength 堆码强度	(4)
standard deviation 标准方差	(15)
stand-up pouch 直立袋	(8)
starch-based adhesive 淀粉黏结剂	(4)
start from scratch 从零开始	(15)
static compressive strength 静态抗压强度	(3)

shock cushion curve 缓冲特性曲线、最大加速度-静应力曲线 (12)
shock fragility 冲击脆值 (14)
shock isolator 隔振器 (14)
shock machine 冲击试验机 (12)
shock programmer 冲击程序装置 (14)
shock pulse 冲击脉冲 (11)
shock testing machine 冲击试验机 (14)
shoot [ʃut] n. 滑道、滑槽 (17)
shrink plastic 收缩塑料 (7)
shrink-wrapped 收缩裹包的 (10)
shroud [ʃraʊd] n. 护罩、保护罩 (17)
shunting speed 调车速度 (10)
shutter [ˈʃʌtə] n. 底门 (17)
SI unit 公制、国际单位制 (8)
side panel （纸箱）侧板 (4)
silica [ˈsilikə] n. 二氧化硅、硅土 (6)
silicone oxides （SiO_x）硅氧化物 (8)
simply put 简单地说 (15)
simulated shipping 仿真运输过程 (20)
sine [sain] n. 正弦 (14)
single face 单面瓦楞纸板（两层） (4)
single wall 单瓦楞纸板（三层） (4)
single white-lined (SWL) paperboard 单面贴有白色面纸的纸板 (3)
single-face pallet 单面托盘 (10)
single-facer 单面机 (4)
single-service packaging 一次性包装 (1)
sinusoidal [ˈsinəˈsɔidl] adj. 正弦（曲线）的 (12)
size [saiz] v. 涂胶 (3)
skin packaging 贴体包装 (18)
slat [slæt] n. 板条、狭板 (17)
slat counter 板条式计数装置 (17)
sled [sled] n. 翘板 (4)
sleek [sliːk] adj. 圆滑的、井然有序 (5)
sleeve [sliv] n. 套筒、套管 (16)
slide boxes 滑盖盒（箱） (20)
slip cover 滑盖 (5)
slip sheet 滑托板、薄垫板 (13)
slipcover 套盖 (5)
slippage [ˈslipidʒ] n. 移动、滑动 (4)
slot [slɔt] v. 开槽 (4)

static stress 静应力	(10)
steady state vibration 稳态振动	(12)
steel banding 钢带捆扎	(13)
step acceleration test 步进式加速度测试	(14)
step velocity test 步进式速度测试	(14)
sterile ['stɛrəl] adj. 无菌的	(17)
sterility [stə'rilətɪ] n. 无菌性	(2)
sterilize ['stɛrəlaiz] v. 消毒、杀菌	(2)
steroid ['stiərɔid] n. 类固醇	(19)
still liquids 不含气液体、静态液体	(16)
stirring blades 搅拌叶片	(17)
stock-picking 选货	(2)
stopper ['stɔpə] n. 塞子	(9)
straight-line filler 直列式灌装机	(16)
strapping ['stræpiŋ] n. 捆扎	(13)
stray volatile 游离挥发物	(2)
stress concentrator 应力集中点	(10)
stress/strain machine 应力/应变机	(4)
stretch [stretʃ] v. 伸展	(8)
stretch-wrapped 拉伸裹包的	(10)
stringer ['striŋə] n. 纵梁、长梁	(10)
strip pack 条式包装	(9)
stroboscope ['strəubəskəup] n. 频闪观测器、频闪仪	(14)
stroke [strəuk] n. 行程	(7)
sturdy ['stə:di] adj. 强健的、坚定的	(10)
styrenics ['staiə,riniks] n. 苯乙烯	(18)
styrofoam ['stairə,fɔm] n. 泡沫聚苯乙烯	(7)
subjecting ... to 受……管制、受到	(13)
sublimate ['sʌblimeit] v. 升华	(2)
substrate ['sʌbstreit] n. 底层、基材	(18)
suck-back 倒吸	(5)
sulfuric acid 硫酸	(3)
supplier [sə'plaiə] n. 供应商、供货方	(1)
supply tank 供料缸	(16)
surface tension 表面张力	(16)
surgical ['sə:dʒikəl] adj. 外科的、手术上的	(19)
surmise [sə'maiz] v. 猜测、推测	(7)
susceptor [sə'septə] n. 感受器、基座	(8)
synthetic [sin'θetik] adj. 合成的、人造的	(7)

syringe [si'rin(d)ʒ] *n.* 注射器 (9)
syrup ['sirəp] *n.* 糖浆、果汁 (2)

T

tabulate ['tæbjuleit] *vt.* 把……制成表格 (15)
tailgate ['teilgeit] *n.* （卡车等的）后挡板 (10)
take-up factor 收缩率 (4)
talcum ['tælkəm] *n.* 滑石 (5)
tamp [tæmp] *v.* 夯实 (17)
tamper evidence 显窃启 (8)
tamperer ['tæmpərə] *n.* 偷换者 (9)
tamper-evident closure 显窃启盖 (1)
tamper ['tæmpə] *n.* 偷换、窜改 (2)
tamperproof ['tæmpəpru:f] *adj.* 防偷换的 (9)
tamper-resistant 阻抗偷换的 (9)
tan [tæn] *n.* 棕色、褐色 (3)
TAPPI 纸浆与造纸工业技术协会 (1)
teflon ['tefla:n] *n.* 聚四氟乙烯 (16)
telescope ['teliˌskop] *v.* 伸缩 (17)
telescope box 套合型纸箱 (4)
temper [tempə] *n.* 韧度 (8)
tenfold ['tenfəuld] *adj.* 十倍的 (16)
tensile or burst strength 抗张或耐破强度 (8)
tensile propertie 拉伸性能 (4)
tensile strength 抗张强度 (3)
thawing [θɔ:iŋ] *n.* 解冻、融化 (2)
the Dark Ages 欧洲中世纪、黑暗时代 (1)
the Grocery Manufacturers of America 美国食品杂货制造商 (13)
the International Safe Transit Association（ISTA） 国际安全运输协会 (13)
the trial-and-error cycle 反复试验过程 (13)
thermal history 受热历程 (7)
thermoform [θə:məuˌfɔ:m] *v.* 加热成型 (2)
thermoformability 热成型 (3)
thermophilic [ˌθə:məu'filik] *adj.* 适温的、喜温的 (2)
thermoplastic [ˌθə:mə'plæstik] *n.* 热塑性塑料 (1)
thermoset [ˈθə:məuset] *n.* 热固性 (7)
thread profile 螺纹牙形 (6)
three-piece can 三片罐 (5)
three-sided-seal pouch 三边封袋 (8)
thumper ['θʌmpə] *n.* 重击物、重锤 (17)